CLIMATE FORCING OF GEOLOGICAL HAZARDS

CLIMATE FORCING OF GEOLOGICAL HAZARDS

EDITED BY
BILL McGUIRE

AON BENFIELD UCL HAZARD CENTRE, DEPARTMENT OF EARTH
SCIENCES, UNIVERSITY COLLEGE LONDON, LONDON, UK

AND

MARK MASLIN

DEPARTMENT OF GEOGRAPHY, UNIVERSITY COLLEGE LONDON,
LONDON, UK

Originating from a Theme Issue published in *Philosophical Transactions A:
Mathematical, Physical and Engineering Sciences*

A John Wiley & Sons, Ltd., Publication

Registered office: John Wiley & Sons, Ltd, The Atrium, Southern Gate, Chichester, West Sussex, PO19 8SQ, UK

Editorial offices: 9600 Garsington Road, Oxford, OX4 2DQ, UK
The Atrium, Southern Gate, Chichester, West Sussex, PO19 8SQ, UK
111 River Street, Hoboken, NJ 07030-5774, USA

For details of our global editorial offices, for customer services and for information about how to apply for permission to reuse the copyright material in this book please see our website at www.wiley.com/wiley-blackwell.

Library of Congress Cataloging-in-Publication Data
Climate forcing of geological hazards / edited by Bill McGuire and Mark Maslin.
    p. cm.
  "Originating from a theme issue published in Philosophical transactions A: mathematical, physical and enginering sciences".
  Includes bibliographical references and index.
    ISBN 978-0-470-65865-9 (cloth)
1. Climatic changes. 2. Natural disasters. I. McGuire, Bill, 1954– II. Maslin, Mark.
III. Philosophical transactions. Series A, Mathematical, physical, and engineering sciences.
  QC903.C566 2013
  363.34'1–dc23
                        2012030742

Cover image: Volcan Eyjafjallajökull in South Iceland. © iStockphoto.com/JochenScheffi
Cover design: Nicki Averill Design & Illustration

Set in 10/12.5 pt Minion by Toppan Best-set Premedia Limited
Printed and bound in Malaysia by Vivar Printing Sdn Bhd

1   2013

Contents

List of contributors

Fabien Albino Nordic Volcanological Centre, Institute of Earth Sciences, University of Iceland, Reykjavik, Iceland

Simon Allen Climate and Environmental Physics, Physics Institute, University of Bern, Switzerland

Richard A. Betts Met Office Hadley Centre, Exeter, UK

Simon Day Aon Benfield UCL Hazard Centre, Department of Earth Sciences, University College London, London, UK

Kim Deeming School of Earth, Atmospheric and Environmental Sciences, University of Manchester, Manchester and Aon Benfield UCL Hazard Centre, Department of Earth Sciences, University College London, Gower Street, London, UK

Tom Dunkley Jones School of Geography, Earth and Environmental Sciences, University of Birmingham, Birmingham, UK

Rachel Flecker School of Geographical Sciences, University of Bristol, Bristol, UK

Serge Guillas Department of Statistical Science and Aon Benfield UCL Hazard Centre, University College London, London, UK

Andrea Hampel Institut für Geologie, Leibniz-Universität Hannover, Hannover, Germany

Stephan Harrison College of Life and Environmental Sciences, University of Exeter, Penryn, UK

Paul Harrop Tessella plc, Abingdon, Oxfordshire, UK

Ralf Hetzel Institut für Geologie und Paläontologie, Westfälische Wilhelms-Universität Münster, Germany

Andrew Hooper Delft University of Technology, Delft, The Netherlands

Christian Huggel Department of Geography, University of Zurich, Switzerland

Ruža F. Ivanović School of Geographical Sciences, University of Bristol, Bristol, UK

Jasper Knight School of Geography, Archaeology and Environmental Studies, University of the Witwatersrand, Johannesburg, South Africa

Margreth Keiler Geographical Institute, University of Bern, Bern, Switzerland

Felicity Liggins Met Office Hadley Centre, Exeter, UK

Björn Lund Department of Earth Sciences, Uppsala University, Sweden

Daniel J. Lunt School of Geographical Sciences, University of Bristol, Bristol, UK

Bill McGuire Aon Benfield UCL Hazard Centre, Department of Earth Sciences, University College London, London, UK

Georgios Maniatis Institut für Geologie, Leibniz-Universität Hannover, Hannover, Germany

Mark Maslin Department of Geography, University College London, London, UK

Matthew Owen Department of Geography, University College London, London, UK

Carolina Pagli School of Earth and Environment, University of Leeds, UK

Virginie Pinel ISTerre, IRD R219, CNRS, Université de Savoie, Le Bourget du Lac, France

Andrew Ridgwell Department of Geography, Bristol University, Bristol

Nadine Salzmann Department of Geography, University of Zurich, and Department of Geosciences, University of Fribourg, Switzerland

Peter Schmidt Department of Earth Sciences, Uppsala University, Sweden

Freysteinn Sigmundsson Nordic Volcanological Centre, Institute of Earth Sciences, University of Iceland, Reykjavik, Iceland

David R. Tappin British Geological Survey, Nottingham, UK

Hugh Tuffen Lancaster Environment Centre, Lancaster University, Lancaster, UK

Paul J. Valdes School of Geographical Sciences, University of Bristol, Bristol, UK

Preface

Since the Last Glacial Maximum, some 20,000 years ago, our world has experienced an extraordinary metamorphosis: flipping – in the blink of an eye, geologically speaking – from a frigid wasteland into the temperate world upon which our civilisation has grown and thrived. Over this period, a staggering 52 million cubic kilometres of water were redistributed about the planet, as the great continental ice sheets melted and previously depleted global sea levels rose 130 m to compensate. Rapid global warming, of around 6°C, resulted in atmospheric circulation patterns changing to accommodate broadly warmer, wetter conditions, leading to a modification of major wind trends and a rearrangement of climatic zones, but these were not the only consequences of our world's dramatic post-glacial transformation. The solid Earth was involved too, as the lithosphere (the brittle, outer layer of our planet that comprises the crust and uppermost mantle) underwent major readjustments in response to the massive changes in water and ice load. Outcomes included major earthquakes in formerly ice-covered regions at high latitudes and a spectacular rise in the level of volcanic activity in Iceland.

The 12 chapters that make up this book together address the many and varied ways in which dramatic climate change, such as that which characterised post-glacial times, is able to 'force' a reaction from the solid Earth, or Geosphere. Of critical importance from a societal point of view, the prospects for anthropogenic climate change driving a hazardous response are also examined and evaluated. The book builds on presentations and dialogue at the Third Johnston–Lavis Colloquium held at University College London in September 2009. The meeting brought together delegates from the UK, Europe and the USA to address the issue of climate forcing of geological and geomorphological hazards, with a particular focus on examining possibilities for a geospheric response to anthropogenic climate change. The chapters that form this volume are a reflection of new research and critical reviews presented in sessions on: climates of the past and future; climate forcing of volcanism and volcanic activity; and climate as a driver of seismic, mass movement and tsunami hazards.

Two introductory papers set the scene. In the first, Bill McGuire summarizes evidence for periods of exceptional past climate change eliciting a dynamic response from the Earth's lithosphere, involving enhanced levels of potentially hazardous geological and geomorphological activity. The response, McGuire notes, is expressed mainly through the triggering, adjustment or modulation of a range of crustal and surface processes that include gas-hydrate destabilisation, submarine and subaerial landslide formation, debris flow occurrence and glacial outburst flooding, and volcanic and seismic activity. Adopting a uniformitarian approach, and acknowledging potential differences in both rate and scale from the period of post-glacial warming, he goes on to examine potential influences of anthropogenic climate change in relation to an array of geological and geomorphological hazards across an assortment of environmental settings. In a second and complementary review paper, Felicity Liggins, and others, evaluate climate change projections from both global and regional climate models in the context of geological and geomorphological hazards. The authors observe that, in assessing potential for a geospheric response, it seems prudent to consider that regional levels of warming of 2°C are unavoidable, with high-end projections associated with unmitigated emissions potentially leading to a global average temperature rise in excess of 4°C, and far greater warming in some regions. Importantly, they note that significant uncertainties exist, not only in relation to climate projections, but also in regard to links between climate change and geospheric responses.

Between them, the following two chapters examine the ways and means whereby rapid climate change has, in the past, increased levels of volcanic activity and the destabilisation of volcanic edifices and promoted magma production, and look ahead to possible ramifications for volcanic landscapes of contemporary climate change. In Chapter 3, Kim Deeming and her co-authors explore the phenomenon of volcano lateral collapse (the large-scale failure and collapse of part of a volcano's flank) in response to a changing climate. The authors present the results of a cosmic ray exposure dating campaign at Mount Etna in Sicily, which constrains the timing and nature of collapse of the Valle del Bove – a major volcanic landslide scar on the eastern flank of the volcano. The authors link pluvial (wet) conditions during the Early Holocene to the formation of a high-energy surface drainage system and to its truncation by a catastrophic lateral collapse event, about 7500 years ago, which opened the Valle del Bove. A possible mechanism is proposed whereby magma emplacement into a water-saturated edifice caused the thermal pressurization of pore-water, leading to a reduction in sliding resistance and subsequent large-scale slope failure. Deeming and her colleagues showcase the mechanism as one possible driver of future lateral collapse at ice-capped volcanoes and at those located in regions predicted to experience enhanced precipitation.

Following on from this, Hugh Tuffen provides a general evaluation of the impact of a changing climate on glaciated volcanoes – looking ahead to how the melting of ice caps on active volcanoes may influence volcanic hazards in the twenty-first century. In reviewing the evidence for current melting of ice increasing the frequency or size of future eruptions, he notes that much remains to be understood in relation to ice loss and increased eruptive activity. In particular, uncertainty surrounds the sensitivity of volcanoes to small changes in ice thickness and how rapidly volcanic systems respond to deglaciation. Nevertheless, Tuffen expects an increase in explosive eruptions at glaciated volcanoes that experience significant ice thinning, and a greater frequency of lateral collapse at glaciated stratovolcanoes in response to anthropogenic warming. On the positive side, deglaciation may ultimately reduce the threat from volcanic debris flows (lahars) and melt-water floods from volcanoes that currently support ice caps.

There is strong evidence for a lithospheric response to the rapidly changing post-glacial climate being elicited by load changes, either as a consequence of unloading at high latitudes and high altitudes due to ice-mass wastage, or as a result of the loading of ocean basins and continental margins in response to a ≥100 m rise in global sea level. In Chapter 5, Freysteinn Sigmundsson and his co-workers evaluate the influence of climate-driven ice loading and unloading on volcanism, focusing on Iceland and, in particular, the Vatnajökull Ice Cap. They note that ice wastage on Icelandic volcanoes reduces pressure at the surface and causes stress changes in magmatic systems. This in turn is capable of promoting an increase in the generation of magma in the uppermost mantle, raising the potential for the 'capture' of magma in the crust – as opposed to its eruption at the surface – and modifying the conditions required for the walls of a magma reservoir to fail. The authors demonstrate that, although pressure-release melting in the mantle may generate an amount of magma comparable with that arising from plate tectonic processes, at least part of this will never reach the surface. Perhaps somewhat surprisingly, Sigmundsson and his colleagues show that long-term ice wastage at Katla volcano may actually reduce the likelihood of eruption, because more magma is needed in the chamber to cause failure, compared with times when ice cover is greater.

Continuing the loading–unloading theme in Chapter 6, Andrea Hampel and her fellow researchers examine how active faults have responded to variations in ice and water volumes as a consequence of past climate change. Using numerical models, the authors demonstrate that climate-driven changes in ice and water volume are able to affect the slip evolution of both thrust and normal faults, with – in general – both the slip rate and the seismicity of a fault increasing with unloading and decreasing with loading. Adopting a case-study approach, Hampel and colleagues provide evidence for a widespread, post-glacial, seismic response on faults located beneath decaying ice sheets or glacial lakes. Looking ahead, the authors point to the implications of their results for ice-mass loss at high latitudes, and speculate that shrinkage of the Greenland and Antarctic ice sheets as a consequence of anthropogenic warming could result in a rise in the frequency of earthquakes in these regions.

In the next chapter, Serge Guillas and others provide a contemporary slant to loading and unloading effects by presenting the results of a statistical analysis of a putative correlation between recent variations in the El Niño–Southern Oscillation (ENSO) and the occurrence of earthquakes on the East Pacific Rise (EPR). The authors observe a significant (95% confidence level) positive influence of the Southern Oscillation Index (SOI) on seismicity, and propose that increased seismicity on the EPR arises due to the reduced sea levels in the eastern Pacific that precede El Niño events, which can be explained in terms of the reduction in ocean-bottom pressure over the EPR by a few kilopascals. Guillas and co-authors note that this provides an example of how variations in the atmosphere and hydrosphere can drive very small changes in environmental conditions which, in turn, are able to trigger a response from the solid Earth. Perhaps most significantly, they speculate that, in a warmer world, comparable and larger changes associated with ocean loading due to global sea level rise, or unloading associated with the passage of more intense storms, may trigger more significant earthquake activity at fault systems in the marine and coastal environments that are in a critical state.

Staying in the marine environment, Dave Tappin provides, in Chapter 8, a comprehensive review of the role of climate in promoting submarine mass failures (SMFs) that may source tsunamis. Tappin highlights the importance of climate in 'preconditioning' sediment so as to promote instability and failure, including its influence on sediment type, deposition rate and post-depositional modification. The author also notes that climate may play a role in triggering SMFs via earthquake or cyclic loading associated with tides or storm waves. Tappin makes the important point that, in the past, climate influence on SMFs appears to have been greatest at high latitudes and associated with glaciation–deglaciation cycles, which had a significant influence on sedimentation, preconditioning and triggering. In fact many of these current geohazards are due to the continued isostatic rebound as the land recovers from ice sheet loading more than 10,000 years ago. As a corollary, Tappin observes that, as the Earth warms, increased understanding of the influence of climate will help to underpin forecasting of tsunami-sourcing SMFs, in particular at high latitudes where climate change is occurring most rapidly.

The theme of slope destabilisation and failure, this time in a subaerial setting, is continued in the next chapter by Christian Huggel and his co-researchers, who examine recent large slope failures in the light of short-term, extreme warming events. Huggel and colleagues demonstrate a link between large slope failures in Alaska, New Zealand and the European Alps, and preceding, anomalously warm episodes. The authors present evidence supporting the view that triggering of large slope failures in temperature-sensitive high mountains is primarily a function of reduced slope strength due to increased production of meltwater from snow and ice, and rapid thaw processes. Looking ahead they expect more frequent episodes of extreme temperature to result in a rise in the number of large slope failures in elevated terrain and warn of potentially serious consequences for mountain communities.

In Chapter 10, Jasper Knight and others give a regional perspective on the slope failure and flood hazard in mountainous terrain, focusing on the influence of contemporary climate change on a broad spectrum of geomorphological hazards in the eastern European Alps, including landslides, rock falls, debris flows, avalanches and floods. In the context of the pan-continental 2003 heat wave and the 2005 central European floods, the authors demonstrate how physical processes and human activity are linked in climatically sensitive alpine regions that are prone to the effects of anthropogenic climate change. Importantly, Knight and colleagues note that, although the European Alps, alongside other glaciated mountain ranges, are being disproportionately impacted upon by climate change, this is further exacerbated by regional factors including local climatology and long-term decay of glaciers and permafrost. The authors conclude that future climate changes are likely to drive rises in the incidence of mountain hazards and, consequently, increase their impact on Alpine communities.

The two concluding chapters centre on gas hydrates (or clathrates) and their sensitivity to a rapidly changing climate. In both marine and continental settings, gas-hydrate deposits have long captured interest, both in relation to their potential role in past episodes of sudden warming, such as during the Paleocene–Eocene Thermal Maximum (PETM), some 55 million years ago, and in the context of future anthropogenic warming. In the first, Mark Maslin and his fellow researchers review the current state of the science as it relates to the hazard potential of gas hydrates. Maslin and colleagues note that gas hydrates may present a serious threat as the world warms, primarily through the release into the atmosphere of large quantities of methane, which is an extremely effective greenhouse gas, resulting in accelerated global warming. In addition, they observe that the explosive release of methane from gas hydrates may also promote submarine slope failure and the consequent generation of potentially destructive tsunamis. The authors also stress, however, that, although the destabilisation of gas hydrates in permafrost terrains can be robustly linked to projected temperature increases at high latitudes, it remains to be determined whether or not future ocean warming will lead to significant methane release from marine hydrates.

In a second paper, and the last chapter of the volume, Tom Dunkley Jones and others look back to the PETM, the most prominent, transient global warming event during the Cenozoic, in order to evaluate the effects of the rapid release of thousands of gigatonnes of greenhouse gases on the planet's climate, ocean-atmosphere chemistry and biota, for which the PETM provides perhaps the best available analogue. Dunkley Jones and his co-workers support the view that, although gas hydrate release was probably not responsible for an initial, rapid, CO_2-driven warming, the as yet unknown event responsible for this subsequently triggered the large-scale dissociation of gas hydrates, which contributed to further warming as a positive feedback mechanism. As the authors note, this somewhat equivocal situation ensures that the question of what role – if any – gas hydrates may play in future anthropogenic warming, remains to be answered.

We feel that this book provides a valuable new insight into how climate change may force geological and geomorphological phenomena, ultimately increasing the risk of natural hazards in a warmer world. Taken together, the chapters build a panorama of a field of research that is only now becoming recognized as important in the context of the likely impacts and implications of anthropogenic climate change. We are keen for this volume to provide a marker that reinforces the idea that anthropogenic climate change does not simply involve the atmosphere and hydrosphere, but can also elicit a response from the Earth beneath our feet. In this regard we are hopeful that it will encourage further research into those mechanisms by which climate change may drive potentially hazardous geological and geomorphological activity, and into the future ramifications for society and economy.

Bill McGuire and Mark Maslin
London, UK

This book was originally published as an issue of the *Philosophical Transactions A: Mathematical, Physical and Engineering Sciences* (volume 368, issue 1919) but has been materially changed and updated.

Hazardous responses of the solid Earth to a changing climate

Bill McGuire

Aon Benfield UCL Hazard Centre, Department of Earth Sciences, University College London, UK

Summary

Periods of exceptional climate change in Earth's history are associated with a dynamic response from the geosphere, involving enhanced levels of potentially hazardous geological and geomorphological activity. The response is expressed through the adjustment, modulation or triggering of a broad range of surface and crustal phenomena, including volcanic and seismic activity, submarine and subaerial landslides, tsunamis and landslide 'splash' waves, glacial outburst and rock-dam failure floods, debris flows and gas-hydrate destabilisation. In relation to anthropogenic climate change, modelling studies and projection of current trends point towards increased risk in relation to a spectrum of geological and geomorphological hazards in a warmer world, whereas observations suggest that the ongoing rise in global average temperatures may already be eliciting a hazardous response from the geosphere. Here, the potential influences of anthropogenic warming are reviewed in relation to an array of geological and geomorphological hazards across a range of environmental settings. A programme of focused research is advocated in order to: (1) better understand those mechanisms by which contemporary climate change may drive hazardous geological and geomorphological activity; (2) delineate those parts of the world that are most susceptible; and (3) provide a more robust appreciation of potential impacts for society and infrastructure.

Introduction

Concern over anthropogenic climate change driving hazardous geological and geomorphological activity is justified on the basis of four lines of evidence: (1) periods of exceptional climate change in Earth's history are associated with a dynamic response from the geosphere; (2) small changes in environmental conditions provide a means whereby physical phenomena involving the atmosphere and hydrosphere can elicit a reaction from the Earth's crust and sometimes at deeper levels; (3) modelling studies and projection of current trends point towards increased risk in relation to a range of geological and geomorphological hazards in a warmer world; and (4) observations suggest that the ongoing rise in global average temperatures may already be eliciting a hazardous response from the geosphere.

A link between past climate change and enhanced levels of potentially hazardous geological and geomorphological activity is well established, with supporting evidence coming mostly, although not exclusively, from the period following the end of the Last Glacial Maximum (LGM) around 20 ka BP (20 thousands of years before present). During the latest Pleistocene and the Holocene, the atmosphere and hydrosphere underwent dramatic transformations. Rapid planetary warming promoted a major reorganisation of the global water budget as continental ice sheets melted to replenish depleted ocean volumes, resulting in a cumulative sea-level rise of about 130 m. Contemporaneously, atmospheric circulation patterns changed to accommodate broadly warmer, wetter conditions, leading to modification of major wind trends and a rearrangement of climatic zones.

The nature of the geospheric response to transitions from glacial to interglacial periods provides the context for evaluating the potential of current greenhouse gas (GHG)-related warming to influence the frequency and incidence of geological and geomorphological hazards. Critically, however, differences in the timescale, degree and rate of contemporary environmental change may result in a different hazardous response. The key question, therefore, is: to what extent does the post-glacial period provide an analogue for climate-change driven hazards in the twenty-first century and beyond.

Climate change as a driver of geological and geomorphological hazards at glacial–interglacial transitions

At the broadest of scales, modification of the global pattern of stress and strain, due to a major redistribution of planetary water, may influence geological and geomorphological activity at times of glacial–interglacial transition (Matthews, 1969; Podolskiy, 2008). As noted in Liggins et al. (2010), however, a more targeted geospheric response to planetary warming and hydrological adjustment during these times is associated with ice-mass loss, rapid sea-level rise and greater availability of liquid water, in the form of either ice melt or increased precipitation levels. These environmental transformations in turn drive load pressure changes and increases in pore-water pressure which, together, act to promote hazardous geological and geomorphological activity. Notably, variations in ice and water load have been linked to fault rupture (Hampel et al., 2007, 2010), magma production and eruption (McNutt & Beavan, 1987; McNutt, 1999; Pagli & Sigmundsson, 2008; Sigmundsson et al., 2010), and submarine mass movements (Lee, 2009; Tappin,

2010). Elevated pore-water pressures are routinely implicated in the formation of subaerial and marine landslides (Pratt et al., 2002; Tappin 2010).

The geospheric response to such changes in environmental conditions at times of glacial–interglacial transition is expressed through the adjustment, modulation or triggering of a wide range of surface and crustal phenomena, including volcanic (e.g. Chappel, 1975; Kennett & Thunell, 1975; Rampino et al., 1979; Hall, 1982; Wallmann et al., 1988; Nakada & Yokose, 1992; Sigvaldason et al., 1992; Jull & McKenzie, 1996; McGuire et al., 1997; Zielinksi et al., 1997; Glazner et al., 1999; Maclennan et al., 2002; Bay et al., 2004; Jellinek et al 2004; Nowell et al., 2006; Licciardi et al., 2007; Bigg et al., 2008; Carrivick et al., 2009a; Huybers & Langmuir, 2009) and seismic (e.g. Anderson, 1974; Davenport et al., 1989; Costain & Bollinger, 1996; Luttrell & Sandwell, 2010; Wu et al., 1999; Wu & Johnston, 2000; Hetzel & Hampel, 2005) activity, marine (e.g. Maslin et al., 1998, 2004; Day et al., 1999, 2000; Masson et al., 2002; Keating & McGuire, 2004; McMurtry et al., 2004; Vanneste et al., 2006; Quidelleur et al., 2008; Lee, 2009; Tappin, 2010) and sub-aerial (Lateltin et al., 1997; Friele & Clague, 2004; Capra, 2006) landslides, tsunamis (McMurtry et al., 2004; Lee, 2009) and landslide 'splash' waves (Hermanns et al., 2004), glacial outburst (Alho et al., 2005) and rock-dam failure (Hermanns et al., 2004), floods, debris flows (Keefer et al., 2000) and gas-hydrate destabilisation (e.g. Henriet & Mienert, 1998; Maslin et al., 2004; Beget & Addison 2007; Grozic, 2009).

The degree to which comparable responses to projected future climate changes could modify the risk of geological and geomorphological hazards is likely to be significantly dependent on the scale and rate of future climate change. The scale of changes in key environmental conditions in post-glacial times was considerable, with the rapid loss of continental ice sheets after the LGM, leading to cumulative load-pressure reductions on the crust of a few tens of megapascals, and sea levels in excess of 100 m higher increasing the total load on the crust by approximately 1 MPa. Rates of change were also dramatic, with annual vertical mass wastage of between 10 and 50 m (corresponding to a load reduction of 10–50 kPa) reported for the Wisconsin Laurentide Ice Sheet (Andrews, 1973). The rate of global eustatic sea level rise may have approached 5 m a century at times, with annual rates of more than 45 mm (Blanchon & Shaw, 1995), resulting in an annual load-pressure increase on the crust of about 1 kPa. Given the scale of absolute changes and the very high rates involved, it is unsurprising that imposition of such stresses within the crustal domain elicited a significant geological and geomorphological response. Although the post-LGM climate of the latest Pleistocene and the Holocene was characterised by considerable variability (Mayewski et al., 2004), the transition from 'ice-world' to 'water-world' broadly altered the moisture balance in favour of a far greater incidence of warm and wet conditions, e.g. during the African Humid Period from about 14,800 to 5500 years BP (Hély et al., 2009) and during the Early Holocene across much of the Mediterranean (Frisia et al., 2006). Higher levels of precipitation raised the potential for higher pore-water pressures in unstable volumes of rock and debris, e.g. promoting landslide formation. Pratt et al. (2002), speculate that enhanced monsoon rainfall during the Early Holocene raised pore-water pressures in the Nepal Himalayas, resulting in an increase in landslide frequency. Similarly, Capra (2006) invokes more humid Holocene conditions to explain an apparent increase in the incidence of lateral collapse of volcanic edifices. In relation to the formation of submarine landslides in the post-glacial period, a range of potential environmental triggers are proposed, most notably elevated levels of seismicity associated with isostatic rebound of previously ice-covered crust, or ocean-loading due to rapid sea-level rise, but also elevated sediment pore pressures and gas hydrate destabilisation (see Lee, 2009 for more comprehensive discussion of these factors).

Projected future climate changes and the potential for a geospheric response

Since the LGM, about 20 ka BP, global average temperatures have risen by around 6°C, with a rise of close to 0.8°C occurring in the last 100 years (Figure 1.1). Without a major change in energy policy, GHG emissions are projected to rise substantially, increasing the global mean temperature by between 1.6°C and 6.9°C relative to pre-industrial times, driving long-term rises in temperature and global sea level and possible changes to the Atlantic Meridional Overturning Circulation (MOC) (IPCC 2007) (Figure 1.2). As noted by Liggins et al. (2010), physical inertia in the climate system ensures that the full effect of past anthropogenic forcing remains to be realised. Considering that global GHG emissions are still on an upward trend, and with no binding agreement in place to reduce this, it is highly likely that warming will result in regional temperature rises of at least 2°C above the pre-industrial period. Under the high-end Intergovernmental Panel on Climate Change (IPCC) SRES (Special Report on Emissions Scenarios) emissions scenario (A1F1), Betts et al. (2011) show that global average temperatures are likely to reach 4°C relative to pre-industrial times by the 2070s, and perhaps as early as 2060. Under the lowest of the main emissions scenarios (B1), the central estimate of warming is projected to be 2.3°C relative to the pre-industrial period. These projections indicate that the current episode of GHG-driven warming is exceptional. As observed in the IPCC AR4 (IPCC 2007), if temperatures rise about 5°C by 2100, the Earth will have experienced approximately the same amount of warming in a few centuries as it did over several thousand years after the LGM. This rate of warming is not matched by any comparable global average temperature rise in the last 50 million years. Furthermore, high latitudes, where most residual ice now resides, are expected to warm even more rapidly. Christensen et al. (2007), for example, propose that under the A1B scenario the Arctic (north of 60° N) could warm by between 2.8°C and 7.8°C by 2080–2099 (relative to 1980–1999). 'High-end' projections under the A2 scenario suggest that surface temperatures across much of the Arctic could increase by 15°C by the 2090s (Sanderson et al., 2011).

Figure 1.1 Records of Northern Hemisphere temperature variation during the last 1300 years. (a) Annual mean instrumental temperature records; (b) reconstructions using multiple climate proxy records; and (c) overlap of the published multi-decadal timescale uncertainty ranges of temperature reconstructions. The HadCRUT2v instrumental temperature record is shown in black. All series have been smoothed with a gaussian-weighted filter to remove fluctuations on timescales <30 years; smoothed values are obtained up to both ends of each record by extending the records with the mean of the adjacent existing values. All temperatures represent anomalies (°C) from the 1961 to 1990 mean. Reproduced from IPCC, 2007 Climate Change 2007: The Physical Science Basis. Contribution of Working Group 1 to the Fourth Assessment Report of the Intergovernmental Panel on Climate Change. Solomon et al. Cambridge University Press, UK & USA. USA Figure 6.10 with permission.

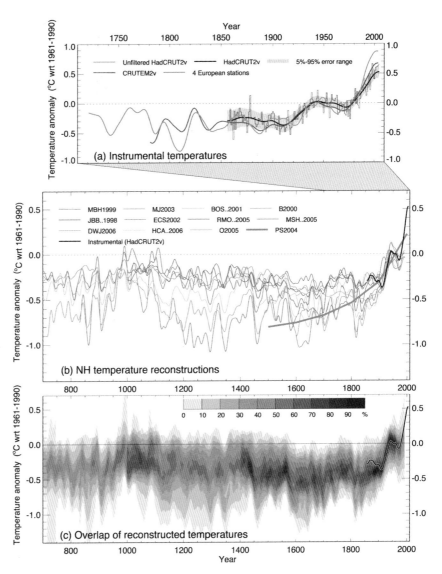

Global temperature rises are driving increases in ocean volume due to thermal expansion of seawater and via melting of glaciers, ice caps, and the Greenland and West Antarctic ice sheets. Dependent on the scenario, annual thermosteric sea-level rise by 2100 could lie between 1.9 ± 1.0 (B1 scenario) and 3.8 ± 1.3 (A2) (Meehl et al., 2007). Total global mean sea-level rise by the end of the century is projected in the IPCC AR4 (IPCC 2007) to be between 0.18 and 0.59 m. Even the high end of these projections may, however, be an underestimate. Rahmstorf (2007), for example, argues for a rise of 0.5–1.4 m by the end of the century, whereas Pfeffer et al. (2008) estimate an upper bound of 2 m by 2100.

Figure 1.2 Projected changes in (a) atmospheric CO_2, (b) global mean surface warming, (c) sea-level rise from thermal expansion and (d) Atlantic Meridional Overturning Circulation (MOC) calculated by eight Earth system models of intermediate complexity (EMICs) for the SRES A1B scenario and stable radiative forcing after 2100, showing long-term commitment after stabilisation. Coloured lines are results from EMICs, grey lines indicate AOGCM results where available for comparison. Anomalies in (b) and (c) are given relative to the year (2000). Vertical bars indicate ±2 standard deviation uncertainties due to ocean parameter perturbations in the C-GOLDSTEIN model. The MOC shuts down in the BERN2.5CC model, leading to an additional contribution to sea-level rise. Individual EMICs treat the effect from non-CO_2 greenhouse gases and the direct and indirect aerosol effects on radiative forcing differently. Despite similar atmospheric CO_2 concentrations, radiative forcing among EMICs can thus differ within the uncertainty ranges currently available for present-day radiative forcing. (Reproduced from IPCC, 2007 Climate Change 2007: The Physical Science Basis. Contribution of Working Group 1 to the Fourth Assessment Report of the Intergovernmental Panel on Climate Change. Solomon et al, Cambridge University Press, UK and USA. USA Figure 10.34.)

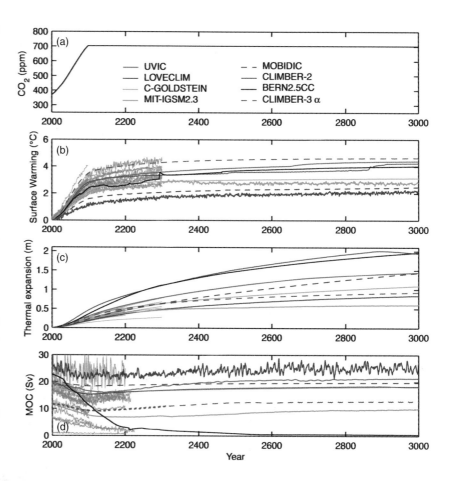

Projected changes in other climate quantities are also relevant in relation to influencing potentially hazardous crustal and surface processes, most notably variations in patterns of precipitation. Under the A1B scenario, for example, the Arctic (north of 60° N) is projected to see a 28% increase in precipitation by 2080–2099 relative to 1980–1999 (Christensen et al., 2007), and a similar rise is expected for Alaska and Kamchatka (Liggins et al., 2010). Ocean warming may also be important, and while this is projected to progress more slowly than over land masses (Meehl et al., 2007), it is expected to be greatest at high latitudes, where it may play a role in accelerating ice wastage and in contributing towards gas hydrate disassociation.

Projected rising temperatures and sea levels, and changes in precipitation, are capable of initiating load changes and elevated pore-water pressures that exceed levels that have been shown to drive a range of geological and geomorphological processes that have hazardous potential. Small ice masses are already experiencing serious wastage, with surging and thinning of some glaciers resulting in vertical mass reduction of tens to hundreds of metres (Doser et al., 2007), leading to load pressure declines on basement rocks of ≥0.5 MPa (Sauber et al., 2000). Comparable load pressure falls may be expected in relation to the Greenland and West

Antarctic ice sheets if increased melting accelerates ice loss and glaciers surge and thin. The projected 0.18- to 0.59-m rise in sea level by the end of the century (IPCC 2007) would result in an increased load on the crust of 1.8–5.9 kPa. For a rise of up to 1.4 m (Rahmstorf, 2007), the load change rises to 14 kPa, and to 20 kPa for the upper bound 2.0 m rise of Pfeffer et al. (2008). Recently (1993–2003), annual sea level rise has been on the order of 2.4–3.8 mm/year (IPCC 2007), which translates broadly to an increased load pressure of 0.1 kPa every 3 years.

Although the load pressure changes associated with GHG-driven ice wastage and sea-level rise are generally small, in terms of both absolute values and rates, increasing evidence supports the view that they may be sufficient to trigger a geospheric response. Mounting evidence makes a convincing case for the modulation or triggering of seismic, volcanic and landslide activity as a consequence of small changes in environmental parameters such as solid Earth and ocean tides, and atmospheric temperature and pressure, as well as in response to specific geophysical events such as typhoons or torrential precipitation (Table 1.1).

Table 1.1 Examples of environmental drivers of seismicity, landslide slip and eruptive activity described in the literature, together with associated driving pressures

Process	Environmental driver	Driving pressures	References
Contemporary observations			
Seismicity (south-east Germany)	Precipitation	<1 kPa	Hainzl et al. (2006)
Seismicity (San Andreas Fault, California)	Precipitation	About 2 kPa	Christiansen et al. (2007)
Seismic tremor/slow fault slip (Japan/Cascadia subduction zone)	Ocean loading	15 kPa	Rubinstein et al. (2008)
Seismicity (Japan)	Snow loading	'A few kPa'	Heki (2003)
Daily slip (Slumgullion landslide, Colorado)	Atmospheric pressure variation	<1 kPa	Schulz et al. (2009)
Eruptive activity (Pavlof, Alaska)	Ocean loading	2 kPa	McNutt & Beavan (1987); McNutt (1999)
Seismicity	Snow unloading and ground-water recharge	>5 kPa	Christiansen et al. (2005)
Seismicity (Juan de Fuca Ridge, Pacific North East)	Ocean loading	30–40 kPa	Wilcock (2001)
Seismicity (East Pacific Rise)	Ocean unloading	1–2 kPa	Guillas et al. (2010)
Climate-change impacts and projections			
Glacier ice-mass loss (Alaska)	Ice unloading	Up to 2 MPa	Sauber & Molnia (2004)
Current 1993–2003 rate of sea level rise	Ocean loading	0.1 kPa every 3 years	IPCC (2007)
Global average sea level rise (by 2100)	Ocean loading	1.8–5.9 kPa	IPCC (2007)
Global average sea level rise (by 2100)	Ocean loading	Up to 14 kPa	Rahmstorf (2007)

Pressure changes related to twentieth-century glacier ice-mass loss in Alaska and to future sea-level rise scenarios are included for comparison.

Based on observations of seismicity from south-east Germany, Hainzl et al. (2006) demonstrate that the crust can sometimes be so close to failure that even tiny (<1 kPa) pore-pressure variations associated with precipitation can trigger earthquakes in the top few kilometres. Christiansen et al. (2007) propose that modulation of seismicity on a creeping section of the San Andreas Fault in the vicinity of Parkfield is linked to the hydrological cycle. The authors suggest that fracturing of critically stressed rocks occurs either as a consequence of pore-pressure diffusion or crustal loading/unloading, and note that hydrologically induced stress perturbations of about 2 kPa may be sufficient to trigger earthquakes on the fault. In volcanic settings, Mastin (1994) relates the violent venting of volcanic gases at Mount St Helen's between 1989 and 1991 to slope instability or accelerated growth of cooling fractures within the lava dome after rainstorms, whereas Matthews et al. (2002) link episodes of intense tropical rainfall with collapses of the Soufriere Hills' lava dome on Montserrat (Caribbean).

Liu et al. (2009) show that slow earthquakes in eastern Taiwan are triggered by stress changes of approximately 2 kPa on faults at depth, associated with atmospheric pressure falls caused by passing tropical cyclones. Rubinstein et al. (2008) have been able to correlate episodes of slow fault slip and accompanying seismic tremor at subduction zones in Cascadia (Pacific North West) and Japan with the rise and fall of ocean tides, which involve peak-to-peak load pressure changes (for Cascadia) of 15 kPa. Heki (2003) demonstrates that snow load seasonally influences the seismicity of Japan through increasing compression on active faults and reducing the Coulomb failure stress by a few kilopascals. Schulz et al. (2009) show that diurnal tidal variations in atmospheric pressure amounting to <1 kPa modulate daily slip on the Slumgullion landslide in south-west Colorado. For volcanoes, Earth tides (Johnston & Mauk, 1972; Hamilton, 1973; Sparks, 1981) and other changing external factors, such as barometric pressure (Neuberg, 2000) or ocean loading (McNutt & Beavan, 1987; McNutt, 1999), have been proposed as having roles in forcing or modulating activity. McNutt and Beavan (1987) and McNutt (1999), for example, suggest that eruptions of the Pavlof (Alaska) volcano, from the early 1970s to the late 1990s, were modulated by ocean loading involving yearly, non-tidal, variations in local sea level as small as 20 cm, which translates to a load pressure change on the crust of 2 kPa. On a geographically broader scale, Bettinelli et al. (2008) explain seasonal variations in the seismicity of the Himalayas in terms of changes in surface hydrology, whereas Christiansen et al. (2005) link shallow (<3 km) seasonal seismicity at large calderas and stratovolcanoes across the western USA with stress changes of >5 kPa associated with snow unloading and ground-water recharge. Guillas et al. (2010) argue for reduced sea level in the eastern Pacific before the development of El Niño conditions, and approximating to a 1- to 2-kPa sea-bed load reduction, triggering increased levels of seismicity on the East Pacific Rise. At the global level, Mason et al. (2004) present evidence from the last 300 years in support of a seasonal signal in volcanic activity. This they attribute to fluctuations across a range of environmental conditions associated with the deformation of the Earth in response to the annual hydrological cycle, including reduced sea levels, millimetre-scale motion of the Earth's crust and falls in regional atmospheric pressure. Although far from established, Podol-skiy (2008) makes a case for an increase in global seismicity in recent decades, citing climate change as one potential driver.

Climate forcing of hazards in the geosphere

In light of the above, the potential is addressed for enhanced responses to changing environmental conditions so as to increase the risk of geological and geomorphological hazards in a GHG-warmed world. In the context of rising atmospheric and ocean temperatures, ice-mass wastage and changing patterns of precipitation, possible implications are examined for high-latitude regions, ocean basins and margins, mountainous terrain and volcanic landscapes (Table 1.2 and Figure 1.3).

Table 1.2 Potential geological and geomorphological hazards in the context of projected future climate changes

Potential hazard	Mechanism/potential relationship with climate change	Relevant climate drivers	Environmental settings	References
Subaerial landslides and debris flows	Permafrost thaw; pore-water pressurisation; intense rainfall destabilising regolith	Temperature rise; ice-mass loss; intense precipitation	Mountainous terrain; volcanic landscapes	Deeming et al. (2010); Huggel et al. (2010); Keiler et al. (2010); Tuffen (2009)
Glacial outburst floods (GLOFs)	Glacier retreat; accumulation of meltwater in pro-glacial lakes	Temperature rise; ice-mass loss	High latitudes; mountainous terrain; glaciated volcanic landscapes	Keiler et al. (2010); Tuffen (2009)
Earthquakes	Ice-sheet and glacier wastage; ocean island and ocean margin loading due to sea-level rise	Temperature rise; ice-mass loss; ocean volume increase	High latitudes; glaciated terrain at mid-to-low latitudes; ocean basins and margins	Hampel et al. (2010); Guillas et al. (2010)
Volcanic activity	Unloading due to ice-sheet and glacier wastage; loading due to sea-level rise; pore-water pressurisation; intense rainfall destabilising regolith	Temperature rise; ice-mass loss; intense precipitation; ocean volume increase	Volcanic landscapes at all latitudes	Deeming et al. (2010); Sigmundsson et al. (2010); Tuffen (2009)
Tsunamis	Submarine and sub-aerial slope failures and volcano lateral collapses; gas-hydrate breakdown; ocean load-related earthquakes; ice quakes	Ocean temperature rise; ocean volume increase; intense precipitation	Ocean basins and margins	Day & Maslin (2010); Maslin et al. (2010); Dunkley Jones et al. (2010); Tappin (2010)

Columns show responsible mechanisms and relationship to climate change, relevant climate drivers, and most susceptible environmental settings.

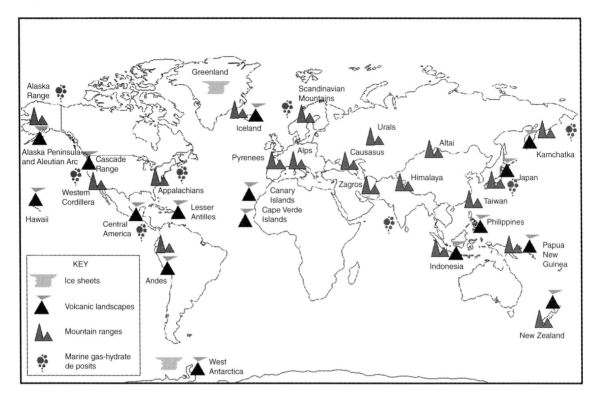

Figure 1.3 Notable high-latitude ice sheets, areas of mountainous terrain, active volcanoes and gas-hydrate concentrations susceptible to the impacts of rising temperatures, ice-mass loss, increasing ocean volume and higher levels of precipitation due to anthropogenic climate change. Consequent potential geological and geomorphological hazards are summarised in Table 1.2.

High latitude regions

The effects of anthropogenic climate change will be greater and more rapidly apparent at high latitudes. The potential for triggering geological and geomorphological hazards is also elevated, most notably as ice-mass is lost from the great ice sheets, smaller ice caps and individual glaciers and ice fields. In Greenland and Antarctica, isostatic rebound as ice mass is reduced may result in increased seismicity (Turpeinen et al., 2008; Hampel et al., 2010), which may in turn trigger submarine landslides that could be tsunamigenic (Tappin, 2010). In Iceland, Kamchatka and Alaska, melting of ice in volcanically and tectonically active terrains may herald a rise in the frequency of volcanic activity (Pagli & Sigmundsson, 2008; Sigmundsson et al., 2010) and earthquakes (Sauber & Molnia, 2004; Sauber & Ruppert, 2008).

During post-glacial times, the melting of major continental ice sheets, such as the Laurentian and Fennoscandian, triggered intense seismic activity associated with isostatic rebound of the crust (e.g. Wu, 1999; Wu et al., 1999; Muir-Wood, 2000). For a 1-km ice load, the rebound may have totalled hundred of metres, with associated stresses totalling several megapascals, comparable with plate-driving stresses (Stewart et al., 2000). Ice thicknesses at Greenland and Antarctica currently exceed 3 km, providing potential for an ultimate rebound of more than 1 km should all the ice melt. Although this is an extreme scenario, smaller-scale ice loss may also trigger a potentially hazardous seismic response as high-latitude temperatures climb. Turpeinen et al. (2008) use finite-element modelling in support of the idea (e.g. Johnston, 1987) that current low levels of seismicity in regions such as Greenland and Antarctica are a consequence of ice-sheet load, and speculate that future deglaciation of these regions may result in a pronounced increase in seismicity.

Song (2009) highlights an additional potential threat from 'ice quakes' (Ekstrom et al., 2003) associated with a future break-up of the Greenland and West Antarctic ice sheets. The author calculates that impulse energies from glacial earthquakes in both Greenland and West Antarctica are capable of generating significantly more powerful tsunamis than submarine earthquakes of similar magnitude, and notes that this may pose a threat to high-latitude regions such as Chile, New Zealand and Newfoundland (Canada).

Maslin et al. (2010) note that isostatic rebound of Greenland and Antarctica may also involve the adjacent continental slope, thereby reducing pressure on any gas hydrates contained in slope sediments, raising the chances of hydrate break-down and the related threat of tsunamigenic submarine landslides. Notwithstanding a gas hydrate trigger, increased numbers of earthquakes may themselves be capable of triggering landsliding of piles of glacial sediment accumulated around the margins of the Greenland and Antarctic land masses. Such a mechanism has been shown to be important in triggering major submarine landslides in the post-glacial period, best known of which is the Storegga Slide, formed off the coast of Norway 8100 ± 250 years BP (Lee, 2009; Tappin 2010). With a volume of between 2500 and 3500 km^3, the Storegga Slide is one of the world's largest landslides, and is widely regarded to have been triggered by a strong earthquake associated with the isostatic rebound of Fennoscandia (Bryn et al., 2005). From the perspective of future hazard potential, it is noteworthy that the Storegga event generated a major tsunami (Tappin, 2010) with tsunami deposits identified at heights above estimated contemporary sea level of 10–12 m on the Norwegian coast, more than 20 m in the Shetland Islands and 3–6 m on the coast of north-east Scotland (Bondevik et al., 2005). A range of mechanisms capable of being driven by anthropogenic climate change is presented by Tappin (2010) as having the potential to trigger the formation of submarine mass failures, including earthquakes and cyclic loading due to storms or tides. Pore-fluid pressurization and gas hydrate instability are held up by the author as possible contributors to slope destabilization, but are thought unlikely to play a triggering role in the failure process.

Projected temperature rises for high latitudes will affect smaller ice caps, ice fields and glaciers more rapidly than the major ice sheets. Of these, the Vatnajökull ice cap (area about 8000 km^2) (Figure 1.4) in Iceland presents the greatest threat

Figure 1.4 Iceland's Vatnajökull ice cap captured by the moderate resolution imaging spectroradiometer (MODIS) on the Terra satellite in November 2004. Pagli and Sigmundsson (2008) predict that reduced ice load due to future climate change will result in an additional 1.4 km^3 of melt being produced in the underlying mantle every century – comparable to an eruption equivalent in size to the 1996 Gjálp eruption beneath Vatnajökull, every 30 years. (Reproduced courtesy of NASA.)

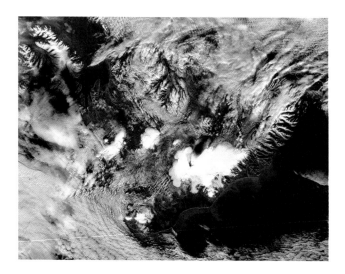

in relation to the resultant triggering of a potentially hazardous geospheric response. As reported in Pagli et al. (2007), mass-balance measurements show that the ice cap is thinning at a current rate of about 0.5 m/year, and lost about 435 km^3 between 1890 and 2003 – about 10% of the total volume. In post-glacial times, the reduction in vertical load associated with an annual ice-thinning rate of about 2 m, across a much larger ice cap (180 km diameter compared with 50 km today) (Pagli et al., 2007), was instrumental in triggering a significant increase in the frequency of volcanic eruptions. Furthermore, Jull and McKenzie (1996) showed that removal of the countrywide ice load reduced pressure on the underlying mantle to such a degree that melt production jumped by a factor of 30. The smaller size of the current Vatnajökull ice cap, and slower thinning rate support a more measured reaction from the crust and mantle to contemporary warming. Nevertheless, Pagli and Sigmundsson (2008) predict, on the basis of finite-element modelling, that the reduced ice load will result in an additional 1.4 km^3 of melt being produced in the underlying mantle every century – comparable to an eruption equivalent in size to the 1996 Gjálp eruption beneath Vatnajökull, every 30 years. The authors also speculate that stress changes in the crust, in response to ice-mass loss, may already be contributing towards elevated levels of seismicity with 'unusual' focal mechanisms in the north west of the region. From a future seismic risk viewpoint, it is worth observing that Hampel et al. (2007) demonstrate a clear seismic response to deglaciation of the 16,500 km^2 area Yellowstone ice cap (north-west USA).

Although the direct effects of increased levels of volcanic eruptions in Iceland may impinge on relatively small populations, large events that are explosive or release significant volumes of sulphur gas may have far wider effects. The Laki (Lakagigar) eruption in 1783, for example, generated a tropospheric sulphurous haze that spread south-eastwards over Europe. This resulted in extremely poor

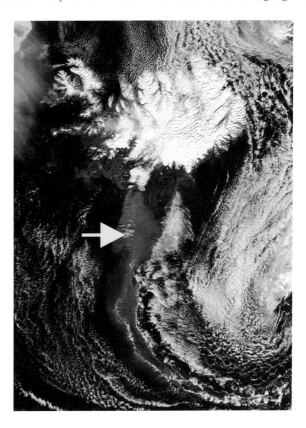

Figure 1.5 The ash plume from Iceland's erupting Eyjafjallajökull volcano on 19 April 2010. Ash in the atmosphere during April and May 2010 resulted in major disruption to air traffic across the UK and Europe. Any future rise in eruptive activity due to the loss of the Vatnajökull ice cap has the potential to cause comparable problems. (Reproduced courtesy of NASA.)

air quality and anomalously high temperatures during the summer months, and dramatically reduced winter temperatures, and led to significant excess deaths in the UK and continental Europe (e.g. Grattan et al., 2005). Furthermore, the 1783 eruption lasted for 6 months; a similar event today would have the potential to cause major disruption to the north polar air transport routes. In this respect, and given the severe impact of the 2010 Eyjafjallajökull eruption (Figure 1.5) on air traffic across the UK and Europe, any potential increase in volcanic activity in Iceland would clearly be unwelcome. The future picture may not, however, be all bad. Sigmundsson et al. (in Chapter 5 of this book), for example, propose that a proportion of the new magma arising from future ice mass loss across Iceland will be 'captured' within the crust rather than erupted at the surface. For the Katla volcano, the recent unrest of which has raised concerns over another Icelandic eruption with air-traffic disrupting potential, such behaviour might actually work to reduce the likelihood of eruption.

Elevated levels of either volcanic or seismic activity on Iceland may also result in the triggering of secondary hazards, most notably glacial floods (*jökulhlaups*) through rapid melting of ice during subglacial eruptions (e.g. Alho et al., 2005) and landslides or snow avalanches caused by ground accelerations during earth-quakes (Saemundsson et al., 2003). *Jökulhlaups* currently pose a periodic threat to settlements and the main coast road, immediately south of Vatnajökull, which could reasonably be expected to increase should the incidence of volcanic activity rise as predicted. *Jökulhlaups* also occur in Greenland, where they may become more common as the climate warms and present a threat to communities and infrastructure (Mernild et al., 2008).

Outside Iceland, at high latitudes, ice-mass wastage is expected to promote a comparable response, leading to increased levels of seismic and volcanic activity. Glacier mass fluctuations in south-central Alaska have been charged with modulating the recent seismic record, and even implicated in the triggering of the 1979 magnitude 7.2 St Elias earthquake (Sauber et al. 2000; Sauber & Molnia, 2004; Sauber & Ruppert, 2008). Rapid ice-mass loss at the many glaciated volcanoes in Alaska and Kamchatka, driven by surface temperature rises that could exceed 15°C by 2100 (Sanderson et al. 2011), has the potential to promote eruptions, either as a consequence of reduced load pressures on magma reservoirs or through increased opportunity for magma–water interaction. In addition, the potential for edifice lateral collapse could be enhanced as a consequence of elevated pore-water pressures arising from meltwater and a significant predicted rise in precipitation (Capra, 2006; Deeming et al., 2010). The potential for both volcanic and non-volcanic landslides may also be promoted by increased availability of water leading to slope destabilisation and failure due to slow cracking, held to be a contributory factor in the formation of *stürtzstroms* (giant, rapidly moving landslides) (e.g. Kilburn & Petley, 2003).

An increase in climate change-driven, non-volcanic, mass movements at high latitudes may already be apparent. Huggel (2009) and Huggel et al. (2008, 2010) speculate that rising temperatures may be behind a recent series of major rock and ice avalanches, with volumes in excess of $10^6\,\mathrm{m}^3$ in Alaska. With atmospheric warming in the state occurring at a rate of 0.03–0.05°C per year (Symon et al. 2005), a continuing trend towards the more frequent formation of large landslides is probable. Generally, a combination of melting permafrost and rising rock temperatures can be expected to increase the incidence of instability development and large-scale mass movement across all regions of elevated terrain at high latitudes.

Ocean basins and margins

Warmer oceans have the potential to influence the stability of gas hydrate deposits in marine sediments and, as a consequence, destabilise submarine slopes. Increased ocean mass, reflected in rising sea levels, may elicit volcanic and seismic responses in coastal and island settings, which, in turn may promote the formation of sub-aerial, volcanic landslides, submarine landslides and tsunamis.

Potentially sensitive to rising ocean temperatures is the stability of gas hydrate deposits contained in marine sediments in many parts of the world (e.g. Henriet & Mienert 1998 and papers therein; Bice & Marotzke, 2002; Day & Maslin, 2010; Maslin et al., 2010). These present a number of prospective hazards, most notably through the release of enormous volumes of methane as a consequence of desta-bilisation, but also through triggering large submarine sediment slides which may in turn generate tsunamis. Gas hydrates are ice-like solids comprising a mixture of water and gas (normally methane), the stability of which is strongly dependent on pressure–temperature conditions. They may become disassociated and release methane gas if ambient temperatures are increased or the pressure reduced. Best estimates of the amount of carbon stored in marine hydrate deposits ranges from 1000 GtC to 3000 GtC (giga-tonnes of carbon), which would have a major impact on planetary warming should all or part of it be released into the atmosphere. In this regard, gas hydrate release on a major scale is believed to have occurred as a consequence of rapid warming during the Palaeogene (the PETM: Palaeocene–Eocene thermal maximum) (Dunkley Jones et al., 2010). Looking ahead, however, the potential for widespread marine hydrate breakdown as a consequence of anthropogenic climate change remains a matter for debate. Although rising ocean temperatures will tend towards destabilising hydrates, increasing load pressures, as a result of rising sea levels, will act in the opposite sense. Maslin et al. (2010) note that even if marine hydrate disassociation is triggered on a large scale it may be that all or much of the methane released will not reach the atmosphere, because (1) thermal penetration of marine sediments to the gas–hydrate interface could be sufficiently tardy to allow a new equilibrium to become established without significant gas release and (2) a fraction of any gas released may be oxidised in the ocean.

Gas–hydrate disassociation has been considered by some (e.g. Kayen & Lee, 1992; Maslin et al. 1998, 2004; Sultan et al., 2004; Owen et al., 2007; Grozic, 2009) as a potential trigger for major submarine sediment slides, through the release of free gas leading to high excess pore pressures and reduction of sediment shear strength. Tappin (2010) cautions, however, that evidence for such a link is largely circumstantial. In addition, from a hazard perspective, current knowledge of the physicochemical properties of hydrates seems to indicate that they are not able to instantaneously dissociate. Lee (2009) points out that few studies demonstrate an unambiguous link between hydrate disassociation and the triggering of a subma-rine landslide, and also notes that the mechanism would seem to be most likely to prevail during glacial periods when sea levels, and consequently load pressure on marine sediments, are reduced.

Modelling studies (Wallmann et al., 1988; Nakada & Yokose, 1992) have dem-onstrated that sea-level changes, about 100 m, are capable of triggering or modu-lating volcanic and tectonic activity. More specifically, Luttrell and Sandwell (2010) have shown that lithospheric flexure due to ocean loading caused by post-glacial sea-level rise was sufficient to 'unclamp' coastal transform faults, such as the San Andreas (California, USA), north Anatolian (Turkey) and Alpine (South Island, New Zealand), thereby promoting failure through the reduction of normal stress. They also demonstrated that, for plate boundary faults at subduction zones, reduced sea levels favoured offshore fault rupture, whereas elevated sea levels promoted landward rupture at greater depths. In a similar vein, Brothers et al. (2011) implicate a combination of loading and pore-pressure increases, associated with periodic Late Holocene flooding of southern California's Salton Sea by the Colorado River, in triggering earthquakes on the southern Andreas Fault. Quidelleur et al. (2008) speculate that erosion and pore-pressure changes associated with rapidly rising sea levels at glacial–interglacial transitions may play a role in major lateral collapse of ocean island volcanoes. McGuire et al. (1997) have linked the incidence of volcanic activity in the Mediterranean region to the rate of sea-level change over the last 80 ka. They note, in particular, a significant increase in intensity of volcanism during times of very rapid Holocene sea-level rise, between 17 and 6 ka BP, broadly coincident with the catastrophic rise events of Blanchon and Shaw (1995), which saw centennial global eustatic sea level rise rates of approximately 5 m. Perhaps most significantly, in relation to the impact of future sea-level rise on volcanic systems, McNutt and Beavan (1987) attribute the modulation of eruptive activity at Pavlof volcano (Alaska) to the development of compressive strain beneath the volcano when adjacent sea levels are elevated, with magma being preferentially squeezed out under these conditions. McGuire et al. (1997) describe finite-element results demonstrating that sea-level rise adjacent to a volcanic body reduces compressive stress within the edifice. They suggest that, during times of rapid sea-level rise, this may result in the triggering of eruptions at 'charged' volcanoes, whereat magma is stored at depths of ≤5 km. The findings of McNutt and Beavan (1987), McNutt (1999) and McGuire et al. (1997) are compatible with ocean loading resulting in a bending moment in the crust at ocean margins, leading to reduced compression at higher levels and increased compression at depth. Progressive bending at ocean margins, as ocean mass increases at the expense of melting glaciers and ice sheets, has the potential to trigger eruptions at 'primed' volcanoes. The volcanic response is likely to occur across a range of timescales dependent on the nature of individual 'plumbing' systems and the availability of magma; the cumulative effect, however, would most probably be an increase in the frequency of eruptions in areas close to the marine environment. Clustering of volcanic eruptions in response to external forcing is addressed in Mason et al. (2004), in the context of the recognized seasonality of eruptions, with a mathematical treatment provided by Jupp et al. (2004).

The numbers of volcanoes potentially susceptible to crustal strain changes associated with future sea-level rise is large. Of the 550 or so volcanoes at which eruptions have been historically documented (Global Volcanism Program, 2010), McGuire et al. (1997) determine that 57% form islands or are coastally located, whereas a further 38% are found within 250 km of a coastline. When or if rising sea levels will result in a recognisable signal in global volcanism remains a matter for debate. It is notable, however, that a 2-m rise by 2100 would result in a cumulative load pressure on the sea floor (20 kPa) that is an order of magnitude greater than that held responsible by McNutt and Beavan (1987) and McNutt (1999) for modulation of Pavlof's eruptive behaviour.

Nakada and Yokose (1992) have demonstrated theoretically that the large (approximately 100 m) sea-level changes associated with glaciation–deglaciation cycles, and resulting in cumulative ocean loading/unloading of about 1 MPa, are capable of triggering tectonic and volcanic activity, particularly at island arcs where the lithosphere is relatively thin. At the other extreme, Rubinstein et al. (2008) have correlated episodes of slow fault slip and accompanying seismic tremor at subduction zones in Cascadia (Pacific North West) and Japan, with the rise and fall of ocean tides, involving peak-to-peak load pressure changes of just 15 kPa. Wilcock (2001) provides convincing evidence for micro-earthquakes on the Endeavour segment of the Juan de Fuca Ridge (Pacific North East) being triggered by the loading effect of ocean tides, which result in vertical stress variations of 30–40 kPa, whereas Guillas et al. (2010) propose that ocean load pressure fluctuations as small as a few kilopascals modulate microseismicity on the East Pacific Rise.

Provided that the crust is sufficiently permeable, increased water load is capable of raising pore-fluid pressure in active fault zones, thereby modulating or triggering seismicity through a reduction in the frictional resistance to fault slip. This mechanism has long been recognised in relation to the filling of reservoirs (Talwani, 1997), and has been held responsible for a lethal earthquake that followed the Koyna reservoir in India in the early 1960s (Simpson et al., 1988). Pore-pressure changes in oceanic or submerged continental crust arising from a 1- to 2-m global sea-level rise this century would be orders of magnitude smaller than those associated with filling of reservoirs. Nevertheless, they must be considered as having the potential to trigger earthquakes on faults that are already critically stressed and, therefore, close to rupture. This in turn provides a means for generating submarine landslides and/or tsunamis, both of which carry threats to coastal communities.

Mountainous terrain

As for elevated topography at high latitudes, the principal impact of climate change in mountainous terrain is expected to be an increase in slope instability, the formation of ice and/or rock avalanches and debris flows and, in the Himalayas in particular, a rise in the number and size of glacial outburst floods. A primary driver is rising rock temperatures and permafrost thaw, the latter being a critical mechanism via which climate is able to control slope stability and natural hazard potential in mountainous terrain (Gruber & Haeberli, 2007). As described in Liggins et al. (2010), the European and New Zealand Alpine ranges, the Pyrenees, Caucasus, Andes and Himalayas are all expected to experience rises in mean temperature, with extreme precipitation and temperature events increasing in both magnitude and frequency.

As mentioned previously, Huggel et al. (2008, 2010) speculate that a recent series of major rock and ice avalanches in Alaska may be a reflection of rising temperatures associated with climate change. These include an ice and rock avalanche with a volume of approximately $50 \times 10^6 \, m^3$ formed from the failure of the summit glacier on Mount Steller in 2005, which travelled 9 km. The authors also note a trend towards increasing slope instability in the Russian Caucasus, reflected in the collapse of part of the Dzhimarai-khokh mountain on to the Kolka glacier

Figure 1.6 On 20 September 2002, collapse of part of the Dzhimarai-Khokh peak on to the Kolka Glacier generated an avalanche of ice and debris that travelled 24 km, buried villages and took more than 100 lives. This image, acquired a week later, shows a long, dark-grey streak running upward through the centre of the scene, marking the position of a gorge now infilled with ice and debris from the avalanche. (Reproduced courtesy of NASA. Image by Robert Simmon and Jesse Allen, based on data from the NASA/GSFC/MITI/ERSDAC/JAROS, and U.S./Japan ASTER Science Team and MODIS Science Team.)

in 2002 (Figure 1.6). The resulting ice and rock avalanche entrained almost the entire glacier, accumulating a total volume of $100 \times 10^6 \, m^3$ travelling at a velocity of up to 80 m/s (Huggel et al., 2005). The avalanche transformed into a debris flow at lower altitudes, which led to the deaths of more than 100 people. Huggel et al. (2008) speculate that warming permafrost may have been implicated in the failure process. During the 20th century, permafrost has warmed by between 0.5 and 0.8°C in the upper tens of metres (Gruber et al., 2004). Permafrost thaw has also been blamed for anomalous rock-fall activity in the Alps during the 2003 heatwave (Gruber & Haeberli, 2007), and can reasonably be expected to drive increased slope destabilisation and failure in frozen terrain at many mountain ranges as the global average temperatures continue to rise. As noted by Keiler et al. (2010), mountain regions are particularly sensitive to climate change as a consequence of strong feedback mechanisms involving snow cover at high elevations and albedo and heat budgets, which act to amplify change. In the European Alps, for example, temperatures have already risen by twice the global average since the late nineteenth century (Keiler et al., 2010).

Increased slope instabilities in some mountainous regions are also likely to be a consequence of a projected rise in the incidence of extreme precipitation events and the large-scale melting of glaciers and ice fields. Episodes of exceptional rainfall across elevated terrain have been shown to be effective at mobilising rock and regoliths to form destructive and lethal debris flows and debris-laden floods. Notable recent events include: the 1999 alluvial fan debris flows, which claimed in excess of 30,000 lives in northern Venezuela (Swiss Re, 2000); the debris flows and landslides in Italy and Switzerland the following year, which resulted in 37 deaths (Swiss Re, 2001); and more than 200 mass movements associated with the major floods in the European Alps in 2005 (Keiler et al., 2010). Chiarle et al. (2007) recognise three types of event that lead to debris flows in the Alps, but which have general application to their formation in mountainous regions: (1) intense and prolonged rainfall, leading to the saturation and failure of debris accumulations; (2) short-duration rainstorms capable of destabilising a glacier drainage system; and (3) glacial lake outbursts or the melting of surface ice fields or buried ice during dry conditions. In a warmer world, all three scenarios are more likely in response to more extreme precipitation events and rising temperatures.

As temperatures rise and glaciers retreat, so a landscape is exposed that is more prone to disturbance from rising temperatures and extreme precipitation events. In the European Alps, for example, close to 50% of the area previously covered by ice has been uncovered since 1850 (Zemp et al., 2006). Keiler et al. (2010) highlight the increased threat from glacial lake floods and from landslides and debris flows originating at steep, water-saturated, slopes that were previously ice covered. In particular, outburst floods from pro-glacial lakes impounded behind ice barriers, terminal moraines or landslides pose an increasing and serious threat in many glaciated regions. Glacial lake outburst floods (GLOFs) have been historically documented from many mountainous regions, including the European Alps (Huggel et al. 2002), the Andes (Reynolds, 1992) and the Himalayas (Watanabe & Rothacher, 1996), and have claimed thousands of lives in the Andes and Himalayas alone (Clague, 2009).

GLOFs are a growing hazard in the Himalayas, presenting an increasing threat to communities and infrastructure in Bangladesh, Bhutan, China, India, Nepal and Pakistan. In Nepal, an outburst flood in 1985 resulted in five 5 and the destruction of a hydropower plant (Horstmann, 2004), whereas another in Bhutan in 1994 took 27 lives (Watanabe & Rothacher, 1996). Nepal appears to be particularly at risk, with 20 potentially dangerous glacial lakes identified, including Tsho Rolpa, the largest moraine-dammed pro-glacial lake in the region. The lake has grown sixfold since the 1950s and is fed by the Tradkarding Glacier, which is retreating at an annual rate of up to 100 m (Rana et al. 2000). A future GLOF from the lake threatens 10,000 people and significant infrastructure, including the 60-MW Khimti hydropower facility (Horstmann, 2004). GLOFs also present a problem elsewhere, including Switzerland where six new pro-glacial lakes were developed as a consequence of retreat of the Grubengletscher Glacier in Wallis Canton. One of these, drained catastrophically in 1968 and again in 1970, resulted in damaging debris flows (Haeberli et al. 2001). GLOFs are also becoming increasingly common in the Andes, where five outbursts occurred in northern Patagonia (Chile) in 2008 and 2009 (Dussaillant et al. 2009). In addition to promoting

GLOFs through glacier melting, future warming may also increase peak discharge of outbursts when they occur. Ng et al. (2007) have demonstrated that peak discharge of GLOFs originating at Merzbacher Lake in Kyrgyzstan is modulated by mean air temperature, which influences the rate of meltwater input to the lake as it drains and the lake-water temperature. The corollary, the authors note, is that future warming can be expected to promote 'higher-impact' GLOFs from pro-glacial lakes worldwide, by increasing the probability of warm weather during their formation.

The moraine dams that impound most pro-glacial lakes are particularly vulnerable to failure because they are composed of loose sediment and debris, and have characteristically steep slopes that are easily destabilised. Failure may be promoted by strong ground motion associated with earthquakes, and such a mechanism has been proposed to explain a 2.5 km^3 volume Late Pleistocene outburst of Lake Zurich (Strasser et al. 2008). Interestingly, the seismic threat to moraine dams may itself be increased as a consequence of the loss of ice mass promoting elevated levels of seismicity Seismic events related to ice-mass loss may also trigger avalanches or landslides capable of displacing pro-glacial lake waters, leading to tsunamis or splash waves, moraine dam overtopping or erosion. Seismogenic landslides may also form natural dams, impounding glacial meltwater and forming new lakes that provide potential sources for future GLOFs. Probably the best known such event occurred in the Gorno-Badakhshan province of Tajikistan in 1911, when an earthquake-triggered landslide blocked the Murghab River, forming Lake Sarez. The lake now has a volume of around 16 km^3, and presents a catastrophic threat to communities downstream should the rock dam become breached.

Volcanic landscapes

The disposition of tectonic plates ensures a non-random distribution of active volcanoes, with large concentrations at high latitudes (Alaska, Kamchatka, Iceland) and in the tropics (Indonesia, Papua New Guinea, the Philippines). High-latitude volcanoes are often glaciated and typically susceptible to deglaciation-related hazards encountered in mountainous terrains of non-volcanic origin. As addressed in detail in Tuffen (2009), a number of questions remain unanswered in relation to the impact of ice-mass loss from glaciated volcanoes. Where volcanoes are buried beneath a significant ice thickness, as at Iceland's Vatnajökull, large-scale melting due to climate change is convincingly predicted to lead to increased mantle melting and eruptive activity (Pagli & Sigmundsson 2008) as a consequence of unloading. In relation to smaller ice volumes capping individual edifices, most notable in Alaska, Kamchatka, the Andes, the Cascades and New Zealand, any unloading effect is, however, likely to be negligible. Tuffen (2009) observes that ice thinning of ≥100 m at volcanoes with ice cover in excess of 150 m, such as Sollipulli (Chile), may promote more explosive eruptions, with increased tephra hazards. Notwithstanding the influence of unloading, other hazardous consequences of ice melting are likely to be significant. These include GLOFs, ice avalanches and lateral collapse events. GLOFs may arise from the catastrophic release of meltwater from the overtopping or breaching of crater lakes or depressions, particularly where the water is impounded by weak pyroclastic material. Such an event occurred at Ruapehu (New Zealand) in 2007, when a tephra 'dam' holding back impounded water in a crater lake was breached, generating a debris-rich flood with an estimated volume of 1.4×10^6 m^3 (Carrivick et al., 2009b). As ice is progressively lost from glaciated volcanoes, however, the GLOF threat is likely to fall. The growing incidence of large volume ice avalanches at active volcanoes, most notably in Alaska (Huggel et al., 2008), can reasonably be expected to increase as ice masses weakened by rising air temperatures are further perturbed by geothermal heat (Huggel 2009). A general rise in edifice instability, leading to greater potential for lateral collapse, may also be promoted by ice-mass loss at glaciated volcanoes. As noted by Tuffen (2009), this may arise due to debuttressing and withdrawal of mechanical support previously supplied by the ice, or as a consequence of meltwater saturation raising pore-water pressures within shallow hydrothermal systems, which may – in turn – promote slip on established planes of weakness (Capra, 2006). Although a progressive loss of ice mass at glaciated volcanoes is likely to result in a reduction in the incidence of GLOF-related debris flows, higher-magnitude intense precipitation events across deglaciated and unglaciated volcanic landscapes may have the opposite effect, particularly at volcanoes in northern mid-latitudes and the southern tropics and sub-tropics, which are already becoming wetter as a consequence of anthropogenic climate change (Zhang et al., 2006). Mobilisation of poorly consolidated debris exposed by retreating ice may, in particular, provide a source for potentially destructive debris flows.

In unglaciated high-relief volcanic regions, including in the Caribbean, Europe, Indonesia, the Philippines and Japan, climate change may drive increased hazardous activity via modified precipitation patterns and, in particular, a rise in the frequency and magnitude of severe rainfall events. The main hazard ramifications are likely to be an increase in debris flow production and an elevated potential for the development of slope instability and landslides due to rises in pore-water pressure. Two recent incidents demonstrate the destructiveness and lethality of precipitation-triggered collapses and debris flows (also know as lahars) in volcanic landscapes. In 1998 at Sarno (Campania, Italy), sustained, extreme rainfall mobilised pyroclastic material derived from the Vesuvius and Campi Flegrei volcanic centres, leading to the formation of approximately 150 debris flows, which resulted in 160 fatalities and extensive damage to Sarno and neighbouring population centres (Brondi & Salvatori, 2003). Later the same year, torrential precipitation associated with Hurricane Mitch triggered a small flank collapse at Casita volcano (Nicaragua) (Scott et al. 2005). The resulting landslide rapidly transformed, first into a watery debris flood and then into a debris flow that inundated two towns and took 2500 lives. Looking ahead, many volcanoes provide a ready source of unconsolidated debris that can be rapidly transformed into potentially hazardous debris flows by extreme precipitation events that are predicted to become broadly more common as a consequence of planetary warming. Notably, as for Casita, many volcanoes occupying coastal, near-coastal or island locations in the tropics are particularly susceptible to torrential rainfall associated with tropical cyclones, which are projected by some to become both more powerful (e.g. Emanuel et al. 2008) and wetter (e.g. Knutson et al. 2008).

Conclusions

Evidence from the study of periods of exceptional climate change, together with contemporary observations, support a robust link between changing climatic conditions and a broad portfolio of potentially hazardous geological and geomorphological processes. Modelling studies and the projection of current trends argue for elevated levels of a range of geological and geomorphological hazards in a warmer world, whereas viable physical mechanisms capable of eliciting a geospheric reaction in response to small changes in environmental conditions are well established. Questions remain, however, most particularly in relation to the timescales over which a geospheric response may be detectable. Although increases in the incidence of climate change-driven, large-volume rock-and-ice avalanches (Huggel et al., 2008, 2010; Huggel, 2009), and the suggested modulation of seismicity in areas of large-scale ice wastage (Sauber & Molnia, 2004; Sauber & Ruppert, 2008), lead to speculation that climate change is already eliciting a crustal response, no increase in the global incidence of either volcanic activity or seismicity has been identified to date, nor has any change in the stability of submarine slopes been detected. It may be the case that modulation of potentially hazardous geological and geomorphological processes due to anthropogenic climate change proves to be too small a signal to extract from the background noise of 'normal' geophysical activity, at least in the short to medium term.

Furthermore, there are few constraints on the timing of a geospheric response, which may well lag significantly behind the warming trend. With respect to ice wastage in Greenland and Antarctica, Turpeinen et al. (2008) and Hampel et al. (2010) suggest that enhanced seismicity may be important on timescales as short as 10–100 years (Figure 1.7). A comparable timescale has been proposed (Gruber et al., 2004; Harris et al. 2009) in relation to the formation of large, deep-rooted, landslides after temperature rise and permafrost thaw in mountain regions. With respect to increased levels of melt production in the mantle beneath Iceland's Vatnajökull ice cap, Sigmundsson et al. (2010) speculate that it could take centuries or longer for fresh magma to reach the surface. There is also considerable uncertainty in relation to the linearity of possible responses, with different elements of the geosphere responding, for example, in a non-linear manner, with thresholds or tipping points resulting in step-like increases in frequency or scale.

In order to improve knowledge and reduce uncertainty, a programme of focused research is advocated so as to: better understand those mechanisms by which contemporary climate change may drive hazardous geological and geomor-

Figure 1.7 A 260 km² km fragment of Greenland's Petermann Glacier broke off in August 2010. Continued ice-mass wastage and consequent lithospheric rebound may ultimately result in an increase in seismic activity in the region.

phological activity; delineate those parts of the world that are most susceptible; and provide a more robust appreciation of potential impacts for society and infrastructure. More specifically, there is a need to (1) better establish potential correlations in the geological (particularly the Quaternary) record, between climate change and significant hazardous events such as large submarine landslides, major volcanic eruptions and ocean island collapses; (2) promote the application of modelling techniques so as to investigate, and more accurately portray, the influences of changing environmental conditions such as ice-mass wastage and ocean loading on potentially hazardous geophysical systems; and (3) encourage monitoring of specific locations perceived already to be demonstrating a climate-change response or that are deemed sensitive enough to do so in the short to medium term. For the first time, the IPCC explicitly addressed the impact of anthropogenic climate change on the geosphere, together with its manifold potentially hazardous consequences, in its 2011 report on climate change and extreme events (IPCC, 2011). It is strongly exhorted to continue to engage with the issue in future assessments and in particular in the IPCC's fifth assessment report, due to be published in 2013 and 2014. This seems to be particularly apposite given the outcome of the 2011 Durban climate conference, which effectively determined that little – if any – effective action to seriously curb global greenhouse gas emissions is likely before 2020.

Acknowledgements

Among a number of colleagues and others whose thoughts and suggestions have helped to improve this paper considerably from earlier versions, Dave Tappin is particularly acknowledged.

References

Alho, P., Russell, A. J., Carrivick, J. L. & Kayhko, J. (2005) Reconstruction of the largest Holocene jokulhlaup within Jokulsa a Fjollum, NE Iceland. *Quaternary Science Reviews* **24**, 2319–2334.

Anderson, D. C. (1974) Earthquakes and the rotation of the Earth. *Science* **186**, 49–50.

Andrews, J. T. (1973) The Wisconsin Laurentide Ice Sheet: dispersal centres, problems of rates of retreat, and climatic implications. *Arctic and Alpine Research* **5**, 185–199.

Bay, R., Bramall, N. & Price, P. (2004) Bipolar correlation of volcanism with millennial climate change. *Proceedings of the National Academy of Sciences of the United States of America* **101**, 6341–6345.

Beget, J. E. & Addison, J. A. (2007) Methane gas release from the Storegga submarine landslide linked to early-Holocene climate changes: a speculative hypothesis. *Holocene* **17**, 291–295.

Betts, R. A., Collins, M., Hemming, D. H., Jones, C. D., Lowe, J. A. & Sanderson, M. (2011) When could global warming reach 4°C? *Philosophical Transactions of the Royal Society of London A* **369**, 67–84.

Bettinelli, P., Avouacb, J.-P., Flouzata, M., et al. (2008) Seasonal variations of seismicity and geodetic strain in the Himalaya induced by surface hydrology. *Earth and Planetary Science Letters* **266**, 332–344.

Bice, K. L. & Marotzke, J. (2002) Could changing ocean circulation have destabilized methane hydrate at the Paleocene/Eocene boundary? *Paleoceanography* **17**, article number 1018.

Bigg, G., Clark, C. & Hughes, A. (2008) A last glacial ice sheet on the Pacific Russian coast and catastrophic change arising from coupled ice-volcanic interaction. *Earth and Planetary Science Letters* **265**, 559–570.

Blanchon, P. & Shaw, J. (1995) Reef drowning during the last glaciation: evidence for catastrophic sea-level rise and ice-sheet collapse. *Geology* **23**, 4–8.

Bondevik, S., Lovholt, F., Harvitz, C., Mangerud, J., Dawson, A. & Svendsen, J. I. (2005) The Storegga slide tsunami – comparing field observations with numerical simulations. *Marine and Petroleum Geology* **22**, 195–208.

Brondi, F. & Salvatori, L. (2003) The 5–6 May (1998) mudflows in Campania, Italy. In: Hervás, J. (ed.), *Lessons learned from landslide disasters in Europe*. European Commission Joint Research Centre, 5–16.

Brothers, D., Kilb, D., Luttrell, K., Driscoll, N. & Kent, G. (2011) Loading of the San Andreas fault by flood-induced rupture of faults beneath the Salton Sea. *Nature Geoscience* **4**, 486–492.

Bryn, P., Berg, K., Forsberg, C. F., Solheim, A. & Kvalstad, T. J. (2005) Explaining the Storegga slide. *Marine and Petroleum Geology* **22**, 11–19.

Capra, L. (2006) Abrupt climate changes as triggering mechanisms of massive volcanic collapses. *Journal of Volcanology and Geothermal Research* **155**, 329–333.

Carrivick, J. L., Russell, A. J., Rushmer, E. L., et al. (2009a) Geomorphological evidence towards a deglacial control on volcanism. *Earth Surface Processes and Landforms* **34**, 1164–1178.

Carrivick, J. L., Manville, V. & Cronin, S. J. (2009b) A fluid dynamics approach to modelling the 18th March (2007) lahar at Mt. Ruapehu, New Zealand. *Bulletin of Volcanology* **71**, 153–169.

Chappel, J. (1975) On possible relationships between upper Quaternary glaciations, geomagnetism and vulcanism. *Earth and Planetary Science Letters* **26**, 370–376.

Chiarle, M., Iannotti, S., Mortara, G. & Deline, P. (2007) Recent debris flow occurrences associated with glaciers in the Alps. *Global and Planetary Change* **56**, 123–136.

Christensen, J. H., Hewitson, B., Busuioc, A., et al. (2007) Regional climate predictions. In: Solomon, S., Qin, D., Manning, M., et al. (eds), *Climate Change: The Physical Science Basis*. Contribution of Working Group I to the Fourth Assessment Report of the Intergovernmental Panel on Climate Change. Cambridge: Cambridge University Press, pp 847–940.

Christiansen, L. B., Hurwitz, S., Saar, M. O., Ingebritsen, S. E. & Hsieh, P. A. (2005) Seasonal seismicity at western United States volcanic centers. *Earth and Planetary Science Letters* **240**, 307–321.

Christiansen, L. B., Hurwitz, S. & Ingebritsen, S. (2007) Annual modulation of seismicity along the San Andreas Fault near Parkfield, CA. *Geophysical Research Letters* **34**, L04306.

Clague, J. J. (2009) Climate change and slope instability. In: Sassa, K. & Canuti, P. (eds), *Landslides – Disaster Risk Reduction*. First World Landslide Forum, UN University, Tokyo, pp 557–572.

Costain, J. K. & Bollinger, G. A. (1996) Climatic changes, streamflow, and long-term forecasting of intraplate seismicity. *Journal of Dynamics* **22**, 97–117.

Day, S. J. & Maslin, M. A. (2010) Gas hydrates: a hazard for the 21st century? Report of an open-discussion session at the 3rd Johnston-Lavis Colloquium. *Philosophical Transactions of the Royal Society of London A* **368**, 2579–2584.

Day, S. J., McGuire, W. J., Elsworth, D., Carracedo, J-C. & Guillou, H. (1999) Do giant collapses of volcanoes tend to occur in warm, wet, interglacial periods and if so, why? Abstract volume, week B, IUGG XXII General Assembly, Birmingham, 174.

Day, S. J., Elsworth, D. & Maslin, M. (2000) A possible connection between sea-surface temperature variations, orographic rainfall patterns, water-table fluctuations and giant lateral collapses of the oceanic island volcanoes. Abstract volume, Western Pacific Geophysics meeting, Tokyo, Japan. WP251.

Davenport, C. A., Ringrose, P. S., Becker, A., Hancock, P. & Fenton, C. (1989) In: Gregersen, J. & Basham, P. W. (eds), *Earthquakes at North-Atlantic Passive Margins: Neotectonics and post-glacial rebound*. NATO Advanced Science Institutes Series C, Mathematical & Physical Sciences 266. NATO, 175–194.

Deeming, K. R., McGuire, W. J. & Harrop, P. (2010) Climate forcing of volcano lateral collapse: evidence from Mount Etna, Sicily. *Philosophical Transactions of the Royal Society of London A.*, **368**, 2559–2578.

Doser, D. I., Wiest, K. R. & Sauber, J. (2007) Seismicity of the Bering Glacier region and its relation to tectonic and glacial processes. *Tectonophysics* **439**, 119–127.

Dunkley Jones, T., Ridgwell, A., Lunt, D. J., Maslin, M. A., Schmidt, D. N., & Valdes, P. J. (2010) A Paleogene perspective on climate sensitivity and methane hydrate instability. *Philosophical Transactions of the Royal Society of London A* **368**, 2395–2416.

Dussaillant, A., Benito, G., Buytaert, W., Carling, P., Meier, C. & Espinoza, F. (2009) Repeated glacial-lake outburst floods in Patagonia: an increasing hazard? *Natural Hazards* **54**, 469–481.

Ekstrom, G., Nettles, M. & Tsai, V. C. (2003) Seasonality and increasing frequency of Greenland glacial earthquakes. *Science* **311**, 1756–1758.

Emanuel, K., Sundararajan, R. & Williams, J. (2008) Hurricanes and global warming. *Bulletin of the American Meteorological Society* **89**, 347–367.

Friele, P. A. & Clague, J. J. (2004) Large Holocene landslides from Pylon Peak, southwestern British Columbia. *Canadian Journal of Earth Sciences* **41**, 165–182.

Frisia, S., Borsato, A., Mangini, A., Spötl, C., Madonia, G. & Sauro, U. (2006) Holocene climate variability in Sicily from a discontinous stalagmite record and the Mesolithic to Neolithic transition. *Quaternary Research* **66**, 388–400.

Glazner, A. F., Manley, C. R., Marron, J. S. & Rojstaczer, S. (1999) Fire or ice: anticorrelation of volcanism and glaciation in California over the past 800,000 years. *Geophysical Research Letters* **26**, 1759–1762.

Global Volcanism Program (2010) *Worldwide Holocene Volcano and Eruption Information*. Washington DC: Smithsonian National Museum of Natural History. Available at: www.volcano.si.edu (accessed January 2010).

Grattan, J., Rabartin, R., Self, S. & Thordarson, T. (2005) Volcanic air pollution and mortality in France 1783–1784. *Comtes Rendus Geoscience* **337**, 641–651.

Grozic, J. L. H. (2009) Interplay between gas hydrates and submarine slope failure, In: Mosher, D. C. Shipp, R. C., Moscardelli, L., et al. (eds), *Submarine Mass Movements and Their Consequences*. Advances in Natural and Technological Hazards Research 28. Dordrecht: Springer, 11–30.

Gruber, S. & Haeberli, W. (2007) Permafrost in steep bedrock slopes and its temperature-related destabilization following climate change. *Journal of Geophysical Research* **112**.

Gruber, S., Hoelzle, M. & Haeberli, W. (2004) Permafrost thaw and destabilization of Alpine rock walls in the hot summer of 2003. *Geophysical Research Letters* **31**, doi: 10.1029/2004GL020051.

Guillas, S., Day, S. J. & McGuire, W. J. (2010) Statistical analysis of ENSO and sea-floor seismicity in the eastern tropical Pacific. *Philosophical Transactions of the Royal Society of London A* **368**, 2481–2500.

Haeberli, W., Kääb, A., Mühll, D. V. & Teysseire, P. (2001) Prevention of outburst floods from periglacial lakes at Grubengletscher, Valais, Swiss Alps. *Journal of Glaciology* **47**, 111–122.

Hainzl, S., Kraft, T., Wassermann, J. & Igel, H. (2006) Evidence for rain-triggered earthquake activity. *Geophysical Research Letters* **33**, L19303.

Hall, K. (1982) Rapid deglaciation as an initiator of volcanic activity: An hypothesis. *Earth Surface Processes Landforms* **206**, 45–51.

Hamilton, W. (1973) Tidal cycles of volcanic eruptions: Fortnightly to yearly periods. *Journal of Geophysical Research* **78**, 3363–3375.

Hampel, A., Hetzel, R. & Densmore, A. L. (2007) Postglacial slip rate increase on the Teton normal fault, northern Basin and Range Province, caused by melting of the Yellowstone ice cap and deglaciation of the Teton Range? *Geology* **35**, 1107–1110.

Hampel, A., Hetzel, R. & Maniatis, G. (2010) Response of faults to climate-driven changes in ice and water volumes on Earth's surface. *Philosophical Transactions of the Royal Society of London A* **368**, 2501–2518.

Harris, C., Arenson, L. U., Christiansen, H. H., et al. (2009) Permafrost and climate in Europe: Monitoring and modelling thermal, geomorphological and geotechnical responses. *Earth Science Reviews* **92**, 117–171.

Heki, K. (2003) Snow load and seasonal variation of earthquake occurrence in Japan. *Earth and Planetary Science Letters* **207**, 159–164.

Hély, C., Braccanot, P., Watrin, J. & Zheng, W. (2009) Climate and vegetation: simulating the African humid period. *Comptes Rendu Geoscience* **341**, 671–688.

Henriet, J-P. & Mienert, J., eds (1998) *Gas Hydrates: Relevance to world margin stability and climate change.* London: Geological Society Special Publication No. 137, 338.

Hermanns, R. L., Niedermann, S., Ivy-Ochs, S. & Kubik, P. W. (2004) Rock avalanching into a landslide-dammed lake causing multiple dam failure in Las Conchas valley (NW Argentina) – evidence from surface exposure dating and stratigraphic analyses. *Landslides* **1**, 113–122.

Hetzel, R. & Hampel, A. (2005) Slip rate variations on normal faults during glacial-interglacial changes in surface loads. *Nature* **435**, 81–84.

Horstmann, B. (2004) *Glacial lake outburst floods in Nepal and Switzerland.* Bonn: Germanwatch, 12pp.

Huggel, C. (2009) Recent extreme slope failures in glacial environments: effects of thermal perturbation. *Quaternary Science Reviews* **28**, 1119–1130.

Huggel, C., Kääb, A., Haeberli, W., Teysseire, P. & Paul, F. (2002) Remote sensing based assessment of hazards from glacier lake outbursts: a case study in the Swiss Alps. *Canadian Geotechnical Journal* **39**, 316–330.

Huggel, C., Zgraggen-Oswald, S., Haeberli, W., et al. (2005) The 2002 rock/ice avalanche at Kolka/Karmadan, Russian Caucasus: assessment of extraordinary avalanche formation and mobility, and application of QuickBird satellite imagery. *Natural Hazards and Earth Systems Sciences* **5**, 173–187.

Huggel, C., Caplan-Auerbach, J. & Wessels, R. (2008) Recent extreme avalanches: triggered by climate change? *EOS, Transactions of the AGU* **89**, 469–470.

Huggel, C., Salzmann, N., Allen, S., et al. (2010) Attributing warm extreme events to high-mountain slope failures: an analysis of worldwide large slope failures. *Philosophical Transactions of the Royal Society of London A* **368**, 2435–2460.

Huybers, P. & Langmuir, C. (2009) Feedback between deglaciation, volcanism and atmospheric CO_2. *Earth and Planetary Science Letters* **286**, 479–491.

Intergovernmental Panel on Climate Change (IPCC) (2007) In: Solomon, S., Qin, D., Manning, M., et al. (eds), *Climate Change 2007: The physical science basis.* Contribution of Working Group I to the Fourth Assessment Report of the Intergovernmental Panel on Climate Change. Cambridge: Cambridge University Press, 996pp.

IPCC (2011) *Managing the Extreme Events and Disasters to Advance Climate Change Adaptation* (SREX). Available at: http://ipcc-wg2.gov/SREX (accessed March 2011).

Jellinek, A., Manga, M. & Saar, M. (2004) Did melting glaciers cause volcanic eruptions in eastern California? Probing the mechanics of dike formation. *Journal of Geophysical Research* **109**,

Johnston, A. C. (1987) Suppression of earthquakes by large continental ice sheets. *Nature* **330**, 467–469.

Johnston, M. & Mauk, F. (1972) Earth tides and the triggering of eruptions from Mount Stromboli. *Nature* **239**, 266–267.

Jull, M. & McKenzie, D. (1996) The effect of deglaciation on mantle melting beneath Iceland. *Journal of Geophysical Research* **101**, 815–828.

Jupp, T., Pyle, D., Mason, B. & Dade, W. (2004) A statistical model for the timing of earthquakes and volcanic eruptions influenced by periodicprocesses. *Journal of Geophysical Research* **109**.

Kayen, R. E. & Lee, H. J. (1992) Pleistocene slope instability of gas hydrate-laden sediment on the Beaufort Sea margin. *Marine Geotechnology* **10**, 125–142.

Keating, B. H. & McGuire, W. J. (2004) Instability and structural failure at volcanic ocean islands and the climate dimension. *Advances in Geophysics* **47**, 175–271.

Keefer, D. K., Moseley, M. E. & de France, S. D. (2000) A 38,000-year record of floods and debris flows in the IIo region of southern Peru and its relation to El Niño events and great earthquakes. *Palaeogeography, Palaeoclimatology, Palaeoecology* **194**, 41–77.

Keiler, M., Knight, J. & Harrison, S. (2010) Climate change and geomorphological hazards in the eastern European Alps. *Philosophical Transactions of the Royal Society of London A* **368**, 2461–2480.

Kennett, J. & Thunell, R. (1975) Global increase in Quaternary explosive volcanism. *Science* **187**, 497–503.

Kilburn, C. R. J. & Petley, D. N. (2003) Forecasting giant, catastrophic slope collapse: lessons from Vajont, northern Italy. *Geomorphology* **54**, 21–32.

Knutson, T., Sirutis, J., Garner, S., Vecchi, G. & Held, I. (2008) Simulated reduction in Atlantic hurricane frequency under twenty-first century warming conditions. *Nature Geoscience* **1**, 359–364.

Lee, H. J. (2009) Timing of occurrence of large submarine landslides on the Atlantic Ocean margin. *Marine Geology* **264**, 53–64.

Lateltin, O., Beer, C., Raetzo, H. & Caron, C. (1997) Landslides in flysch terranes of Switzerland: causal factors and climate change. *Ecologae, Geologicae, Helvetiae* **90**, 401–406.

Licciardi, J., Kurz, M. & Curtice, J. (2007) Glacial and volcanic history of Icelandic table mountains from cosmogenic 3He exposure ages. *Quaternary Science Reviews* **26**, 1529–1546.

Liggins, F., Betts, R. & McGuire, W. J. (2010) Projected climate changes in the context of geospheric responses. *Philosophical Transactions of the Royal Society of London A* **368**, 2347–2368.

Liu, C., Linde, A. T. & Sacks, I. S. (2009) Slow earthquakes triggered by typhoons. *Nature* **459**, 833–836.

Luttrell, K. & Sandwell, D. (2010) Ocean loading effects on stress at near shore plate boundary fault systems. *Journal of Geophysical Research* **115**, B08411.

McGuire, W. J., Howarth, R. J., Firth, C. R., et al. (1997) Correlation between rate of sea level change and frequency of explosive volcanism in the Mediterranean. *Nature* **389**, 473–476.

Maclennan, J., Jull, M., McKenzie, D., Slater, L. & Grönvold, K. (2002) The link between volcanism and deglaciation in Iceland. *Geochemistry, Geophysics, Geosystems* **3**, 1062.

McMurtry, G. M., Watts, P., Fryer, G. J., Smith, J. R. & Imamura, F. (2004) Giant landslides, mega-tsunamis, and palaeo-sea level in the Hawaiian Islands. *Marine Geology* **203**, 219–233.

McNutt, S. (1999) Eruptions of Pavlof Volcano, Alaska, and their possible modulation by ocean load and tectonic stresses: Re-evaluation of the hypothesis based on new data from 1984–1998 *Purea and Applied Geophysics* **155**, 701–712.

McNutt, S. & Beavan, R. (1987) Eruptions of pavlof volcano and their possible modulation by ocean load and tectonic stresses. *Journal of Geophysical Research* **92**, 11509–11523.

Mason, B., Pyle, D., Dade, W. & Jupp, T. (2004) Seasonality of volcanic eruptions. *Journal of Geophysical Research* **109**.

Masson, D. G., Watts, A. B., Gee, M. J. R., et al. (2002) Slope failures on the flanks of the western Canary Islands. *Earth Science Reviews* **57**, 1–35.

Maslin, M. A., Mikkelsen, N., Vilela, C. & Haq, B. (1998) Sea-level- and gas-hydrate controlled catastrophic sediment failures of the Amazon Fan. *Geology* **26**, 1107–1110.

Maslin, M. A., Owen, M., Day, S. & Long, D. (2004) Linking continental slope failure to climate change: testing the Clathrate Gun Hypothesis. *Geology* **32**, 53–56.

Maslin, M. A., Owen, M., Betts, R., Day, S. J., Dunkley Jones, T. & Ridgewell, A. (2010) Gas hydrates: past and future geohazard? *Philosophical Transactions of the Royal Society of London A* **368**, 2369–2394.

Mastin, L. G. (1994) Explosive tephra emissions of Mount St. Helens 1989–1991: The violent escape of magmatic gas following storms? *Geological Society of America Bulletin* **106**, 175–185.

Matthews, A. J., Barclay, J., Carn, S., et al. (2002) Rainfall-induced volcanic activity on Montserrat. *Geophysical Research Letters* **29**, 2211–2214.

Matthews, R. K. (1969) Tectonic implications of glacio-eustatic sea level fluctuations. *Earth and Planetary Science Letters* **5**, 459–462.

Mayewski, P. A., Rohling, E. E., Stager, J. C., et al. (2004) Holocene climate variability. *Quaternary Research* **62**, 243–255.

Meehl, G. A., Stocker, T. F., Collins, W. D., et al. (2007) Global climate projections. In: Solomon, S., Qin, D., Manning, M., et al. (eds), *Climate Change 2007: The physical science basis.* Contribution of Working Group I to the Fourth Assessment Report of the Intergovernmental Panel on Climate Change. Cambridge: Cambridge University Press, pp 747–846.

Mernild, S. H., Hasholt, B., Kane, D. L. & Tidwell, A. C. (2008) Jökulhlaup observed at Greenland ice sheet. *Eos* **99**, 321–322.

Muir-Wood, R. (2000) Deglaciation seismotectonics: a principal influence on intraplate seismogenesis at high latitudes. *Quaternary Science Reviews* **19**, 1399–1411.

Nakada, M. & Yokose, H. (1992) Ice age as a trigger of active Quaternary volcanism and tectonism. *Tectonophysics* **212**, 321–329.

Neuberg, J. (2000) External modulation of volcanic activity. *Geophysical Journal International* **142**, 232–240.

Ng, F., Liu, S., Mavlyudov, B. & Wang, Y. (2007) Climatic control on the peak discharge of glacier outburst floods. *Geophysical Research Letters* **34**, L21503.

Nowell, D., Jones, C. & Pyle, D. (2006) Episodic Quaternary volcanism in France and Germany. *Journal of Quaternary Science* 645–675.

Owen, M., Day, S. & Maslin, M. (2007) Late Pleistocene submarine mass movements: occurrence and causes. *Quaternary Science Reviews* **26**, 958–978.

Pagli, C. & Sigmundsson, F. (2008) Will present day glacier retreat increase volcanic activity? Stress induced by recent glacier retreat and its effect on magmatism at the Vatnajökull ice cap, Iceland. *Geophysical Research Letters* **35**, L09304.

Pagli, C., Sigmundsson, F., Lund, B., et al. (2007) Glacio-isostatic deformation around the Vatnajökull ice cap, Iceland, induced by recent climate warming: GPS observations and finite element modeling. *Journal of Geophysical Research* **112**, B08405.

Pfeffer, W. T., Harper, J. T. & O'Neill, S. (2008) Kinematic constraints on glacier contributions to 21st-century sea-level rise. *Science* **321**, 1340–1343.

Pratt, B., Burbank, D. W., Heimsath, A. & Ojha, T. (2002) Impulsive alluviation during early Holocene strengthened monsoons, central Nepal Himalaya. *Geology* **30**, 911–914.

Podolskiy, E. A. (2008) Effects of recent environmental changes on global seismicity and volcanism. *Earth Interactions* **13**, 1–14.

Quidelleur, X., Hildenbrand, A. & Samper, A. (2008) Causal link between Quaternary palaeoclimatic changes and volcanic islands evolution. *Geophysical Research Letters* **35**: L02303.

Rahmstorf, S. (2007) A semi-empirical approach to projecting future sea-level rise. *Science* **315**, 368–370.

Rampino, M., Self, S. & Fairbridge, R. (1979) Can rapid climate change cause volcanic eruptions? *Science* **206**, 826–828.

Rana, B., Shrestha, A. B., Reynolds, J. M., Aryal, R., Pokhrel, A. P. & Budhathoki, K. P. (2000) Hazard assessment of the Tsho Rolpa Glacier Lake and ongoing remediation measures. *Journal of Nepal Geological Society* **22**, 563–570.

Reynolds, J. M. (1992) The identification and mitigation of glacier-related hazards: examples from the Cordillera Blanca, Peru. In: McCall, G. J. H., Lamming, D. J. C. & Scott, S. C. (eds), *Geohazards*. London: Chapman & Hall, pp 143–157.

Rubinstein, J. L., La Rocca, M., Vidale, J. E., Creager, K. C. & Wech, A. G. (2008) Tidal modulation of nonvolcanic tremor. *Science* **319**, 186–189.

Saemundsson, T., Petursson, H. G. & Decaulne, A. (2003) Triggering factors for rapid mass movements in Iceland. In: D. Rickenmann and C. Chen (eds) *Debris-flow Hazards Mitigation, Mechanics, Prediction and Assessment*. New York: American Society of Civil Engineers, pp 167–178.

Sanderson, M. G., Hemming, D. L. & Betts, R. A. (2011) Regional temperature and precipitation changes under high-end (≥4°C) global warming. *Philosophical Transactions of the Royal Society of London A* **369**, 85 – 98.

Sauber, J. & Molnia, B. F. (2004) Glacier ice mass fluctuations and fault instability in tectonically active Southern Alaska. *Global Planetary Change* **42**, 279–293.

Sauber, J. & Ruppert, N. A. (2008) Rapid ice mass loss: does it have an influence on earthquake occurrence in southern Alaska? In: Freymüller, J. T., Haeussler, P. J., Wesson, R. L., & Ekström, G. (eds), *Active Tectonics and Seismic Potential of Alaska*. Geophysics Monograph Series 179. Washington DC: American Geophysical Union.

Sauber, J., Plafker, G., Molnia, B. F. & Bryant, M. A. (2000) Crustal deformation associated with glacial fluctuations in the eastern Chugach Mountains, Alaska. *Journal of Geophysical Research* **105**, 8055–8077.

Schulz, W. H., Kean, J. W. & Wang, G. (2009) Landslide movement in southwest Colorado triggered by atmospheric tides. *Nature Geoscience* **2**, 863–866.

Scott, K. M., Vallance, J. W., Kerle, N., Macías, J. L. Strauch, W. & Devoli, G. (2005) Catastrophic precipitation-triggered lahar at Casita volcano, Nicaragua: occurrence, bulking and transformation. *Earth Surface Processes and Landforms* **30**, 59–79.

Sigmundsson, F., Pinel, V., Lund, B., et al. (2010) Climate effects on volcanism: Influence on magmatic systems of loading and unloading from ice mass variations with examples from Iceland. *Philosophical Transactions of the Royal Society of London A* **368**, 2519–2534.

Sigvaldason, G., Annertz, K. & Nilsson, M. (1992) Effect of glacier loading/ deloading on volcanism: Postglacial volcanic eruption rate of the Dyngjufjoll area, central Iceland. *Bulletin of Volcanology* **54**, 385–392.

Simpson, D., Leith, W. & Scholz, C. (1988) Two types of reservoir-induced seismicity. *Bulletin of the Seismological Society of America* **78**, 2025–2040.

Song, T. (2009) Glacial earthquake tsunami – a consequence of climate change. Proceedings, Third Johnston-Lavis Colloquium: Climate Forcing of Geological and Geomorphological Hazards. University College London. (Abstract.)

Sparks, R. (1981) Triggering of volcanic eruptions by earth tides. *Nature* **290**, 448.

Stewart, I. S., Sauber, J. & Rose, J. (2000) Glacio-seismotectonics: ice sheets, crustal deformation and seismicity. *Quaternary Science Reviews* **19**, 1367–1389.

Strasser, M., Schindler, C. & Anselmetti, F. S. (2008) Late Pleistocene earthquake-triggered moraine dam failure and outburst of Lake Zurich, Switzerland. *Journal of Geophysical Research* **113**.

Sultan, N., Cochonat, P., Foucher, J-P. & Mienert, J. (2004) Effect of gas hydrates melting on seafloor slope instability. *Marine Geology* **213**, 379–401.

Swiss Re (2000) *Natural Catastrophes and Man-made Disasters in 1999: Storms and earthquakes lead to the second-highest losses in insurance history*. Sigma. No 2/2000. Swiss Re, 35pp.

Swiss Re (2001) *Natural Catastrophes and Man-made Sisasters in 2000: Fewer insured losses despite huge floods*. Sigma. No 2/2001. Swiss Re, 20pp.

Symon, C., Arris, L. & Heal, B., eds (2005) *Arctic Climate Impact Assessment: Scientific Report*. New York: Cambridge University Press, 1042pp.

Talwani, P. (1997) On the nature of reservoir-induced seismicity. *Pure and Applied Geophysics* **150**, 473–492.

Tappin, D. R. (2010) Submarine mass failures as tsunami sources: their climate control. *Philosophical Transactions of the Royal Society A* **368**, 2417–2434.

Tuffen, H. (2009) How will melting of ice affect volcanic hazards in the 21st century? *Philosophical Transactions of the Royal Society of London A* **368**, 2535–2558.

Turpeinen, H., Hampel, A., Karow, T. & Maniatis, G. (2008) Effect of ice sheet growth and melting on the slip evolution of thrust faults. *Earth and Planetary Science Letters* **269**, 230–241.

Vanneste, M., Mienert, J. & Bunz, S. (2006) The Hinlopen Slide: a giant submarine slope failure on the northern Svalbard margin, Arctic Ocean. *Earth and Planetary Science Letters* **245**, 373–388.

Wallmann, P. C., Mahood, G. A. & Pollard, D. D. (1988) Mechanical models for correlation of ring-fracture eruptions at Pantelleria, Strait of Sicily, with glacial sea-level drawdown. *Bulletin of Volcanology* **50**, 327–339.

Watanabe, T. & Rothacher, D. (1996) The 1994 Lugge Tsho glacial lake outburst flood, Bhutan Himalaya. *Mountain Research and Development* **16**, 77–81.

Wilcock, W. (2001) Tidal triggering of micro earthquakes on the Juan de Fuca Ridge. *Geophysical Research Letters* **28**, 3999–4002.

Wu, P. (1999) Intraplate earthquakes and postglacial rebound in eastern Canada and Northern Europe. In: Wu, P. (ed.), *Dynamics of the Ice Age Earth, A Modern Perspective*. Zurich: Trans Tech Publications, 443–458.

Wu, P. & Johnston, P. (2000) Can deglaciation trigger earthquakes in N. America? *Geophysical Research Letters* **27**, 1323–1326.

Wu, P., Johnston, P. & Lambeck, K. (1999) Postglacial rebound and fault instability in Fennoscandia, *Geophysical Journal International* **139**, 657–670

Zemp, M., Haeberli, W., Hoelzle, M. & Paul, F. (2006) Alpine glaciers to disappear within decades? *Geophysical Research Letters* **33**, L13504.

Zhang, X. et al. (2006) Detection of human influence on twentieth century precipitation trends. *Nature* **448**, 461–465.

Zielinksi, G., Mayewksi, P., Meeker, L., et al. (1997) Volcanic aerosol records and tephrochronology of the Summit, Greenland, ice cores. *Journal of Geophysical Research* **102**, 625–626, 640.

2 Projected future climate changes in the context of geological and geomorphological hazards

Felicity Liggins[1], Richard A. Betts[1] and Bill McGuire[2]

[1] Met Office Hadley Centre, Exeter, UK
[2] Aon Benfield UCL Hazard Centre, Department of Earth Sciences, University College London, UK

Summary

On palaeoclimate timescales, enhanced levels of geological and geomorphological activity have been linked to climatic factors, including examples of processes that are expected to be important in current and future anthropogenic climate change. Planetary warming leading to increased rainfall, ice-mass loss and rising sea levels is potentially relevant to geospheric responses in many geologically diverse regions. Anthropogenic climate change, therefore, has the potential to alter the risk of geological and geomorphological hazards through the twenty-first century and beyond. Here, we review climate change projections from both global and regional climate models in the context of geohazards. In assessing the potential for geospheric responses to climate change, it appears prudent to consider regional levels of warming of 2°C above average pre-industrial temperature as being potentially unavoidable as an influence on processes requiring a human adaptation response within this century. At the other end of the scale, when considering changes that could be avoided by reduction of emissions, scenarios of unmitigated warming exceeding 4°C in the global average include much greater local warming in some regions. However, considerable further work is required to better understand the uncertainties associated with these projections, uncertainties inherent not only in the climate modelling but also in the linkages between climate change and geospheric responses.

Introduction

Changes in many aspects of geological and geomorphological activity have been linked to climate variability on palaeoclimate timescales. This includes glacial lake outbursts and rock-dam failures, submarine and subaerial landslides, tsunamis and landslide 'splash' waves, seismic and volcanic activity, and gas hydrate destabilization. With human influence on climate now becoming significant and potentially very large, an emerging question is whether this could influence any or all of these processes and hence alter the risk of geological and geomorphological hazards. This field of study is very much in its infancy, but, as part of the background to other research, this chapter assesses climate change projections in the context of climate-related geological and geomorphological processes.

Although changes in the global pattern of stress and strain in the Earth's crust cannot be ignored, the more focused response of the geosphere to post-glacial planetary warming and hydrological adjustment is linked primarily to a relatively limited portfolio of ice-mass loss, rapidly rising sea levels and increased availability of liquid water, the last arising from either ice melt or elevated levels of precipitation. These, in turn, drive a change in geological and geomorphological activity through changes in a small number of key environmental parameters acting mainly within the crust, notably variations in load pressure and increases in pore-water pressure. Although the former are capable of promoting fault rupture and magma migration towards the surface, most notably in areas of ice-mass loss or sea-level rise, the latter are particularly instrumental in promoting the failure of masses of unstable rock or sediment in both marine and subaerial settings. The consequence of these climate-change-forced adjustments in environmental parameters, during post-glacial times, are: increased levels of seismicity (e.g. Davenport et al., 1989; Wu et al., 1999; Wu & Johnston, 2000; Hetzel & Hampel, 2005) and volcanism, both in and adjacent to formerly ice-covered regions (Sigvaldasson et al., 1992; Jull & McKenzie, 1996; Licciardi et al., 2007; Huybers & Langmuir, 2009), and also in the marine environment (McGuire et al., 1997; Huybers & Langmuir, 2009); a higher incidence of large, submarine landslides along continental margins (Lee, 2009); and elevated levels of mass movement in areas of mountainous terrain (Lateltin et al., 1997; Friele & Clague, 2004; Hermanns et al., 2004).

Here, we examine the potential, in a warming world, for such responses to changes in environmental conditions to be accentuated at local, regional and global scales, leading to possible changes – via triggering, adjustment or modulation – in hazardous geological and geomorphological activity. Principal relevant climate quantities will be rising atmospheric and ocean temperatures, ice-mass loss, ocean volume increase and a greater incidence of extreme precipitation. Environmental settings that are most susceptible appear to be (1) regions of ice loss at high latitudes and high altitudes, (2) ocean margins, (3) mountainous terrain and (4) regions predicted to experience significantly increased levels of precipitation (McGuire, 2010).

Climate change research: informing mitigation and adaptation

The Earth's climate has always changed over time, to a varying degree and on a range of timescales. For most of geological history, these changes were driven by natural forcings, most notably the Milankovitch and solar cycles and degassing in volcanic environments. Since the middle of the nineteenth century, increasingly detailed analysis of our weather and climate has revealed unusual changes that cannot be explained by these natural factors alone. The observed rise in global temperatures over recent decades is very likely to be due to the influence of anthropogenic changes in the concentrations of greenhouse gases (GHGs) such as CO_2, CH_4 and N_2O (Intergovernmental Panel on Climate Change or IPCC, 2007a). Anthropogenic CO_2 emissions show a general increase over time (Carbon Dioxide Information Analysis Center or CDIAC – see http://cdiac.ornl.gov) and, in the absence of major changes in energy policy, this increase is expected to continue for some decades (Nakicenovic et al., 2000), and is potentially limited only by the still large availability of fossil fuel reserves. Under such scenarios, increases in global mean temperature between 1.6°C and 6.9°C relative to the pre-industrial period are projected (IPCC, 2007a).

Long response times in the climate system mean that the full effect of past anthropogenic climate forcing has not yet been realized, and hence the global climate is already committed to further change even if anthropogenic GHG emissions were to cease immediately. Moreover, given that GHG emissions continue to show an overall upward trend, and the international policy has yet to achieve a binding agreement on a workable mechanism for reduction of emissions, further increases in year-on-year emissions also seem inevitable, and limiting long-term cumulative emissions also appears to be increasingly challenging. In practice, then, the level of 'committed' warming appears likely to be above the level suggested by the physical inertia of the climate system alone. There are therefore two reasons to assess the potential impacts of future climate change under a range of different scenarios. The first reason is to inform mitigation action: how much, and how soon, do GHG emissions need to be reduced in order to avoid harmful impacts? Clearly, this issue requires information on harmful impacts, which could potentially include geological and geomorphological hazards. The second reason is to assess the impacts of climate change that are now unavoidable, in order to inform adaptation planning to help minimize these impacts. Again, if the risks of geological and geomorphological hazards are altered by anthropogenic climate change even within the horizon of committed change, then this is important information for managing exposure to those risks.

One approach to the 'mitigation question' is to refer to logical ultimate consequences of ongoing climate warming, such as the loss of major ice sheets (Hansen, 2003). Should all land ice melt, global sea levels would rise by up to approximately 70 m – this represents an extreme scenario that could be the ultimate, long-term response of the Earth's system to a doubling of atmospheric carbon dioxide levels above pre-industrial levels (Hansen et al., 1997). Such loss of all land ice would require disintegration of the East Antarctic ice sheet, a potentially catastrophic event, but one that is not of immediate concern based on current climate modelling. However, there is evidence that lower levels of warming could lead to an eventual loss of the Greenland and West Antarctic ice sheets, resulting in a possible commitment to several more metres of sea-level rise beyond 2100.

Insight into these large impacts can be gained from palaeoclimate evidence, e.g. during the Middle Pliocene, around 3 million years BP (before present), sea level was $25 \pm 10\,\text{m}$ above today's, in long-term equilibrium with a global average temperature just 2–3°C higher than that of the present day (Dowsett et al., 1994). However, the palaeoclimate record of natural climate variability provides no direct analogues for the full range of changes that may occur as a consequence of anthropogenic forcing of climate, owing to the different processes and timescales involved. Therefore, it is valid to use numerical models of the Earth's climate system to assess potential future changes and their impacts.

Modelling the climate

The past, present and future influences of natural and anthropogenic driving forces on the climate system can be assessed on a planetary scale using a range of models, from simple climate models (SCMs) that focus on key planetary-scale processes such as the global energy balance and carbon cycle to general circulation models (GCMs) that simulate the atmospheric circulation and other atmospheric processes as mechanistically as possible (still requiring a large number of approximations). For a more detailed evaluation at higher temporal and spatial resolutions, regional climate models (RCMs) are 'nested' within GCM projections to downscale the results of larger, coarse resolution models in the context of finer-scale influences such as orography and land use.

Uncertainties in climate projections arise from a number of sources, with the influence of each of these varying through time (Meehl et al., 2007; Hawkins & Sutton, 2009). The first area of uncertainty is natural variability – in other words, those fluctuations in our climate that occur on seasonal, annual and decadal timescales. In near-term climate projections, the greatest uncertainties arise from these regional climate responses. The second area of uncertainty arises from the fact that models are only approximations of the real world and hence are not perfect predictors of future change. This uncertainty is extremely difficult to quantify because the limitations of the approximations may not be known. However, one approach beginning to account for this uncertainty is to use a large number of models that deal with the need for approximations in different ways. Under the same driving conditions, individual climate models can give different results depending on their structural nature. This area of uncertainty is important to consider in the short term, alongside natural variability, and its influence increases over time.

In the long term, a third area of uncertainty becomes important, namely the effect of uncertainties in the anthropogenic emissions of GHGs. In the near term, the projected climate responses to the various plausible emission scenarios are not significantly different in comparison with the uncertainties arising from natural climate variability and uncertainties in the climate response to any given emission scenario. This is partly due to the long response timescales of the climate system described above – some change is already locked in, so will occur whichever emission scenario is followed, and the full response of any given emission scenario will take several decades to emerge. However, later in the twenty-first century, the projections due to different emission scenarios start to differ, and the uncertainty in future emissions becomes a major factor. Nevertheless, uncertainties in the climate response to any given emission scenario still remain very significant, especially at regional scales and in precipitation as opposed to temperature.

Emission scenarios

The consequences of different potential future changes in anthropogenic emissions are examined through the use of emission scenarios. These are typically based on internally consistent storylines of future population growth, technology change, economic state and the character of the geopolitical system. The most widely used emission scenarios so far are from the IPCC's special report on emissions scenarios (SRES; Nakicenovic et al., 2000). These consider a wide range of future socioeconomic pathways, but do not consider policies explicitly intended to reduce anthropogenic influences on climate. Therefore, these scenarios can be used to assess the potential consequences of not implementing reduction of emissions. Although these scenarios were developed in the 1990s and can therefore now be compared with actual emissions, it is still regarded as too early to reliably assess whether any of the SRES scenarios are emerging as more realistic than any others (Leggett & Logan, 2008; Van Vuuren et al., 2008). More recently, new scenarios accounting for such climate policy have begun to be used in climate modelling studies, in order to assess the level of 'residual' climate change that would still occur with reduction of these emissions.

Climate change projections

The SRES scenarios have been used in both SCMs and GCMs to assess the potential range of climate change that could arise from socioeconomic futures that do not include an action to reduce anthropogenic influence on climate. These suggest a likely range of global mean temperature increase of 1.1–6.4°C by 2090–2099 relative to 1980–1999 (IPCC, 2007a), which equates to approximately 1.6–6.9°C relative to pre-industrial levels. Under the highest emission scenario of the main SRES group, known as A1FI (where 'FI' denotes 'fossil intensive'), a warming of 4°C relative to pre-industrial level would most likely be reached in approximately the 2070s, with a possible earlier date of approximately 2060 (Betts et al., 2012). The lowest of the main emission scenarios, known as B1, results in a central estimate of warming of 1.8°C by the 2090s relative to 1980–1999, which is 2.3°C relative to pre-industrial level.

At present, the most comprehensive set of climate projections with GCMs is that produced for the IPCC's fourth assessment report (AR4), commonly known as the 'AR4 multi-model ensemble'. Members of this ensemble of GCMs were used to simulate twenty-first century climate with a subset of the SRES scenarios. This subset included the 'low' B1 scenario described above, the A1B scenario (many similarities to the A1FI scenario but with a balanced growth between fossil fuel and other energy sources) and the A2 scenario, which assumes high population growth but results in lower emissions than A1FI from the 2020s onwards. The A1FI scenario was not used in the AR4 ensemble.

The full multi-model ensemble included 23 GCMs, but not all models were used for all scenarios. The most comprehensively studied scenario using all 23 models was A1B, with the central estimate of global warming from the GCMs being 2.8°C by the 2090s relative to 1980–1999. The 'likely range' of this warming for A1B from SCMs and expert judgement was 1.7–4.4°C.

Scenarios including global policies to reduce emissions have only recently begun to be used in GCM studies, but SCMs have been used to assess the implications of these for minimizing the level of future global warming. As with the projections using the SRES scenarios, these 'mitigation' scenarios produce a range of possible trajectories of global warming that reflect uncertainties in the climate response. Scenarios giving a significant probability of remaining below 2°C warming relative to pre-industrial level feature a peak in global emissions approximately around the year 2020 (IPCC, 2007b).

Using climate projections to inform mitigation and adaptation

The available climate projections are relevant for informing both mitigation and adaptation, but different aspects of the projections are relevant to these different issues. A key distinction is the time horizons over which the scenarios remain similar, and at which they begin to diverge, because this distinguishes the avoidable climate change relevant to mitigation from the unavoidable climate change to which the world will need to adapt. Using the SRES-based GCM ensemble, the climate change projected through the 2020s and into the 2030s is insensitive to the range of non-mitigation emission scenarios used in IPCC AR4, with the projections only diverging by 0.05°C from each other in the period 2011–2030 (Figure 2.1) (Meehl et al., 2007). It is only after the 2030s that the projections begin to diverge (Meehl et al., 2007). A similar time horizon of divergence is seen for mitigation scenarios assuming deep emission cuts.

This implies that projections after the 2030s are relevant to the question of which impacts of climate change could be avoided by implementing mitigation policies. The SRES-based GCM projections, and other projections using SCMs, can be used to assess the potential impacts of unmitigated climate change as part of the evidence to inform international climate policy. Before the 2030s, the projected climate change is not affected by reduction of emissions and the uncertainty arises only from the climate system itself. This means that climate change impacts, identified as taking place before this time horizon, may be addressed only through adaptation.

Figure 2.1 Past changes in global mean temperature (black curve), and projected future changes resulting from the IPCC's special report on emissions scenarios (SRES) marker scenarios of greenhouse gas (GHG) and aerosol emissions (coloured curves and grey bars), relative to the 1980–2000 mean (Meehl et al., 2007). Climate changes under the A2, A1B and B1 scenarios were projected with general circulation models or GCMs (red, green and blue lines, with plumes showing 5–95% range of model projections without uncertainties in climate–carbon cycle feedbacks). The full set of marker scenarios, including a range of strengths of climate–carbon cycle feedbacks, were examined with simple climate models (SCMs). Grey bars show the 'likely range' of warming for each scenario, from expert assessment based on all available evidence from GCMs, SCMs and observational constraints. Red line, A2; green line, A1B; blue line, B1; orange line, year 2000 constant concentrations; black line, twentieth century. (Adapted from Climate Change 2007: The Physical Science Basis. Working Group 1 Contribution to the Fourth Assessment Report of the Intergovernmental Panel on Climate Change, Figure SPM.5, Cambridge University Press, UK with permission.)

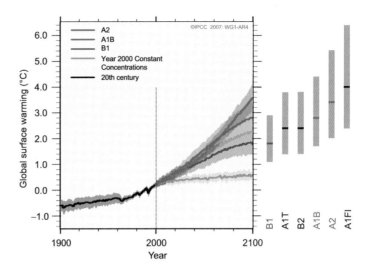

A further issue regarding adaptation concerns the residual climate change after the 2030s under mitigation scenarios. This relates to the commitment to some ongoing climate change even if GHG concentrations are stabilized. To illustrate this, some of the AR4 GCMs were used to simulate climate with GHG concentrations stabilized at the year 2000. About half of the projected rise in global mean temperature from the present day to the 2030s was attributed to the 'committed climate change'. Fixing GHG concentrations at 2100 for the B1 and A1B scenarios still led to a further ongoing rise in projected global mean temperatures of 0.4–0.8°C by 2300. Moreover, mitigation scenarios with SCMs project that global temperatures could remain near 2°C above pre-industrial level for the next century, even with rapid and deep emission cuts (Met Office, 2007). Furthermore, even with mitigation measures being implemented within the twenty-first century, thermal expansion of the oceans will continue (as in the 2100 stabilization scenario modelling of Figure 2.2).

Regional climate change

This chapter uses regional climate change assessments from AR4, which provide plausible ranges for temperature and precipitation averages by the end of the twenty-first century (2080–2099). They are derived from a set of global models using the SRES A1B emission scenario. Each projection presents the maximum and minimum possible change, alongside the median, although it must be emphasised that these projections do not encompass all possible futures. The focus on the A1B scenario for regional climate uncertainty estimates meant that the AR4 regional climate change assessment did not examine the state of the climate for higher levels of global warming that would be expected if a scenario of high emis-

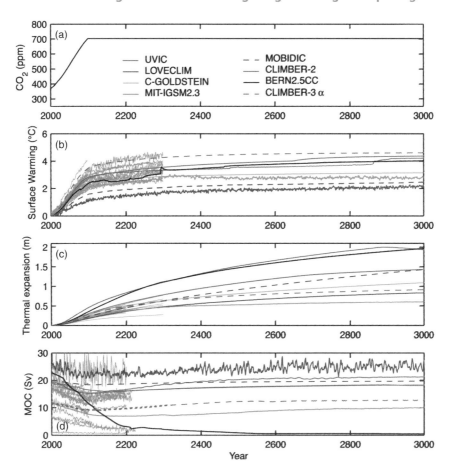

Figure 2.2 (a) Atmospheric CO_2, (b) global mean surface warming, (c) sea-level rise from thermal expansion, and (d) Atlantic meridional overturning circulation (MOC, sverdrups) calculated by eight Earth system models of intermediate complexity (EMIC) for the special report on emissions scenarios (SRES) A1B scenario and stable radiative forcing after 2100, showing long-term commitment after stabilization. Coloured lines are results from EMICs and grey lines indicate atmosphere–ocean general circulation model (GCM) results where available for comparison. Anomalies in (b) and (c) are given relative to the year 2000. Vertical bars indicate ±2 SD uncertainties due to ocean parameter perturbations in the C-GOLDSTEIN model. The MOC shuts down in the BERN2.5CC model, leading to an additional contribution to sea-level rise. Individual EMICs treat the effect from non-CO_2 GHGs and from the direct and indirect aerosol effects on radiative forcing differently. Despite similar atmospheric CO_2 concentrations, radiative forcing among EMICs can thus differ within the uncertainty ranges currently available for present-day radiative forcing. Red line, UVIC; dark blue line, LOVECLIM; light blue line, C-GOLDSTEIN; green line, MIT-IGSM2.3; dashed dark blue line, MOBIDIC; pink line, CLIMBER-2; black line, BERN2.5CC; dashed pink line, CLIMBER-3a. (Adapted from Climate Change 2007: The Physical Science Basis. Working Group 1 Contribution to the Fourth Assessment Report of the Intergovernmental Panel on Climate Change, Figure 10.34. Cambridge University Press with permission.)

sions is followed and/or if feedbacks in the climate system are stronger than expected in the standard models. This is an important omission in the context of geological and geomorphological hazards and is particularly important to consider for high-latitude regions. Some limited regional analysis is available for projections of higher levels of global warming, e.g. Sanderson et al. (2012) assess the 17 members of the AR4 multi-model ensemble that were used to project climate change under the A2 scenario, and distinguish a subset of nine that project warming of ≥4°C by the 2090s relative to the pre-industrial level. Sanderson et al. (2012) refer to these as 'high-end' projections, and this name is also used here.

Similarly, there is little information available as yet on the regional impacts of mitigation scenarios, for informing long-term assessment of committed change. The regional climate projections presented here provide some context for other work on potential changes in geological and geomorphological hazards, but considerable further work is required in order to assess the potentially larger changes in regional climate change that may further increase the risks of hazardous geospheric responses, and indeed the implications of committed climate change for these potential hazards.

Climate forcing of hazards in the geosphere

The geological and geomorphological hazards introduced in Chapter 1 are here evaluated in the context of regional climate change. Table 2.1 summarizes these hazards alongside relevant climate quantities, projected climate changes, and susceptible region and environmental settings.

Global oceans

Warmer oceans and increased ocean mass due to input of land ice meltwater have the potential to influence the stability of gas-hydrate deposits in marine sediments and, as a consequence, alter the stability of submarine slopes. Increased ocean mass may also elicit volcanic and seismic responses in coastal and island settings. These, in turn, may contribute to the formation of subaerial, volcanic landslides, submarine landslides and tsunamis.

During the twenty-first century, global climate models project non-uniform increases in temperature across the surface of the Earth (Figure 2.3). The spatial pattern of projected surface temperature change, presented in IPCC AR4, suggests warming occurring to a greater extent over the land than the sea (Meehl et al., 2007). Observations of land–sea temperatures throughout the twentieth and into the twenty-first century reinforce the existence of a land–sea contrast (e.g. Dommenget, 2009). This contrast is, in part, due to the thermal inertia of the oceans in comparison with that of the land when responding to the driving force of anthropogenic climate change. However, other studies have demonstrated that

Table 2.1 Projected climate changes of greatest relevance to geological and geomorphological hazards, adapted from Climate Change 2007: The Physical Science Basis. Working Group 1 Contribution to the Fourth Assessment Report of the Intergovernmental Panel on Climate Change, Cambridge University Press with permission.)

Environmental setting	Example region	Relevant climate quantities	Projected changes in climate by 2080–2099, relative to 1980–1999		Potential hazards
			Standard AR4 GCM ensemble, A1B scenario	Full range of SRES-based projections	
Globe		Global mean temperature		1.1–6.4°C	
Global oceans	Global marine gas-hydrate deposits; coastal and island volcanoes and ocean margin seismogenic faults	Ocean temperatures; sea level	Varied increase in annual mean temperature dependent on region; 0.21–0.48 m rise in sea level without rapid changes in ice dynamics	Varied increase in annual mean temperature dependent on region; 0.18–0.59 m rise in global mean sea level without rapid changes in ice dynamics	Submarine landslides; tsunamis; volcanic eruptions; earthquakes
High latitudes	Arctic, including Greenland, Iceland, Alaska and Kamchatka	Temperature and ice mass	2.8–7.8°C rise in annual mean temperature; 10–28% change in precipitation	15°C rise in temperature across much of Arctic in 'high-end' projections (>4°C by 2090s)	Glacial outburst floods; ice-quakes; earthquakes; submarine and subaerial landslides; tsunamis; volcanic eruptions; debris flows
	Antarctica	Temperature and ice-mass	1.4–5.0°C rise in annual mean temperature; −2% to 35% change in precipitation	Not yet assessed	Glacial outburst floods; volcanic activity, earthquakes
Non-volcanic mountain regions	European Alps; Pyrenees; Caucasus; New Zealand Alps (and other glaciated non-volcanic mountain chains)	Temperature and ice-mass; precipitation	2.2–5.1°C rise in annual mean temperature and −27% to −4% change in precipitation in southern Europe	Not yet assessed	Glacial outburst floods; subaerial landslides and ice and rock avalanches; debris flows; earthquakes
	Himalayas	Temperature and ice mass; precipitation		7–8°C rise in temperature in parts of Himalayas in 'high-end' projections (>4°C by 2090s)	

(Continued)

Table 2.1 (*Continued*)

Environmental setting	Example region	Relevant climate quantities	Projected changes in climate by 2080–2099, relative to 1980–1999		Potential hazards
			Standard AR4 GCM ensemble, A1B scenario	Full range of SRES-based projections	
Volcanic landscapes	Andes; Cascades (USA)	Temperature and ice-mass; precipitation	2.1–5.7°C rise in annual mean temperature and −3% to 14% change in precipitation in western North America	Not yet assessed	Subaerial landslides and rock and ice avalanches; debris flows; volcanic eruptions
	Central America, eastern Caribbean, Indonesia, Papua New Guinea	Precipitation	1.8–5.0°C rise in annual mean temperature and −48% to 9% change in precipitation in Central America	Not yet assessed	
	Hawaii, Canary Islands, Cape Verde Islands	Precipitation	1.5–3.7°C rise in annual mean temperature and −2% to 15% change in precipitation in south-east Asia; 0.18–0.59 m rise in sea level	Not yet assessed	
	Japan and Taiwan	Precipitation		Not yet assessed	
	Coastal and island concentrations of volcanoes	Sea level		Not yet assessed	

Column 4 shows regional projections from the multi-model ensemble of atmosphere–ocean general circulation models (GCMs) driven by prescribed CO_2 and other greenhouse gas (GHG) concentrations for the A1B emission scenario. This experimental design does not account for other scenarios or uncertainties in climate–carbon cycle feedbacks – the CO_2 concentration projections used in AR4 are from a model in the middle of the range of projected feedback strengths. Column 5 shows projections from the full range of emissions used by IPCC (2007a); although global-scale information is available for all these projections, regional information is more limited.

SRES, special report on emissions scenarios.

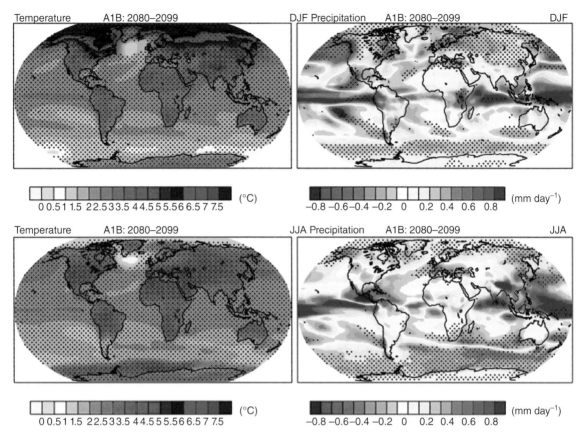

Figure 2.3 Multi-model mean changes in surface air temperature (°C, left) and precipitation (mmd −1, right) for boreal winter (December–January–February, top) and summer (June–July–August, bottom). Changes are given for the special report on emissions scenarios (SRES) A1B scenario, for the period 2080–2099 relative to 1980–1999. Stippling denotes areas where the magnitude of the multi-model ensemble mean exceeds the intermodel standard deviation. (Adapted from Climate Change 2007: The Physical Science Basis. Working Group 1 Contribution to the Fourth Assessment Report of the Intergovernmental Panel on Climate Change, Cambridge University Press UK with permission.)

this contrast is also evident in equilibrium climate models and hence cannot be a function of the differential heat capacity alone. Joshi et al. (2007), Sutton et al. (2007) and most recently Dommenget (2009) all argue that the land–sea contrast seen in both equilibrium and transient climate models is driven mainly by the asymmetrical response and feedbacks of the land surface temperatures to the warming oceans.

The greatest ocean surface warming is projected for the Arctic, partly as a result of changes in albedo from melting sea ice and partly due to changes in poleward energy transport. The melting and forming of sea ice also contribute to the meridional overturning circulation of the North Atlantic. Trends in sea-ice cover since the late 1970s have revealed both decreases in the extent of sea-ice coverage of about 0.3×10^6 km^2 per decade (Cavalieri et al., 2003; Stroeve et al., 2005) and a thinning of the icepack by 40% over the 30-year period from 1966, especially in areas where sea ice was initially thickest (Rothrock et al., 1999, 2003). Research shows how the natural drivers of variations in ice extent and thickness alone cannot explain the decreases observed in recent decades – it is only when the influence of anthropogenic GHG emissions is included that the observed changes can be fully understood (Min et al., 2008). Equatorial regions also see large increases in ocean surface temperatures in the AR4 multi-model ensemble, particularly in the eastern Pacific (e.g. DiNezio et al., 2009).

Conversely, some regions must, by definition, see less warming than the global average – one region, in particular, projected to warm less rapidly is the North Atlantic between Greenland and the UK (Figure 2.3). It is suggested that this is a result of a slowdown of the meridional overturning circulation (Meehl et al., 2007). Looking at zonal changes within the oceans, modelling suggests that the warming is limited to the mixed layer through the 2020s, with warming penetrating into the deeper ocean occurring later in the twenty-first century.

The increases in global ocean temperatures outlined above result in an increase in seawater volume, and hence thermosteric sea-level rise. By the end of the century, this thermal expansion could be 1.9 ± 1.0, 2.9 ± 1.4 and 3.8 ± 1.3 mm/year under the scenarios SRES B1, A1B and A2, respectively, using the IPCC AR4 global climate model ensemble (Meehl et al., 2007). According to IPCC AR4, through the twenty-first century, thermal expansion contributes between 0.10 and 0.41 m to sea-level rise, a large proportion of the projected total global mean sea-level rise of 0.18–0.59 m under the full range of SRES scenarios including A1FI. The remainder is from the melting of glaciers, ice caps and the Greenland and West Antarctic ice sheets, which increases the total mass of water in the oceans. The IPCC acknowledges that the contribution from melting ice sheets is largely unconstrained and could increase the upper end of the sea-level rise and ocean-mass increase estimates. The rate of global mean sea-level rise has increased from a few centimetres per century in recent millennia, to a few tens of centimetres per century in recent decades (Milne et al., 2009). Some studies forecast a global sea-level rise by 2100 that is significantly higher than that projected in the IPCC AR4. Notably, Rahmstorf (2007) proposes a rise of between 0.5 and 1.4 m, whereas Pfeffer et al. (2008) estimates an upper bound of 2 m by the end of the century. These studies remain controversial because they rely on major approximations, but they do highlight the urgent need for better understanding and quantification of the processes of ice dynamics, which are of huge importance to global sea-level rise projections.

High-latitude regions

The effects of anthropogenic climate change may be greater and occur more rapidly at high latitudes. The polar regions exhibit large interannual and decadal variability in their climates. This, alongside difficulties in representing the tele-connections and the interactions of the atmosphere–land–cryosphere–ocean–ecosystem feedback models at high latitudes within global and regional climate, results in a degree of uncertainty in the projections, especially for the Antarctic, which is greater than that ascribed to many other regions (Christensen et al., 2007).

Climate change projections in the polar regions, particularly the Arctic, are sensitive to changes in sea-ice extent due to the ice–albedo feedback mechanism (Ridley et al., 2008). Projections for the twenty-first century suggest a pattern of major declines in summer sea-ice extent around the Arctic and, to a lesser degree, declines in winter extent. Projected changes to sea-ice thickness are largely uncon-strained. As discussed, the restrictions in modelling capabilities in the Arctic and Antarctic regions result in a wide spread of possible futures.

In 2007, the IPCC AR4 concluded that the Arctic (north of 60° N) is very likely (>90% probability) to warm over the twenty-first century, with the annual mean warming very likely to exceed the global mean warming, with some seasonal vari-ation. Alongside rises in precipitation of up to 28%, temperature increases in this region under the A1B scenario could be between 2.8°C and 7.8°C, with a median of 4.9°C, by 2080–2099, relative to 1980–1999 (Christensen et al., 2007). Similar temperature projections under A1B for Alaska and northern Asia (including Kam-chatka) are 3.0–7.4°C, with a median of 4.5°C, and 2.7–6.4°C, with a median of 4.3°C, respectively. Precipitation in these regions is projected to increase by up to a quarter. In 'high-end' projections under the A2 scenario, projections of global warming of ≥4°C suggest that surface temperatures in much of the Arctic could rise by 15°C by the 2090s (Sanderson et al., 2012).

Not only is warming intimately related to the sea-ice cover in the region, but it would also have a direct impact on the extent, depth and timing of snow cover across the Arctic, and glacial mass in the mountainous areas of Alaska and Kam-chatka, in addition to influencing the thaw depth of permafrost areas. Stendel & Christensen (2002) project the thickness of the permafrost active layer to increase by 30–40% across the Northern Hemisphere towards the end of the twenty-first century, with the extent of permafrost retreating polewards. However, modelling the large-scale responses of permafrost to climate change remains a considerable challenge.

When considering ice-mass loss in Greenland, recent observations and studies have reinforced the findings of IPCC AR4. It stated that the margins of the ice sheet are thinning and that, despite inland increases in mass, the overall balance of the ice sheet is negative and is therefore contributing to current, and future, sea-level rise (Lemke et al., 2007; Allison et al., 2009). Antarctica is also undergo-ing some ice-mass loss in the West Antarctic ice sheet, although there is significant variation across the continent and the much larger East Antarctic ice sheet is actu-ally increasing in mass (Lemke et al., 2007). Throughout the twenty-first century, the projected changes in precipitation in Antarctica (from −2% to +35%) and the increases in temperature (1.4–5.0°C) and sea-level rise may further influence the disintegration of ice shelves and the contribution of the West Antarctic ice sheet to rising sea levels.

Mountain regions

Mountain regions susceptible to geospheric responses to climate change include the European and New Zealand Alpine ranges, the Pyrenees, Caucasus, Andes and the Himalayas.

In all of these regions, mean temperatures are projected to rise, with extreme precipitation and temperature events increasing in both magnitude and frequency. In southern Europe (as defined by Christensen et al., 2007) including the mountainous regions of the Alps and Pyrenees, annual mean temperatures are projected to rise by between 2.2°C and 5.1°C (median of 3.5°C) by the end of the twenty-first century under the A1B scenario. The greatest increases in seasonal mean temperature are projected to occur in summer; this is a key point because the impacts of warming may be more important in some seasons than others, particularly in relation to the seasonality of temperature fluctuations across the freezing point and consequent effects on glacier mass balance and freeze–thaw weakening of slopes. The New Zealand Alps are projected to see smaller rises in average temperature, whereas increases in the high-altitude regions of Asia are projected to be comparable to those of southern Europe (Christensen et al., 2007). In the 'high-end' subset of simulations with the A2 scenario, the multi-model average warming projected over the Himalayas reached 8°C or more in December–January–February, and 7°C in June–July–August (Sanderson et al., 2012).

A warming climate is projected to influence precipitation patterns across these mountainous regions. Most global climate models and RCMs are not, however, of sufficiently high resolution to represent the complex topography of these regions, leading to inadequate representation of contemporary precipitation; consequently, many regional projections for the twenty-first century are often regarded as unreliable for use in detailed impact investigations. Nevertheless, analysis of large-scale circulation patterns within these climate models can inform general precipitation projections. Within the European Alps, studies (e.g. Christensen & Christensen, 2007; Déqué et al., 2007) have shown that precipitation is expected to intensify over the summer months. Annually, however, precipitation in this region, in common with New Zealand, is projected to decrease. Across Europe, the snow season could be shortened, with a 50–100% decrease in snow depth by 2100, depending on altitude, and a shift in precipitation from snow to rain as temperatures increase. This could be accompanied by a rise of the snow line by an average of 150 m for every degree Celsius increase (Christensen et al., 2007). Glaciers are also expected to retreat in many regions, such as the Himalayas. However, the snow pack and ice mass of higher-altitude regions and some of the very cold, northern areas of Scandinavia and Russia could be less sensitive to rising temperatures than the lower latitudes and altitudes of Europe where temperatures are already generally nearer freezing point.

Volcanic landscapes

Both glaciated and non-glaciated volcanic landscapes are susceptible to the afore-mentioned global and regional projections of climate change, most notably as a consequence of temperature rises and changes in extreme precipitation. The Cascade Range volcanoes of western North America, for example, could see increases in annual mean temperature of 2.1–5.7°C (with a median of 3.4°C), together with a −3% to +14% change in precipitation. For the Andes, warming by the end of the century could reach 4.5–5°C (Vuille et al., 2008), under an SRES A2 emission scenario. As noted by McGuire (2010), close to 60% of all active volcanoes are coastally located or form islands, whereas most of the balance occurs within 250 km of the coast. As discussed earlier, this means that most active volcanic systems have the potential to be influenced by changes in crustal stress and strain associated with ocean loading caused by future sea-level rise. The disposition of tectonic plates ensures a non-random distribution of active volcanoes, with large concentrations at high latitudes (e.g. Alaska, Kamchatka and Iceland) and in the tropics (including Indonesia, Papua New Guinea and the Philippines).

High-latitude volcanoes are often glaciated and consequently susceptible to many of the same climate-change-driven, deglaciation-related hazards as mountainous terrains of non-volcanic origin.

Non-glaciated, high-relief volcanic regions, including in the Caribbean, Europe, Indonesia, the Philippines and Japan, could also be affected by climate change, principally due to modified precipitation patterns and especially a rise in the frequency and magnitude of severe rainfall events. Of particular concern in many of these regions could be changes to the frequency and intensity of tropical cyclones. Following the active hurricane seasons in the North Atlantic in 2004 and 2005, the World Meteorological Organisation (WMO) made a statement that 'No individual tropical cyclone can be directly attributed to climate change' (IWTC, 2006). As an active area of research, a number of studies have been conducted in recent years examining the influence of the warming climate on tropical cyclones around the world (Shepherd & Knutson, 2007). Within the IPCC's AR4, the conclusion was reached from a study of coarse-resolution climate models (50–100 km grid spacing) alongside finer-scale projections (down to approximately 9 km grid spacing) that climate change could lead to future tropical cyclones having increased peak wind intensities and more intense precipitation, particularly towards the centres of the storms (Meehl et al., 2007). However, many of these studies also show fewer tropical storms occurring globally, and with varying degrees of change to cyclone intensities (Sugi et al., 2002; McDonald et al., 2005; Chauvin et al., 2006; Oouchi et al., 2006; Yoshimura et al., 2006; Bengtsson et al., 2007; Caron & Jones, 2008).

Conclusions

On palaeoclimate timescales, enhanced levels of geological and geomorphological activity have been linked to climatic factors, including examples of processes that are expected to be important in current and future anthropogenic climate change. Planetary warming leading to increased rainfall, ice-mass loss and rising sea levels is potentially relevant to geospheric responses in many geologically diverse regions. Anthropogenic climate change therefore has the potential to alter the risk of geological and geomorphological hazards through the twenty-first century and beyond. Such changes in risk have not yet been systematically assessed.

An appropriate application of climate model projections over a range of timescales, combined with an understanding of geohazards, can go some way towards improving understanding of these changing risks. This requires quantification of possible changes in relevant climate quantities in the regions of interest, considering both the potential climate changes under unmitigated emissions that could still be avoided, and the residual climate changes that would not be avoidable by mitigation and to which adaptation will therefore be necessary. Within these two categories, there are large uncertainties in global and regional climate responses.

Scenarios of unmitigated climate change project global warming of between 1.6°C and 6.9°C by 2100 relative to pre-industrial level. If GHG concentrations are stabilized in the lower part of this range by the end of this century, warming would continue by approximately a further 0.4–0.8°C by 2300 for low- and medium-emission scenarios – the ongoing commitment to warming has not been assessed for higher warming scenarios, but the warming commitment appears to increase with the temperature at which stabilization occurs. Political aims of keeping global warming <2°C above pre-industrial level may be achievable if global emissions peak during the 2010s and decline by several per cent per year thereafter, but longer delays in peaking emissions further decrease the likelihood of achieving this ambition.

Regional climate changes have been assessed in detail for only a limited number of projections that do not cover the full range of projected global changes, but certain key messages can still be drawn from the available studies, e.g. much of the ocean surface is expected to warm by less than the global average, although the Arctic Ocean is expected to warm by more than the global average. Over land, most regions are projected to warm by more than the global average.

Under the unmitigated scenarios, global mean sea level is projected by IPCC to rise by 0.18–0.59 m by 2100 relative to present day, including thermal expansion and ice loss from Greenland and Antarctica but without future rapid dynamical changes in ice flow. IPCC recognized high uncertainty in dynamic ice processes and did not provide a best estimate or upper bound on sea-level rise projections – more recent work has attempted to quantify the implications of rapid ice flow changes, but projections of global sea-level rise of up to 2 m by 2100 remain controversial. Future committed sea-level rise is already significant and stabilization of GHG concentrations by 2100 could lead to further committed rise for the next millennium. A significant factor in this committed rise would be whether major ice sheets become committed to irreversible decline.

Future precipitation changes are subject to even higher uncertainty and, in the case of annual and seasonal mean precipitation, even the sign of the change cannot be confidently given in many cases. The majority of climate models agree on projections of increased precipitation in the high latitudes, but projections in other regions vary in sign. However, precipitation events are expected to become more intense.

When assessing the potential for geospheric responses to climate change, it appears prudent to consider regional levels of warming of 2°C above pre-industrial level as being potentially unavoidable as an influence on processes requiring a human adaptation response within this century. At the other end of the scale when considering changes that could be avoided by reduction of emissions, scenarios of unmitigated warming exceeding 4°C in the global average include much greater local warming in some regions, such as 8°C in the Himalayas and 15°C in the Arctic. However, considerable further work is required to better understand the uncertainties associated with these projections, uncertainties inherent not only in the climate modelling but also in the linkages between climate change and geospheric responses.

Acknowledgements

The work of RAB and FL was supported by the Joint DECC and Defra Integrated Climate Programme – DECC/Defra (GA01101)

References

Allison, I., Alley, R. B., Fricker, H. A., Thomas, R. H. & Warner, R. C. (2009) Ice sheet mass balance and sea level. *Antarctic Science* **21**, 413–426.

Bengtsson, L., Hodges, K. I., Esch, M., et al. (2007) How may tropical cyclones change in a warmer climate? *Tellus A* **59**, 539–561.

Betts, R. A., Collins, M., Hemming, D. H., Jones, C. D., Lowe, J. A. & Sanderson, M. (2012) When could global warming reach 4°C? *Philosophical Transactions of the Royal Society of London* **369**, 67–84.

Caron, L-P. & Jones, C. G. (2008) Analysing present, past and future tropical cyclone activity as inferred from an ensemble of coupled global climate models. *Tellus A* **60**, 80–96.

Cavalieri, D. J., Parkinson, C. L. & Vinnikov, K. Y. (2003) 30-Year satellite record reveals contrasting Arctic and Antarctic decadal sea ice variability. *Geophysical Research Letters* **30**, 1970.

Chauvin, F., Royer, J-F. & Déqué, M. (2006) Response of hurricane-type vortices to global warming as simulated by ARPEGE-Climat at high resolution. *Climate Dynamics* **27**, 377–399.

Christensen, J. H. & Christensen, O. B. (2007) A summary of the PRUDENCE model projections of changes in European climate by the end of this century. *Climate Change* **81**, S7–S30.

Christensen, J. H., Hewitson, B., Busuioc, A., et al. (2007) Regional climate predictions. In: Solomon, S., Qin, D., Manning, M., et al. (eds), *Climate Change: The Physical Science Basis. Contribution of Working Group I to the Fourth Assessment Report of the Intergovernmental Panel on Climate Change.* Cambridge: Cambridge University Press, pp 847–940.

Davenport, C. A., Ringrose, P. S., Becker, A., Hancock, P. & Fenton, C. (1989) In: Gregersen, S. & Basham, P. W. (eds), Earthquakes at North-Atlantic passive margins: neotectonics and postglacial rebound). NATO Advanced Science Institutes Series, Series C. Mathematics and Physical Sciences, vol. 266. Dordrecht, The Netherlands: Kluwer, pp 175–194.

Déqué, M., Collins, M., Hemming, D. H., Jones, C. D., Lowe, J. A. & Sanderson, M. (2007) An intercomparison of regional climate simulations for Europe: assessing uncertainties in model projections. *Climate Change* **81**, S53–S70.

DiNezio, P. N., Clement, A. C., Vecchi, G. A., Soden, B. J., Kirtman, B. P. & Lee, S-K. (2009) Climate response of the equatorial Pacific to global warming. *Journal of Climate* **22**, 4873–4892.

Dommenget, D. (2009) The ocean's role in continental climate variability and change. *Journal of Climate* **22**, 4939–4952.

Dowsett, H. J., Thompson, R., Barron, J., et al. (1994) Joint investigations of the Middle Pliocene climate. *Global Planetary Change* **9**, 169–195.

Friele, P. A. & Clague, J. J. (2004) Large Holocene landslides from Pylon Peak, southwestern British Columbia. *Canadian Journal of Earth Sciences* **41**, 165–182.

Hansen, J. (2003) Can we defuse The Global Warming Time Bomb? *Natural Science.* Available at: http://naturalscience.com/ns/articles/01-16/ns_jeh.html (accessed 15 February (2010).

Hansen, J., Johnson, D., Lacis, A., et al. (1997) Climate impacts of increasing atmospheric carbon dioxide. In: Goudie, A. (ed.), *The Human Impact Reader.* Oxford: Blackwell, pp 230–247.

Hawkins, E. & Sutton, R. (2009) The potential to narrow uncertainty in regional climate predictions. *Bulletin of the American Meteorological Society* **90**, 1095–1107.

Hermanns, R. L., Niedermann, S., Ivy-Ochs, S. & Kubik, P. W. (2004) Rock avalanching into a landslide-dammed lake causing multiple dam failure in Las Conchas valley (NW Argentina) – evidence from surface exposure dating and stratigraphic analyses. *Landslides* **1**, 113–122.

Hetzel, R. & Hampel, A. (2005) Slip rate variations on normal faults during glacial–interglacial changes in surface loads. *Nature* **435**, 81–84.

Huybers, P. & Langmuir, C. (2009) Feedback between deglaciation, volcanism and atmospheric CO_2. *Earth and Planetary Science Letters* **286**, 479–491.

Intergovernmental Panel on Climate Change (IPCC), Solomon, S., Qin, D., Manning, M., et al., eds (2007a) *Climate Change 2007: The physical science basis.* Contribution of Working Group I to the Fourth Assessment Report of the Intergovernmental Panel on Climate Change. Cambridge: Cambridge University Press.

IPCC, Metz, B., Davidson, O. R., Bosch, P. R., Dave, R. & Meyer, L. A., eds (2007b) *Climate change 2007: Mitigation of climate change.* Contribution of Working Group III to the Fourth Assessment Report of the Intergovernmental Panel on Climate Change. Cambridge: Cambridge University Press.

IWTC (2006) WMO Tropical Meteorology Research, International Workshop on Tropical Cyclones. Statement on tropical cyclones and climate change. Available at: www.wmo.ch/pages/prog/arep/tmrp/documents/iwtc_statement.pdf (accessed February 2010).

Joshi, M. M., Gregory, J. M., Webb, M. J., Sexton, D. M. H. & Johns, T. C. (2007) Mechanisms for the land/sea warming contrast exhibited by simulations of climate change. *Climate Dynamics* **30**, 455–465.

Jull, M. & McKenzie, D. (1996) The effect of deglaciation on mantle melting beneath Iceland. *Journal of Geophysical Research* **101**, 815–828.

Lateltin, O., Beer, C., Raetzo, H. & Caron, C. (1997) Landslides in flysch terranes of Switzerland: causal factors and climate change. *Eclogae Geologica Helvetica* **90**, 401–406.

Lee, H. J. (2009) Timing of occurrence of large submarine landslides on the Atlantic Ocean margin. *Marine Geology* **264**, 53–64.

Leggett, J. A. & Logan, J. (2008) Are carbon dioxide emissions rising more rapidly than expected? CRS Report RS22970, Congressional Research Service, Washington, DC (17 October 2008 version).

Lemke, P., Ren, J., Alley, R. B., et al. (2007) Observations: changes in snow, ice and frozen ground. In: Solomon, S., Qin, D., Manning, M., et al. (eds), *Climate Change 2007: The physical science basis.* Contribution of working group I to the fourth assessment report of the Intergovernmental Panel on Climate Change. Cambridge: Cambridge University Press, pp 337–383.

Licciardi, J., Kurz, M. & Curtice, J. (2007) Glacial and volcanic history of Icelandic table mountains from cosmogenic ^{3}He exposure ages. *Quaternary Science Reviews* **26**, 1529–1546.

McDonald, R. E., Bleaken, D. G., Cresswell, D. R., Pope, V. D. & Senior, C. A. (2005) Tropical storms: representation and diagnosis in climate models and the impacts of climate change. *Climate Dynamics* **25**, 19–36.

McGuire, B. (2010) Potential for a hazardous geospheric response to projected future climate changes. *Philosophical Transactions of the Royal Society of London A* **368**, 2317–2345.

McGuire, W. J., Howarth, R. J., Firth, C. R., et al. (1997) Correlation between rate of sea level change and frequency of explosive volcanism in the Mediterranean. *Nature* **389**, 473–476.

Meehl, G. A. et al. (2007) Global climate projections. In: Solomon, S., Qin, D., Manning, M., et al. (eds), *Climate Change 2007: The physical science basis.* Contribution of working group I to the fourth assessment report of the Intergovernmental Panel on Climate Change. Cambridge: Cambridge University Press, pp 747–846.

Met Office (2007) Climate research at the Met Office Hadley Centre: informing government policy in the future. Exeter, UK: Met Office. Available at: www.metoffice.gov.uk/publications/brochures/clim_res_had_fut_pol.pdf (accessed February 2010).

Milne, G. A., Gehrels, W. R., Hughes, C. W. & Tamisiea, M. W. (2009) Identifying the causes of sea level change. *Nature Geoscience* **2**, 471–478.

Min, S-K., Zhang, X., Zwiers, F. W. & Agnew, T. (2008) Human influence on Arctic sea ice detectable from early (1990)s onwards. *Geophysical Research Letters* **35**, L21701.

Nakicenovic, N. et al. (2000) *IPCC Special Report on Emissions Scenarios.* Cambridge: Cambridge University Press.

Oouchi, K., Yoshimura, J., Yoshimura, H., Mizuta, R., Kusunoki, S. & Noda, A. (2006) Tropical cyclone climatology in a global-warming climate as simulated in a 20 km-mesh global atmospheric model: frequency and wind intensity analyses. *Journal of the Meteorological Society of Japan* **84**, 259–276.

Pfeffer, W. T., Harper, J. T. & O'Neill, S. (2008) Kinematic constraints on glacier contributions to 21st-century sea-level rise. *Science* **321**, 1340–1343.

Rahmstorf, S. (2007) A semi-empirical approach to projecting future sea-level rise. *Science* **315**, 368–370.

Ridley, J., Lowe, J. & Simonin, D. (2008) The demise of Arctic sea ice during stabilisation at high greenhouse gas concentrations. *Climate Dynamics* **30**, 333–341.

Rothrock, D. A., Yu, Y. & Maykut, G. A. (1999) Thinning of the Arctic sea ice cover. *Geophysical Research Letters* **26**, 3469.

Rothrock, D. A., Zhang, J. & Yu, Y. (2003) The Arctic ice thickness anomaly of the 1990s: a consistent view from observations and models. *Journal of Geophysical Research* **108**, 3083.

Sanderson, M. G., Hemming, D. L. & Betts, R. A. (2012) Regional temperature and precipitation changes under high-end (>4°C) global warming. *Philosophical Transactions of the Royal Society of London* **369**, 85–98.

Shepherd, J. M. & Knutson, T. (2007) The current debate on the linkage between global warming and hurricanes. *Geographical Compass* **1**, 1–24.

Sigvaldasson, G., Annertz, K. & Nilsson, M. (1992) Effect of glacier loading/deloading on volcanism: postglacial volcanic eruption rate of the Dyngjufjoll area, central Iceland. *Bulletin of Volcanology* **54**, 385–392.

Stendel, M. & Christensen, J. H. (2002) Impact of global warming on permafrost conditions in a coupled GCM. *Geophysical Research Letters* **29**, 1632.

Stroeve, J. C., Serreze, M. C., Fetterer, F., et al. (2005) Tracking the Arctic's shrinking ice cover: another extreme September minimum in 2004. *Geophysical Research Letters* **32**, L04501.

Sugi, M., Noda, A. & Sato, N. (2002) Influence of the global warming on tropical cyclone climatology: an experiment with the JMA global model. *Journal of the Meteorological Society of Japan* **80**, 249–272.

Sutton, R. T., Dong, B. & Gregory, J. M. (2007) Land/sea warming ratio in response to climate change: IPCC AR4 model results and comparison with observations. *Geophysical Research Letters* **34**, L02701.

van Vuuren, D. P., de Vries, B., Beusen, A. & Heuberger, P. S. C. (2008) Conditional probabilistic estimates of 21st century greenhouse gas emissions based on the storylines of the IPCC-SRES scenarios. *Global Environmental Change* **18**, 635–654.

Vuille, M., Francou, B., Wagnon, P., et al. (2008) Climate change and tropical Andean glaciers – past, present and future. *Earth-Science Reviews* **89**, 79–96.

Wu, P. & Johnston, P. (2000) Can deglaciation trigger earthquakes in N. America? *Geophysical Research Letters* **27**, 1323–1326.

Wu, P., Johnston, P. & Lambeck, K. (1999) Postglacial rebound and fault instability in Fennoscandia. *Geophysical Journal International* **139**, 657–670.

Yoshimura, J., Sugi, M. & Noda, A. (2006) Influence of greenhouse warming on tropical cyclone frequency. *Journal of the Meteorological Society of Japan* **84**, 405–428.

3 Climate change and collapsing volcanoes: evidence from Mount Etna, Sicily

Kim Deeming[1], Bill McGuire[1] and Paul Harrop[2]

[1] Aon Benfield UCL Hazard Centre, Department of Earth Sciences, University College London, UK
[2] Tessella plc., Abington, UK

Summary

Similar to many volcanoes, Sicily's Mount Etna underwent major collapse of its flanks during pre-history. Here we present evidence for Early Holocene climate conditions providing circumstances favourable to major lateral collapse at Mount Etna, Sicily. The volcano's most notable topographic feature is the Valle del Bove, a 5×8 km cliff-bounded amphitheatre excavated from the eastern flank of the volcano. Its origin due to prehistoric lateral collapse is corroborated by *stürtzstrom* deposits adjacent to the amphitheatre's downslope outlet, but the age, nature and cause of amphitheatre excavation remain matters for debate. Cosmogenic helium-3 (^{3}He) exposure ages determined for eroded surfaces within an abandoned watershed flanking the Valle del Bove support channel abandonment at about 7.5 ka BP (thousands of years before present), as a consequence of its excavation in a catastrophic collapse event. Watershed development was largely dictated by pluvial conditions during the Early Holocene, which are also implicated in slope failure. A viable trigger is magma emplacement into rift zones in the eastern flank of a water-saturated edifice, leading to the development of excess pore pressures, consequent reduction in sliding resistance, detachment and collapse. Such a mechanism is presented as one potential driver of future lateral collapse in volcanic landscapes forecast to experience increased precipitation or melting of ice cover as a consequence of anthropogenic warming.

Introduction

The start of the Holocene, some tens of thousands of years before the present (10 ka BP), marks a dramatic transformation in the Earth's system and its component parts – the atmosphere, hydrosphere and geosphere (e.g. Roberts, 1998;

Figure 3.1 General view of Mount Etna; looking north from the south flank town of Nicolosi.

Walker & Bell, 2004). Rapid planetary warming promoted a major reorganisation of the global water budget as continental ice sheets melted to replenish depleted ocean volumes. Contemporaneously, atmospheric circulation patterns changed to accommodate broadly warmer, wetter conditions, leading to modification of major wind trends and a rearrangement of climatic zones. A volcanic response to post-glacial environmental change is recognised (e.g. Zielinski et al., 1997), attributable to ice unloading in glaciated volcanic terrains (e.g. Jull & McKenzie, 1996; Licciardi et al., 2007; Carrivick et al., 2009), to ocean loading associated with global sea-level rise (McGuire et al., 1997a) and to increased precipitation arising from warmer, wetter conditions (Capra, 2006).

Here we present evidence, from Mount Etna in Sicily (Figure 3.1), for Holocene environmental change driving catastrophic failure of the volcano's eastern flank. This resulted in the excavation of the Valle del Bove amphitheatre, the volcano's most pronounced topographic feature (Figure 3.2). Catastrophic lateral collapse and ensuing *stürtzstrom* (rapidly emplaced rock avalanche) formation is now recognised as marking a common, transitory stage in the lifecycles of many long-lived volcanoes (Ui, 1983; Siebert, 1984; Holcomb & Searle, 1991; McGuire, 1996, 2006), ranging from stratovolcanoes to basaltic shields; in some cases such behaviour is displayed many times in the history of a single, long-lived volcano. Lateral collapse is a natural consequence of growing gravitational instability, as an edifice increases in mass and volume. Instability is often compounded by steep slopes, a mechanically weak structure, and sometimes by high precipitation rates associated with a topographically elevated location. A destabilised edifice can be induced to collapse in one of a number of ways: notably increased ground accelerations associated with accompanying earthquakes, elevated mechanical stresses due to gravitational loading or dyke injection, and pore-water pressurisation, the latter two being associated with the emplacement of fresh magma. Lateral collapse velocities typically exceed 40 m/s, and may reach >100 m/s (Ward & Day, 2001, 2003), with collapse volumes ranging from <1 km^3 to >10 km^3 at many

Figure 3.2 View of the southern boundary wall of the Valle del Bove flank collapse amphitheatre, taken in 1980. Since this time, much of the amphitheatre's floor has been covered with recent lava flows that reach thicknesses of more than 100 m in places.

continental and subduction zone volcanoes, and to $\geq 1000 \, km^3$ at the great basalt shields of Hawaii (McGuire, 1996). Lateral collapse events are ubiquitous, with approaching 500 recognised at more than 300 volcanoes, and a time-averaged rate of close to 20 collapses a century over the last 500 years (R. Lowe, unpublished data). In relation to the Holocene climate, the lateral failure of volcanoes may be promoted by increased availability of water as glaciated volcanoes lose their ice caps and precipitation levels rise as the world becomes warmer and wetter (Capra, 2006), thereby favouring the development of increased pore-fluid pressures in edifices.

Lateral collapse at Mount Etna

With a current height of 3340 m, and occupying an area of $1250 \, km^2$, Mount Etna is arguably the largest volcano located on continental crust. It has grown and developed over a period of 580 ka (Branca et al., 2004) on the east coast of Sicily, upon a seaward-dipping substrate of sediments and basement rocks. Etna is a polygenetic volcano constructed from several centres of activity which show a broad age progression (oldest to youngest) from east to west (Figure 3.3). The earliest central-vent activity was concentrated in the vicinity of the Valle del Bove, leading to the formation of the Calanna centre, and later the Trifoglietto and associated centres further to the west. At around 80 000 years BP (Gillot et al., 1994; Branca et al., 2004), activity switched to the north-west, building the large central-vent stratovolcano known as the Ellittico. After caldera collapse around 15 000 years BP, activity became focused at the current centre of activity (Mongibello) (see summit craters in Figure 3.3), which is located within the Ellittico caldera. Historical activity is marked by eruptions from the summit vents and also by effusive eruptions lower down on the flanks which are fed by laterally propa-

Figure 3.3 Digital elevation model of Mount Etna showing the location of the Valle del Bove relative to the current summit of the volcano. Also indicated are ancient centres of activity: Ellittico (E), Trifoglietto and associated centres (Tr), and Calanna (C). The extents of the Chiancone fanglomerates (Ch) and the Milo Lahars (ML) are also delineated. Black arrows indicate the positions and directions of dyke propagation of the defunct south-east and east–north-east rift zones. (Tarquini S. et al., 2007. TIN/ITALY/01: a New Triangular Irregular Network of Italy. Ann. Geophys-Italy 50, 407–425 with permission.)

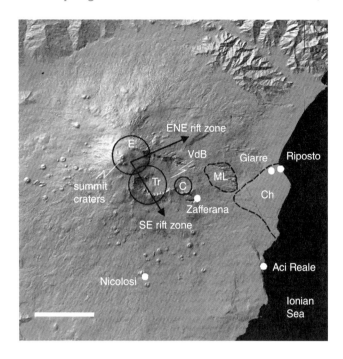

gating dykes. These follow preferential routes that define rift zones, the most important of which trend north and south from the summit (McGuire & Pullen, 1989).

The volcano's most notable topographic feature is the Valle del Bove (see Figures 3.2 and 3.3), a large (5 × 8 km) amphitheatre cut into the eastern flank, bounded by cliffs up to 1 km high and opening towards the Ionian Sea. More than 100 exposed dykes in the cliff walls mark the positions of east–north-east and south-east trending defunct rift zones (McGuire & Pullen, 1989), the intersection of which is bisected by the long axis of the collapse amphitheatre (Figure 3.3). Although an origin due to prehistoric lateral collapse is now corroborated by subaerial and offshore *stürtzstrom* deposits adjacent to the amphitheatre's open, downslope outlet (e.g. Calvari et al., 1998; Del Negro & Napoli, 2002; Pareschi et al., 2006a; Pareschi et al., 2006b), important aspects of the Valle del Bove have remained a matter for debate – most notably the nature and cause of its excavation, and its age. Models for its origin are diverse and include marine (Lyell, 1849) or fluvial (Lyell, 1858) erosion, glacial excavation (Vagliasindi, 1949), gravity-driven progressive landsliding (Guest et al., 1984), edifice spreading (Borgia et al., 1992), eruption-triggered catastrophic failure (Kieffer, 1977; McGuire, 1982), and tectonically or magmatically driven failure after formation of the Ellittico caldera (Calvari et al., 2004).

Age estimates for the collapse are also disparate, ranging from approximately 5 ka BP to 25 ka BP (Kieffer, 1970; McGuire, 1982; Guest et al., 1984; Chester et al.,

1987; Calvari & Groppelli, 1996). Accumulating evidence, however, increasingly supports a Holocene age for Valle del Bove formation. Calvari et al (1998) allocate a carbon-14 (^{14}C) age of 8400 years BP (corresponding to an intercept of radiocarbon age with calibration curve at 6400 BC) to a palaeosol on top of a debris flow unit immediately above a *stürtzstrom* component within a sequence of debris flows (the Milo Lahars – ML in Figure 3.3) exposed immediately east of the Valle del Bove's seaward outlet. The authors link the *stürtzstrom* to an initial excavation phase of the Valle del Bove that they propose, on a petrological basis, involved the northern part of the Valle del Bove only and was linked to lateral collapse of the Ellittico eruptive centre. They recognise, however, that this interpretation may be erroneous, an alternative explanation being that *stürtzstrom* debris derived from the southern flank of the Valle del Bove is not encountered within the exposed fraction of the Milo Lahar sequence. More recently, Calvari et al. (2004) reiterate the link with the Ellittico centre, proposing formation of an initial Valle del Bove due to lateral collapse 'about 10 000 years ago' of the prehistoric Ellittico eruptive centre, with subsequent enlargement occurring through episodes of minor collapse. A date of 5340 ± 60 years BP (Coltelli et al., 2000) for an overlying scoria layer provides an upper limit to the age of the Milo Lahar sequence, which is reasonably explained in terms of the downslope reworking of *stürtzstrom* debris in the millennia immediately following Valle del Bove formation (Calvari et al., 2004).

Additional support for an Early- to Mid-Holocene collapse age comes from the Chiancone (Ch in Figure 3.3), an approximately 300-m thick (Ferrara, 1975) fanglomerate sequence, the exposed units of which are primarily fluviatile, with secondary debris flow and pyroclastic units (Calvari & Groppelli, 1996). The Chiancone crops out immediately east of the Milo Lahars and takes the form of a gently sloping topographic fan, with an area of 40 km^2, extending out into the Ionian Sea. The Chiancone, and its submarine extent, has a volume of at least 14 km^3 (Del Negro & Napoli, 2002) and is accepted (e.g. Rittmann, 1973; McGuire, 1982; Guest et al., 1984; Calvari & Groppelli, 1996; Del Negro & Napoli, 2002; Pareschi et al., 2006a) as representing reworked volcanic products from the Valle del Bove in its upper portion and – in its deeper, unexposed levels – an earlier lateral collapse event, or events, for which the topographic evidence is now lost. Calvari and Groppelli (1996) report a ^{14}C age of 7590 ± 130 years BP from a palaeosol about 1 m above a debris flow unit at the base of the exposed portion of the Chiancone. This they propose provides a minimum age for an 'important eruptive event' that may have been associated with the emplacement of the underlying debris flow unit. The age is comparable to that of the Milo Lahars and very close to that obtained for channel abandonment, and could plausibly reflect the timing of redeposition downslope of *stürtzstrom* material derived from lateral collapse. The approximately 30 m of mainly fluviatile material deposited above this basal (to the exposed part of the Chiancone) debris flow unit is ascribed by Calvari and Groppelli (1996) to post-formation enlargement of the Valle del Bove and continued downslope reworking in episodes of high-energy fluvial transport.

Flank failure and watershed abandonment at Mount Etna

Although effusive activity from the youngest, major eruptive centres (Ellittico 80–15 ka and Mongibello 15 ka BP; Branca et al., 2004) have extensively remodelled and resurfaced the topography of much of the volcano, the outer flanks of the Valle del Bove preserve prominent, deeply incised valleys (Figures 3.4 and 3.5) reminiscent of the parasol drainage pattern distinctive of volcanoes in tropical environments with high rainfall and surface runoff. The valleys are the downslope remnants (Figure 3.5a) of a more vertically extensive channel system which originally drained elevated topography in that part of the volcano now occupied by the Valle del Bove, but is now truncated by its enclosing cliff wall. The channel system, which is inactive today, may have been initiated by meltwater associated with the disintegration of late Pleistocene snow and ice fields that capped the Ellittico eruptive centre at the time of the Last Glacial Maximum (Neri, 2002). The watershed will have been reduced in size after caldera collapse of the Ellittico caldera approximately 15 000 years BP, although it may have benefited from snow melt and runoff associated with a caldera lake or lakes (Del Carlo et al., 2004). Its current form is, however, interpreted primarily as a reflection of erosive drainage

Figure 3.4 Digital elevation model of the Valle del Bove showing the prominent, deeply incised, truncated valleys on the outer northern and southern flanks of the Valle del Bove. AR: Acqua Rocca; AT: Acqua del Turco. Locations of samples from which cosmogenic ^{3}He exposure dates were obtained are also shown (see Table 3.1 for additional information including altitude, longitude, latitude and nature of sampled exposed surface). Samples other than those indicated were all obtained from Acqua Rocca. The location of the baseline test sample (ET00KD1) at Due Monti (Lat: 37.47.04.80; Long: 15.03.06.02) is not shown. (Tarquini S. et al. (2007); TIN/ITALY/01: A New Triangular Irregular Network of Italy. Ann. Geophys-Italy 50, 407–425 with permission.)

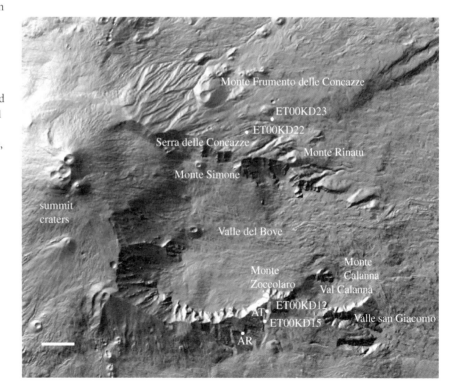

Figure 3.5 The truncated Valle degli Zappini incised into the outer southern flank of the Valle del Bove, with the summit craters of Etna visible in the distance. (a) The Acqua Rocca palaeo-waterfall is obvious in the bottom centre of the image. (b) Close-up view of the 75 m high Acqua Rocca palaeo-waterfall.

events associated with the anomalously (for Etna) warmer and wetter conditions that prevailed in Sicily during the Early Holocene (Frisia et al., 2006). Evidence for highly energetic fluvial flow is provided by the presence of palaeo-waterfalls up to 75 m high (Figure 3.5b), large (>1 m), rounded and polished in-channel boulders and erosive scour features at elevations of up to 3 m above current channel floors.

Here, we seek to better constrain the timing and nature of the Valle del Bove lateral collapse using cosmogenic helium-3 (^{3}He) exposure dating to determine maximum fluvial erosion-surface ages within the drainage channels. These we interpret as providing a best-estimate minimum age for channel system abandonment due to destruction of its watershed by the excavation of the Valle del Bove, the assumption being that this event effectively removed the channel catchments resulting in a cessation of erosive fluvial episodes.

Cosmogenic ^{3}He exposure dating of channel abandonment at Mount Etna

Through determining the duration of surface exposure to extraterrestrial cosmic rays, cosmogenic exposure dating (Kurz, 1986) allows the ages to be established of processes that expose, exhume or extrude rock. High-energy cosmic rays bombard the Earth and, after attenuation by the geomagnetic field and atmosphere, and shielding by topography, they generate cosmogenic nuclides in surface exposed rocks. One of the more common cosmogenic nuclides generated *in situ* by this bombardment is ^{3}He. This is a stable isotope that is retained in the crystal lattices of the minerals garnet, hornblende, olivine and clinopyroxene, the last two of which are common constituents of Etnean volcanic products. As ^{3}He is a particularly rare, naturally occurring isotope on the Earth, small amounts produced by cosmic ray bombardment of the Earth's surface leave a strong signature in surface-exposed minerals.

Cosmogenic exposure dating, based on *in situ* generated nuclides, is an established tool used for providing quantitative estimates of the timing and rate of a range of geomorphological and geological processes including uplift, resurfacing, burial, faulting, fluvial erosion, marine action, glacial retreat and rapid mass movement (see Cockburn & Summerfield, 2004 and references therein for a comprehensive review of geomorphological applications). On volcanoes, the method is increasingly widely utilised to date lava flows and cinder cones, to build and constrain eruptive chronologies, and to shed light on processes and mechanisms such as rift-zone reorganisation (e.g. Kurz et al., 1990; Poreda & Cerling, 1992; Cerling & Craig, 1994; Foeken et al., 2009). Cosmogenic exposure dating has also been successfully utilised to constrain the timing of lateral collapses at Piton de la Fournaise (Réunion Island) (Staudacher & Allègre, 1993), Tenerife (Harrop, 1996) and Fogo (Cape Verde Islands) (Foeken et al., 2009), through dating of appropriate lava flows and faults that indirectly bracket the timing of the collapse.

Palaeo-fluvial features have also been targeted in cosmogenic exposure dating campaigns (e.g. Cerling, 1990; Cerling et al., 1994), and examples of the application of the technique to dating bedrock in fluvial channels can be found in Pratt et al. (2002) and Schaller et al. (2005). Most relevant to this study, Seidl et al. (1997) conducted a cosmogenic exposure dating campaign along the length of a river valley on Kauai island (Hawaii), sampling bedrock, boulders, channel sides and a waterfall lip for multiple cosmogenic nuclide analysis, with the aim of elucidating valley evolution.

In this study at Mount Etna, sample sites (see Figure 3.4) were selected according to availability of exposed rock containing a high percentage of clinopyroxene phenocrysts, and at elevated locations above channel floors to minimise the chances of (1) previous burial by tephra fall, reworked tephra or sediment and (2) erosive influences from any post-truncation flow associated with localised heavy rainfall. Samples for analysis were collected from three truncated channels that showed good preservation of bedrock and other exposed surfaces and were least affected by rock falls (Table 3.1): the Acqua Rocca and Acqua del Turco valleys on the southern flank of the Valle del Bove, and an unnamed truncated valley on its northern flank in the vicinity of the Serra delle Concazze.

In total, 28 samples were collected for the purposes of cosmic ray exposure dating yielding 16 ages (Table 3.1 and see Figure 3.4). The remaining 12 samples were excluded from analysis due to insufficient, suitable, phenocryst material.

In the Acqua Rocca, a 75-m-high palaeo-waterfall located in the Vallone Acqua Rocca degli Zappini, 18 samples were taken from a variety of sites, including in-channel boulders and bedrock surfaces, 11 of which yielded cosmogenic ^{3}He ages. Exposure in the neighbouring Acqua del Turco channel and tributaries, immediately to the east, was significantly poorer. As a result, four samples were collected, one from an in-channel boulder and the others from exposed bedrock. Just two of the samples provided ages. Three exposed bedrock samples were taken in the unnamed north flank valley, two in the vicinity of a palaeo-waterfall, both of which yielded an age. In order to validate the method, three samples (two from in-channel boulders and one from exposed bedrock) were also collected from the actively eroding Due Monti valley on the north flank, which is known to still act as a conduit for rainfall runoff and snow-melt and would consequently be expected to yield a zero age. Only one sample provided an age.

Cosmogenic ^{3}He was extracted from clinopyroxene crystals liberated from bulk rock samples. Crushing of samples released phenocrysts of clinopyroxene from the matrix, which were concentrated and separated using a combination of magnetic and heavy liquid separation techniques, and then thoroughly cleaned. Gas extracted from the separated phenocrysts was purified before analysis in an MAP 215 90° sector noble gas mass spectrometer. Comparison with a well-characterised calibration gas (air enriched with artificially mixed helium) yielded accurate helium isotope ratios and abundances required for exposure age calculation. For individual samples, the exposure age (t) is determined on the basis of the cosmogenic ^{3}He abundance (N) and a given production rate (P) of *stable* cosmogenic nuclides, calculated from $t = N/P$. The production rate is sensitive to variation in altitude, latitude, sample depth, past geomagnetic field variations and topographic shielding, for all of which subsequent corrections are applied. Deeming (2002) provides a detailed discussion of sampling strategies, preparation methods, analytical techniques, correction process and age determination.

Table 3.1 Surface exposure ages determined from abandoned drainage channels on the northern and southern flanks of the Valle del Bove

Sample	Location	Latitude	Longitude	Altitude (m)	Exposed surface	Exposure age (years BP)	Interpretation of age yielded
ET00KD2021	Acqua Rocca	37/42/25.75	12/02/49.49	1500	Bedrock; waterfall top	25 966 ± 6149	Analytical error
ET00KD22	Unnamed valley: north flank	37/45/36.38	15/03/05.96	1925	Bedrock; waterfall base	7419 ± 1256	Minimum channel abandonment age for northern flank valley
ET00KD7	Acqua Rocca	37/42/25.75	15/02/49.49	1400	Bedrock; waterfall base	7222 ± 949	Minimum channel abandonment age for Acqua Rocca
ET00KD15	Acqua Turco	37/42/32.23	15/03/16.05	1475	Elevated chute	5858 ± 1031	Post-abandonment burial and re-exposure
ET99KD21	Acqua Rocca	37/42/25.75	15/02/49.49	1400	Bedrock; waterfall base	4426 ± 636	Post-abandonment burial and re-exposure
ET99KD16	Acqua Rocca	37/42/25.75	15/02/49.49	1500	Bedrock; tor at waterfall top	3787 ± 881	General weathering
ET99KD17	Acqua Rocca	37/42/25.75	15/02/49.49	1500	In-channel boulder	3061 ± 411	Boulder disturbance
ET00KD19	Acqua Rocca	37/42/25.75	15/02/49.49	1500	Bedrock; pothole	3038 ± 407	Post-abandonment burial and re-exposure
ET99KD15	Acqua Rocca	37/42/25.75	15/02/49.49	1500	In-channel boulder	2014 ± 315	Boulder disturbance
ET99KD20	Acqua Rocca	37/42/25.75	15/02/49.49	1500	Bedrock; pothole	1985 ± 489	Post-abandonment erosion
ET00KD23	Unnamed valley: north flank	37/45/44.48	15/03/29.47	1700	Bedrock; waterfall base	1840 ± 576	Post-abandonment erosion
ET00KD18	Acqua Rocca	37/42/25.75	15/02/49.49	1500	Bedrock; scour	1409 ± 434	Post-abandonment burial and re-exposure
ET99KD18	Acqua Rocca	37/42/25.75	15/02/49.49	1500	Bedrock; scour	0 (131) ± 398	Unknown
ET00KD12	Acqua Turco	37/42/42.78	15/03/20.14	1575	Questionable bedrock	0 (−1253) ± −812	Recent rock fall
ET99KD19	Acqua Rocca	37/42/25.75	15/02/49.49	1500	Bedrock; scour	0 (−3217) ± −412	Unknown
ET00KD1	Due Monti	37/47/04.80	15/03/06.02	1675	In-channel boulder	0 (−39) ± −75	Baseline test sample: actively eroding channel

All errors are to one standard deviation. Interpretations of younger ages are tentative and based upon local (to sample) conditions. Ages designated as 'zero' are shown together with the actual determined dates (in brackets).

Results and interpretations

Cosmogenic exposure dating of eroded surfaces within abandoned drainage channels on an active volcano is not straightforward. This is a consequence of the potential of a suite of past and contemporary processes that could operate to deliver cosmogenic exposure ages that post-date the abandonment event. These include periodic tephra accumulation, rare erosive (and limited) streamflow associated with highly localised extreme precipitation, and boulder saltation or movement due to earthquake-related ground shaking, together capable of shielding original surfaces or exposing new ones. Tephra shielding is considered as the most likely source of younger exposure ages from channel samples; the prevailing north-westerly wind direction typically transporting ash and scoriaceous fall material to the south-east of the Mongibello (active since 15 ka BP) summit vents and across the sample sites. During historic times, individual explosive events are likely to have deposited significant thicknesses of tephra in the vicinity of the sample sites. A caldera-forming event in 122 BC, for example, deposited 0.25 m in the city of Catania, 25 km south south-east of the Valle del Bove's southern flank, with estimated thicknesses of 0.5 m in the vicinity of the Acqua Rocca and 0.10–0.25 m across the northern flank of the Valle del Bove (Coltelli et al., 1998). Accumulations of several centimetres of tephra in the sample areas have also been observed by one of the authors (McGuire) on a number of occasions over the last 30 years, associated with moderate explosive events at the summit vents. Although individual tephra burial events may be temporary (months to decades), such is the level of tephra production at the summit vents that cumulative shielding arising from repeated burial over thousands of years would be sufficient to reduce exposure ages significantly.

As a consequence of post-abandonment processes, the expected product of a cosmogenic exposure dating campaign targeting surfaces within such a channel system is an array of ages, the oldest of which will provide a best-estimate *minimum* age for channel abandonment. For Etna, the campaign returned an array of 16 ages (see Table 3.1), 15 of these ranging from zero to 7419 ± 1256 years BP, together with a single outlier (sample ET00KD2021) of 25966 ± 6149 years BP determined from exposed bedrock near the lip of the waterfall in the Acqua Rocca valley. We propose that the outlier is an artefact of the analytical process and address this issue in more detail later.

Discounting this age leaves comparable (within error) maximum ages for each of the abandoned channels sampled: 7419 ± 1256 years BP (ET00KD22) for the unnamed northern flank valley; 7222 ± 949 years BP (ET00KD7) for the Acqua Rocca and 5858 ± 1031 years BP (ET00KD15) for the Acqua Turco. We accept the first two ages as approximating to the best-estimate minimum ages for abandonment of, respectively, the unnamed north flank valley and the Acqua Rocca on the southern flank. Similarity of the ages supports abandonment that was effectively simultaneous on both flanks and occurred approximately 7500 years BP. The somewhat younger maximum age for the Acqua Turco may be an artefact of burial and re-exposure. It is notable that the oldest ages were both delivered by samples taken at elevated positions (2 m for ET00KD7 and 5 m from ET00KD22) above the valley floors and at the bases of palaeo-waterfalls. Such waterfalls are unlikely to have been utilised after destruction of the channel watersheds, thus minimising post-abandonment erosion, whereas elevation of sample sites would minimise the potential for tephra burial. In contrast, the Acqua Turco location was only 0.5 m above the floor and therefore more likely to have undergone burial by tephra for significant periods of time. Although it is possible that a larger sample of dates might deliver an older age, we have confidence that the above dates are close to the true age of channel abandonment. Although some later expansion of the Valle del Bove must be expected due to gravitational instabilities, we argue that the main event was catastrophic, thereby truncating the channels and decapitating the watershed for both flanks simultaneously.

Younger ages in the array are interpreted as representing different exposure histories of the samples (see Table 3.1). Based on local (to sample) conditions, these are most likely to be a consequence of boulder disturbance, perhaps due to: rare high-energy, fluvial action or earthquake-related ground shaking (samples ET99KD17, ET99KD15); variable burial (by tephra, debris, rock falls) and (re-) exposure (samples ET99KD21, ET00KD19, ET00KD18); post-abandonment fluvial erosion (ET99KD20, ET00KD23); or unknown cause (ET99KD18, ET99KD19). A 3787 ± 881 year BP age obtained for sample ET99KD16, taken from an elevated tor within the Acqua Rocca, is probably a consequence of general weathering, whereas the zero age (actually −1253 ± −812 years BP) delivered by sample ET00KD12 seems to reflect doubts in the field about the nature of the sample site, and is probably derived from a block in a recent rock fall. ET00KD1 is a control sample collected from an in-channel boulder within the Due Monti active drainage channel on the north flank of the Valle del Bove. This actively eroding sample was expected to provide a zero exposure age. The determined zero age (actually −39 ± 75 years BP) obtained validates use of the method on Etna and supports its utility in constraining timing of the cessation of high-energy fluvial flow along the sampled abandoned channels. All zero ages determined during the campaign demonstrate that insufficient time had elapsed to accumulate significant abundances of cosmogenic ^3He_c (less than about 300 years) and are recorded as negative values. These negative ages are the product of insufficient ^3He release from the fusion step to account for all the inherited ^3He. This may be the result of the crush $^3He{:}^4He$ ratio being different from the $^3He{:}^4He$ ratio of the trapped He in the fusion step contrary to the assumption in equation (1) below that these values are the same. Such variations may be caused by radiogenic He or sample shielding, e.g.

$$^3He_c = {}^3He_f - {}^4He_f[{}^3He/{}^4He]_i \qquad\qquad (1)$$

whereby the cosmogenic ^3He component (subscript c denotes cosmogenic) can be derived using the inherited component $^3He{:}^4He$ ratio from the crushing (subscript i denotes the inherited component) to correct the ^3He_f (subscript f denotes release by fusion) for any remaining ^3He_i. The crush-released Ne was used to correct the ^4He_i for atmospheric contamination assuming that all the Ne was atmospheric (Cerling et al., 1994).

The single outlier (sample ET00KD2021) age of $25\,966 \pm 6149$ years BP may be explained in one of three ways. First, the anomalously high age could represent the true timing of abandonment of the waterfall and the channel system as a whole. Should this be an accurate exposure age, the campaign would have delivered more ages in the 7000–26 000 year BP range.

Assuming that the data from the mass spectrometric analyses are correct, a second possibility is that this sample could have had a long period(s) of exposure before channel abandonment or could have formed by low-energy water flow slowly cutting through the bedrock a few centimetres in total for a large amount of time (approximately 18 000 years). In this scenario, when water flow subsequently ceased about 7000 years BP, 18 000 years of pre-exposure had already accumulated, producing an anomalously high exposure age. The low steady-state erosion rate of 0.03 m/ka required to generate the 'extra' 18 000 years of exposure before the Valle del Bove collapse, assuming a maximum eruption age for the Trifoglietto lava substrate of 65 ka and an erosion depth of 0.54 m, is not, however, compatible with the high-energy flows expected to be associated, first, with melting of summit snow and ice fields, and, second, with the pluvial conditions of the Early Holocene. Such conditions are attested to by the form of the truncated valleys themselves and by fluvial scour features to a height of 3 m above the present channel floor, indicative of deep, energetic, high-volume flows. Further support comes from the Due Monti sample (ET00KD1) detailed below, taken from a currently but ephemerally active valley to the north of the Valle del Bove. This 'baseline test' sample produced a zero exposure age, indicating periodic, high-energy flows affecting these valleys. This demonstrates that, although such valleys are active, cosmogenic nuclide build-up in the surface is negligible.

A third possibility for the outlier age relates to the analytical process. In the case of ET00KD2021, the fusion-released ^4He was unusually low compared with the fusion-released ^3He, with very little ^4He surviving the crushing process in this sample compared with the others. We suspect that, as a consequence, the mass spectrometer was unable to locate the ^4He peak during the analysis routine. In such circumstances, the routine reverts to a stored mass position that appears not to have been accurate in this case. The measurement could not be repeated due to a paucity of sample material. The anomalously large age could therefore be due to analytical error. In addition, very little ^4He_i survived the crushing process in sample ET00KD2021 compared with the rest of the samples. Thus, given that ET00KD2021 is not a gas poor sample, more fusion-released ^4He should have been expected from this sample. The problem is not with a low abundance of fusion-released ^4He being detectable, because sample ET98KD6 released half the amount of ^4He released by ET00KD2021 and was clearly detected. In light of the above discussion, we conclude that the exposure age outlier is most probably due to incorrect reading of the peak position.

Implications of exposure ages for the formation of the Valle del Bove

The approximately 7500 years BP age of collapse is broadly consistent with those dates obtained from the Milo Lahar and Chiancone sequences, and discussed earlier. Although the 8400 ^{14}C age of Calvari et al. (1998) is about 1000 years older than our age for channel abandonment, it is within error of the 7419 ± 1256 year BP age from the northern flank and close to within error of the Acqua Rocca age of (7222 ± 949 years BP) and is highly likely to mark the same event.

Importantly, it is also worth noting that the single ^{14}C age of Calvari et al. (1998) may not be entirely reliable. Radiocarbon ages determined at active volcanoes are known to be notoriously susceptible to contamination by magmatic CO_2, which has the outcome of reducing the ^{14}C content of plants, thereby increasing the apparent age of sampled material, an effect recognised as being particularly significant in carbon dating of material <10 ka old (Sulerzhitzky, 1969). Notably, Sulerzhitzky demonstrated, for the Kamchatka and Kurile volcanic regions, that contamination with magmatically derived carbon could result in recent plant remains yielding ^{14}C ages as old as 6 ka BP. The carbon contamination problem has previously been recognised at Etna, and has been proposed by Guest et al. (1984) to explain a suite of wide-ranging ^{14}C ages on the upper tephra ashes, exposed in a number of valleys truncated by the southern rim of the Valle del Bove. Consequently, it may be that the age of emplacement of the *stürtzstrom* material within the Milo Lahars is somewhat younger than that proposed by the single 8400 ^{14}C age of Calvari et al. (1998).

In linking channel abandonment with the excavation of the Valle del Bove we assume that the sampled valleys were actively eroding up to the time of watershed decapitation. This is supported by the fact that the Early Holocene northern Sicilian climate remained warm and wet at the time indicated by the maximum exposure ages, before the onset of much drier conditions around 6000 years ago (Frisia et al., 2006). Making the assumption that the channel system was initiated 16 ka BP when melting of ice fields capping the Ellittico volcanic centre commenced (Neri, 2002), maximum erosion rates over the period from initiation to abandonment can be determined. For the Acqua Rocca waterfall on the southern flank and at the north flank location of sample site ET00KD22, which yielded the 7419 ± 1256 year BP exposure age these erosion rates are very similar (0.44 m/ka and 0.38 m/ka respectively) and comparable to incision rates from fluvially driven channel erosion in comparable volcanic terrains, such as at Kauai (Hawaii) (Seidl et al., 1997).

Nature of the collapse mechanism

We propose a collapse model predicated upon the onset of early Holocene pluvial conditions in northern Sicily (Figure 3.6), with a sustained wet climate modifying

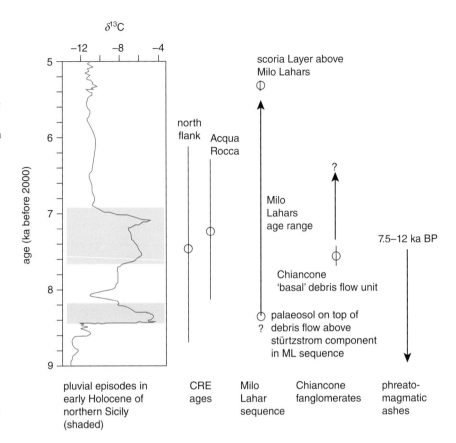

Figure 3.6 Comparison of northern Sicily climate curve and Early-Holocene pluvial episodes (left) (adapted from Frisia et al., 2006) with age brackets of Valle del Bove north and south flank channel abandonment, and age constraints on the Milo Lahars, Chiancone fanglomerates and phreato-magmatic ashes (right). See text for explanation. All ages are indicated by circles with error bars. (Based on data from Frisia, S et al (2006); Holocene climate variability in Sicily from a discontinuous stalagmite record and the Mesolithic to Neolithic transition. Quaternary Res. 66, 388–400 with permission.)

the hydrology of Mount Etna so as to ensure significant elevation of the water table and saturation of the upper part of the edifice. Such conditions are consistent with frequent, high-energy surface run-off and the formation of the deeply incised drainage system now truncated by the Valle del Bove rim. The character of the Milo Lahar and Chiancone sequences, dominated by fluviatile and debris flow lithologies, also support conditions significantly wetter than those prevailing over the last 6000 years, and it is perhaps noteworthy that ages from the top of both sequences coincide with the prevalence of more arid conditions (Figure 3.6). The character of tephra units also attests to water playing a significant role in eruptive activity during the Early Holocene. Del Carlo et al. (2004) (Figure 3.6), for example, report the occurrence of a sequence of phreato-magmatic ashes, ranging in age from 12 000 years BP to 7500 years BP. These they interpret as having been erupted by a series of sub-Plinian eruptions taking place within a lake- or snow-filled Ellittico Caldera, suggesting an elevated water table and saturated edifice at which rising magma and water were able to interact easily. Guest et al. (1984) also speculate that the water table could have been higher at this time, and that the summit caldera could have acted as a trap for snow and meltwater. Although phreato-magmatic eruptions have occurred more recently, most notably at 3150 ± 60 years BP (Del Carlo et al., 2004), such activity does not seem to have been as common during the drier Late Holocene.

Water has long been recognised as playing a critical role in destabilising slopes and triggering landslides, in both non-volcanic and volcanic terrains (e.g. Day, 1996; Voight & Elsworth, 1997). In the latter, the pressurisation of pore water in saturated rock by rising magma has been proposed as an effective trigger for lateral collapse (e.g. Elsworth & Voight, 1995; Elsworth & Voight, 1996; Iverson, 1996; Elsworth & Day, 1999). When magma is emplaced into a saturated, porous material, it mechanically strains the surrounding medium, so resulting in pressurisation of the pore fluid. Magmatic heat will also act to increase pore pressures through thermal expansion of the fluid and by driving groundwater flow capable of spreading excess pore pressures further afield. In concert, the resulting pressurisation can lead to a reduction in the sliding resistance sufficient to initiate lateral collapse. On Etna, opportunity for high-level magma–water interaction, leading to edifice destabilisation and spontaneous collapse, was optimal at a time of maximum utilisation of the drainage channels, when water tables were highest and surface run-off at a level far greater than that observed today.

Lateral collapse under these conditions may have been triggered at Etna through forced dyke emplacement from the central conduit system into intra-volcanic rift zones in the volcano's water-saturated eastern flank – perhaps already seriously weakened by hydrothermal alteration within an elevated water table. Elsworth & Day (1999) show that dykes on the order of 1 m wide and more than around 1 km long are capable of generating mechanical and thermal fluid pressures along a basal décollement and magmastatic pressures at the dyke surface which, together, can trigger failure and lateral collapse. Significantly, deformation monitoring of contemporary dyke intrusion events at Etna reveal a metre or more of horizontal displacement associated with each rifting episode (McGuire et al., 1990). Due to growth of Etna on an uplifting and eastward-tilting basement, high-level magma emplacement at this time occurred preferentially in the form of dykes concentrated along the eastern (east–north-east and south-east) rift zones, the intersection of which is bisected by the long axis of the Valle del Bove, and that today are well exposed in the back wall of the amphitheatre (McGuire et al., 1997b). A contribution to pre-collapse instability may also have been provided by progressive lateral mechanical displacement of the eastern flank associated with repeated dyke emplacement. Whether or not the Valle del Bove lateral collapse triggered a violent explosive eruption as a consequence of the exposure and decompression of magma stored at high levels in the edifice is not known. No particularly distinctive or widespread tephra unit has an age corresponding to the timing of collapse, which may have occurred without significant accompanying eruptive activity. It is also possible that there was no magma involvement at all. If the eastern flank, or a part thereof, was already sliding under its own weight on the tilted basement, then elevated pore pressures alone, or water-driven slow cracking – held to be a contributory factor in the formation of non-volcanic *stürtzstrom* (e.g. Kilburn & Petley, 2003) – may have been sufficient to trigger failure and collapse.

Conclusion

There is strong evidence for lateral collapse at Mount Etna approximately 7500 years BP, coinciding with Early Holocene pluvial conditions in northern Sicily, and for elevated water tables and a saturated edifice being instrumental in promoting instability and initiating collapse. A robust cause-and-effect link remains to be determined, however, justifying further study of both Etna's Valle del Bove and its associated deposits, and potential climate connections with volcano instability and lateral collapse in general. A link between volcano lateral collapse and increased water availability has previously been tentatively proposed. Day et al. (1999, 2000) have argued in favour of a role for elevated pore-fluid pressures in triggering lateral collapses at ocean island volcanoes during warm, wet, interglacial periods. Capra (2006) has also proposed that more humid post-glacial conditions may have promoted failure of unstable volcanic edifices and suggested that future warming may play a similar role. To date, however, ages of collapse events are neither sufficiently numerous, nor well enough constrained, to be able to make a categorical causal link with wetter conditions or other climate parameters (Keating & McGuire, 2004). Notwithstanding this, anthropogenic planetary warming may lead to greater water availability in many volcanic landscapes and may reasonably present increased opportunity for high-level water–magma interaction, together providing greater potential for future lateral collapse. With conditions for elevated water tables promoted through increased meltwater production at ice-capped volcanoes, e.g. in Kamchatka, Alaska, the Andes and Iceland, forecast precipitation increases across volcanic terrains such as Indonesia and Papua New Guinea and at high latitudes (IPCC, 2007, 2011).

Acknowledgements

KRD would like to thank the following: the NERC for making the research possible; Mrs C. Davies for invaluable assistance during chemical procedures and analysis; Mr D. Blagburn and Dr D. Harrison for support during noble gas analysis; Mr B. Clementson and Mr B. Gale for constructing components vital to sample analysis; Dr J. Whitby for help in sample collection; and Dr R. Burgess for advice during thesis and paper compilation.

References

Borgia, A., Ferrari, L. & Pasquarè, G. (1992) Importance of gravitational spreading in the tectonic and volcanic evolution of Mount Etna. *Nature* **357**, 231–235.

Branca, S., Coltelli, M. & Groppelli, G. (2004) Geological evolution of the Etna volcano. In: Bonaccorso, A., Calvari, S., Coltelli, M., Del Negro, C. & Falsaperla,

S. (eds), *Mt. Etna: Volcano Laboratory*. Geophysical Monograph 143. Washington DC: American Geophysical Union, 49–64.

Calvari, S. & Groppelli, G. (1996) Relevance of the Chiancone volcaniclastic deposit in the recent history of Etna volcano, Sicily. *Journal of Volcanology and Geothermal Research* **72**, 239–258.

Calvari, S., Tanner, L. H. & Groppelli, G. (1998) Debris-avalanche deposits of the Milo Lahar sequence and the opening of the Valle del Bove on Etna volcano (Italy). *Journal of Volcanology and Geothermal Research* **87**, 193–209.

Calvari, S., Tanner, L. H., Groppelli, G. & Norini, G. (2004) Valle del Bove, eastern flank of Etna volcano: a comprehensive model for the opening of the depression and implications for future hazards. In: Bonaccorso, A., Calvari, S., Coltelli, M., Del Negro, C. & Falsaperla, S. (eds), *Mt. Etna: Volcano Laboratory*. Geophysical Monograph 143. American Geophysical Union, 65–76.

Capra, L. (2006) Abrupt climate changes as triggering mechanisms of massive volcanic collapses. *Journal of Volcanology and Geothermal Research* **155**, 329–333.

Carrivick, J. L., Russell, A. J., Rushmer, E. L., et al. (2009) Geomorphological evidence towards a deglacial control on volcanism. *Earth Surface Processes and Landforms* **34**, 1164–1178.

Cerling, T. E. (1990) Dating geomorphologic surfaces using cosmogenic ^{3}He. *Quaternary Research* **33**, 148–156.

Cerling, T. E. & Craig, H. (1994) Geomorphology and in-situ cosmogenic isotopes. *Annual Review of Earth and Planetary Sciences* **22**, 273–317.

Cerling, T. E., Poreda, R. J. & Rathbone, S. L. (1994) Cosmogenic ^{3}He and ^{21}Ne ages of the Big Lost River flood, Snake River Plain, Idaho. *Geology* **22**, 227–230.

Chester, D. K., Duncan, A. M. & Guest, J. E. (1987) The pyroclastic deposits of Mount Etna volcano, Sicily. *Geological Journal* **22**, 225–243.

Cockburn, H. A. P. & Summerfield, M. S. (2004) Geomorphological applications of cosmogenic isotope analysis. *Progress in Physical Geography* **28**, 1–42.

Coltelli, M., Del Carlo, P. & Vezzoli, L. (1998) The discovery of a Plinian basaltic eruption of Roman age at Etna volcano (Italy). *Geology* **26**, 1095–1098.

Coltelli, M., Del Carlo, P. & Vezzoli, L. (2000) Stratigraphic constraints for explosive activity for the past 100 ka at Etna volcano, Italy. *International Journal of Earth Sciences* **89**, 665–677.

Day, S. J. (1996) Hydrothermal pore fluid pressure and the stability of porous, permeable volcanoes. In: *Volcano Instability on the Earth and Other Planets* (eds. W. J. McGuire, A. P. Jones and J. Neuberg). Geological Society of London Special Publication **110**, 77–93.

Day, S. J., McGuire, W. J., Elsworth, D., Carracedo, J-C. & Guillou, H. (1999) Do giant collapses of volcanoes tend to occur in warm, wet, interglacial periods and if so, why? Abstract volume, week B, IUGG XXII General Assembly, Birmingham, UK, p174.

Day, S. J., Elsworth, D. & Maslin, M. (2000) A possible connection between sea-surface temperature variations, orographic rainfall patterns, water-table fluctuations and giant lateral collapses of the oceanic island volcanoes. Abstract volume, Western Pacific Geophysics meeting, Tokyo, Japan. WP251.

Deeming, K. R. (2002) Catastrophic lateral collapse at Mount Etna: cosmic ray exposure dating and characterisation. PhD thesis, University of Manchester.

Del Carlo, P., Vezzoli, L. & Coltelli, M. (2004) Last 100 ka tephrostratigraphic record of Mount Etna. In: Bonaccorso, A., Calvari, S., Coltelli, M., Del Negro, C. & Falsaperla, S. (eds), *Mt. Etna: Volcano Laboratory*. Geophysical Monograph 143. American Geophysical Union, 77–89.

Del Negro, C. & Napoli, R. (2002) Ground and marine magnetic surveys of the lower eastern flank of Etna volcano (Italy). *Journal of Volcanology and Geothermal Research* **114**, 357–372.

Elsworth, D. & Day, S. J. (1999) Flank collapse triggered by intrusion: the Canarian and Cape Verde Archipelagoes. *Journal of Volcanology and Geothermal Research* **94**, 323–340.

Elsworth, D. & Voight, B. (1995) Dike intrusion as a trigger for large earthquakes and the failure of volcano flanks. *Journal of Volcanology and Geothermal Research* **100**, 6005–6024.

Elsworth, D. & Voight, B. (1996) Evaluation of volcano flank instability triggered by dyke intrusion. In: McGuire, W. J., Jones, A. P. & Neuberg, J. (eds), *Volcano Instability on the Earth and Other Planets*. Special Publication 110. London: Geological Society of London, 45–53.

Ferrara, V. (1975) Idrogeologia del versante orientale dell'Etna. III Convegno Internationale sulle Acque Sotterranee, Palermo, Sicily, 91–144.

Foeken, J. P. T., Day, S. & Stuart, F. M. (2009) Cosmogenic He-3 exposure dating of the Quaternary basalts from Fogo, Cape Verdes: implications for rift zone and magmatic reorganisation. *Quaternary Geochron* **4**, 37–49.

Frisia, S., Borsato, A., Mangini, A., Spötl, C., Madonia, G. & Sauro, U. (2006) Holocene climate variability in Sicily from a discontinous stalagmite record and the Mesolithic to Neolithic transition. *Quaternary Research* **66**, 388–400.

Gillot, P. Y., Kieffer, G. & Romano, R. (1994) The evolution of Mount Etna in the light of potassium-argon dating. *Acta Vulcanologica* **5**, 81–88.

Guest, J. E., Chester, D. K. & Duncan, A. M. (1984) The Valle del Bove, Mount Etna: its origin and relation to the stratigraphy and structure of the volcano. *Journal of Volcanology and Geothermal Research* **26**, 384–387.

Harrop, P. J. (1996) Dating recent surface processes using cosmic-ray generated ³He in rocks. PhD thesis, University of Manchester.

Holcomb, R. T. & Searle, R. C. (1991) Large landslides from oceanic volcanoes. *Mar. Geotechnol.* **10**, 19–32.

Iverson, R. M. (1996) Can magma-injection and groundwater forces cause massive landslides on Hawaiian volcanoes? *Journal of Volcanology and Geothermal Research* **66**, 295–308.

Intergovernmental Panel on Clinate Cange or IPCC (2007) *Climate Change 2007: The Physical Science Basis*. Cambridge: Cambridge University Press, 996pp.

IPCC (2011) *Managing the Risks of Extreme Events and Disasters to Advance Climate Change Adaptation* (SREX). Available at: http://ipcc-wg2.gov/SREX (accessed March 2011).

Jull, M. & McKenzie, D. (1996) The effect of deglaciation on mantle melting beneath Iceland. *Journal of Geophysical Research* **101**, 815–828.

Keating, B. H. & McGuire, W. J. (2004) Instability and structural failure at volcanic ocean islands and the climate dimension. *Advances in Geophysics* **47**, 175–271.

Kieffer, G. (1970) Les depots detriques et pyroclastiques du versant oriental de l'Etna. *Atti dell'Accademia gioenia di scienze naturali in Catania* **2**, 3–32.

Kieffer, G. (1977) Données nouvelles sur l'origine de la Valle del Bove et sa place dans l'histoire volcanologique de l'Etna. *CR de l'Académie des Sciences Paris* **285**, 1391–1393.

Kilburn, C. R. J. & Petley, D. N. (2003) Forecasting giant, catastrophic slope collapse: lessons from Vajont, northern Italy. *Geomorphology* **54**, 21–32.

Kurz, M. D. (1986) In situ production of terrestrial cosmogenic helium and some applications to geochronology. *Geochimica et Cosmochimica Acta* **50**, 2855–2862.

Kurz, M. D., Colodner, D., Trull, T. W., Moore, R. B. & O'Brien, K. (1990) Cosmic ray exposure dating with in situ produced cosmogenic ^{3}He: results from young Hawaiian lava flows. *Earth and Planetary Science Letters* **97**, 177–89.

Licciardi, J. M., Kurz, M. D. & Curtis, J. M. (2007) Glacial and volcanic history of Icelandic table mountains from cosmogenic He-3 exposure ages. *Quaternary Science Reviews* **26**, 1529–1546.

Lyell, C. (1858) On the structure of lavas which have consolidated on steep slopes, with remarks on the mode of origin of Mount Etna and on the theory of 'Craters of Elevation'. *Philosophical Transactions of the Royal Society of London* **148**, 703–786.

Lyell, C. (1849) On craters of denudation, with observations on the structure and growth of volcanic cones. *Quaterly Journal of the Geological Society of London* **6**, 209–234.

McGuire, W. J. (1982) Evolution of Etna volcano: information from the southern wall of the Valle del Bove caldera. *Journal of Volcanology and Geothermal Research* **13**, 241–271.

McGuire, W. J. (1996) Volcano instability: a review of contemporary themes. In: McGuire, W. J., Jones, A. P. and Neuberg, J. (eds), *Volcano Instability on the Earth and Other Planets.* Geological Society Special Publication **110**, 1–23.

McGuire, W. J. (2006) Lateral collapse and tsunamigenic potential of marine volcanoes. In: *Mechanisms of Activity and Unrest at Large Calderas.* Special Publications 269. London: Geological Society, 121–140.

McGuire, W. J. & Pullen, A. D. (1989) Location and orientation of eruptive fissures and feeder-dykes at Mount Etna: influence of gravitational and regional tectonic stress regimes. *Journal of Volcanology and Geothermal Research* **38**, 325–344.

McGuire, W. J., Pullen, A. D. & Saunders, S. J. (1990) Recent dyke-induced large-scale block movement at Mount Etna and potential slope failure. *Nature* **343**, 357 – 359.

McGuire, W. J., Howarth, R. J., Firth, C. R., et al. (1997a) Correlation between rate of sea-level change and frequency of explosive volcanism in the Mediterranean. *Nature* **389**, 473–476.

McGuire, W. J., Stewart, I. S. and Saunders, S. J. (1997b) Intra-volcanic rifting at Mount Etna in the context of regional tectonics. *Acta Vulcanologica* **9**, 147–156.

Neri, M. (2002) The influence of the Pleistocene glaciers in the morpho-structural evolution of the Etna volcano (Sicily, Italy). *Terra Glacialis* **5**, 9–35.

Pareschi, M. T., Boschi, E., Mazzarini, F. & Favalli, M. (2006a) Large submarine landslides offshore Mt. Etna. *Geophysical Research Letters* **33**.

Pareschi, M. T., Boschi, E., Mazzarini, F. & Favalli, M. (2006b) Lost tsunami. *Geophysical Research Letters* **33**.

Poreda, R. J. & Cerling, T. E. (1992) Cosmogenic neon in recent lavas from the Western United States. *Geophysical Research Letters* **19**, 1863–1866.

Pratt, B., Burbank, D. W., Heimsath, A. & Ojha, T. (2002) Impulsive alleviation during early Holocene strengthened monsoons, central Nepal Himalayas. *Geology* **30**, 911–14.

Rittmann, A. (1973) Structure and evolution of Mount Etna. *Philosophical Transactions of the Royal Society of London* **274**, 5–16.

Roberts, N. (1998) *The Holocene: An environmental history.* Oxford: Wiley-Blackwell, 344pp.

Schaller, M., Hovius, N., Willett, S. D., Ivy-Ochs, S., Synal, H. A. & Chen, M. C. (2005) Fluvial bedrock incision in the active mountain belt of Taiwan from in situ-produced cosmogenic nuclides. *Earth Surface Processes and Landforms* **30**, 955–971.

Seidl, M. A., Finkel, R. C., Caffee, M. W., Bryant Hudson, G. & Dietrich, W. E. (1997) Cosmogenic isotope analyses applied to river longitudinal profile evolution: problems and interpretations. *Earth Surface Processes and Landforms* **22**, 195–209.

Siebert, L. (1984) Large volcanic debris avalanches: characteristics of source areas, deposits, and associated eruptions. *Journal of Volcanology and Geothermal Research* **22**, 163–197.

Staudacher, T. & Allègre, C. J. (1993) Ages of the second caldera of Piton de la Fournaise volcano (Réunion) determined by cosmic ray produced ^{3}He and ^{21}Ne. *Earth and Planetary Science Letters* **119**, 395–404.

Sulerzhitzky, L. D. (1969) Radiocarbon dating of volcanoes. *Bulletin Volcanologique* **35**, 85–94.

Tarquini, S., Isola, I., Favalli, M., et al. (2007) TINITALY/01: A new triangular irregular network of Italy. *Annali di geofisica* **50**, 407–425.

Ui, T. (1983) Volcanic dry avalanches deposits – identification and comparison with non-volcanic debris stream deposits. *Journal of Volcanology and Geothermal Research* **18**, 135–150.

Vagliasindi, C. (1949) L'Etna durante il periodo glaciale e la formazione dela Valle del Bove. *1st Geo-Palaeontol. Univers. Catania Mem.* **2**, 1–80.

Voight, B. & Elsworth, D. (1997) Failure of volcano slopes. *Géotechnique* **47**, 1–31.

Walker, M. & Bell, M. (2004) *Late Quaternary Environmental Change: Physical and human perspectives.* Englewood Cliffs, NJ: Prentice Hall, 376pp.

Ward, S. N. & Day, S. J. (2001) Cumbre Vieja volcano – potential collapse and tsunami at La Palma, Canary Islands. *Geophysical Research Letters* **28**, 397–400.

Ward, S. N. & Day, S. J. (2003) Ritter Island volcano – lateral collapse and the tsunami of 1888. *Geophysical Journal International* **154**, 891–902.

Zielinski, G. A., Mayewski, P. A., Meeker, L. D., et al. (1997) Volcanic aerosol records and tephrochronology of the Summit, Greenland, ice cores. *Journal of Geophysical Research* **102**, 625–640.

4 Melting ice and volcanic hazards in the twenty-first century

Hugh Tuffen

Lancaster Environment Centre, Lancaster University, UK

Summary

Glaciers and ice sheets on many active volcanoes are rapidly receding. There is compelling evidence that melting of ice during the last deglaciation triggered a dramatic acceleration in volcanic activity. Will melting of ice this century, which is associated with climate change, similarly affect volcanic activity and associated hazards? A critical overview is provided here of the evidence for current melting of ice increasing the frequency or size of hazardous volcanic eruptions. Many aspects of the link between ice recession and accelerated volcanic activity remain poorly understood. Key questions include how rapidly volcanic systems react to melting of ice, whether volcanoes are sensitive to small changes in ice thickness, and how recession of ice affects the generation, storage and eruption of magma at stratovolcanoes. A greater frequency of collapse events at glaciated stratovolcanoes can be expected in the near future, and there is strong potential for positive feedbacks between melting of ice and enhanced volcanism. Nevertheless, much further research is required to remove current uncertainties about the implications of climate change for volcanic hazards in the twenty-first century.

Introduction

There is growing evidence that past changes in the thickness of ice covering volcanoes has affected their eruptive activity. Dating of Icelandic lavas has shown that the rate of volcanic activity in Iceland accelerated by a factor of 30–50 after the last deglaciation at about 12 ka (thousands of years) (Maclennan et al., 2002). Analyses of local and global eruption databases have identified a statistically significant correlation between periods of climatic warming associated with recession of ice and an increase in the frequency of eruptions (Jellinek et al., 2004; Nowell et al., 2006; Huybers & Langmuir, 2009). Today the bodies of ice found

Climate Forcing of Geological Hazards, First Edition. Edited by Bill McGuire and Mark Maslin.
© 2013 The Royal Society and John Wiley & Sons, Ltd. Published 2013 by John Wiley & Sons, Ltd.

on many volcanoes are rapidly thinning and receding. These bodies range from extensive ice sheets to small tropical glaciers and thinning is thought to be triggered by contemporary climate change (e.g. Rivera et al., 2006; Björnsson & Pálsson, 2008; Vuille et al., 2008). New data indicate that even modest annual changes in snow thickness may significantly affect the likelihood of eruptions, suggesting that some volcanic systems can be highly sensitive and responsive even to small changes in ice or snow thickness (Albino et al., 2010).

This leads to the following question: will the current ice recession provoke increased volcanic activity and lead to increased exposure to volcanic hazards? In this chapter we analyse our current knowledge of how variations in ice thickness influence volcanism and identify several unresolved issues that currently prevent quantitative assessment of whether activity is likely to accelerate in the coming century. These include the poorly constrained response time of volcanic systems to unloading of ice, uncertainty about how acceleration in volcanic activity scales to the rate and total amount of melting, and the lack of models to simulate how melting of ice on stratovolcanoes may affect magma storage and eruption to the surface. The explosive eruption of the ice-covered Eyjafjallajökull volcano in 2010 is used to illustrate some of these current gaps in our understanding. In conclusion we highlight some of the future research needed for better understanding of how melting of ice may force volcanic activity.

What are hazards for ice- and snow-covered volcanoes, and where are they found?

Many volcanoes are mantled by ice and snow, especially those located at high latitudes or that reach over 4000 m in altitude. Notable examples occur in the Andes, the Cascades, the Aleutian–Kamchatkan arc, Iceland, Antarctica, Japan and New Zealand (Figures 4.1 and 4.2). The nature of ice and snow cover spans a broad spectrum from seasonal snow (Mee et al., 2006), small bodies of ice and firn in summit regions (Houghton et al., 1987; Julio-Miranda et al., 2008), larger Alpine glaciers on volcano flanks (Figures 4.1b and 4.2a) (Rivera et al., 2006; Vuille et al., 2008), thick ice accumulations within summit craters and calderas (e.g. Gilbert et al., 1996; Huggel et al. 2007a), to substantial ice sheets that completely cover volcanic systems (Figure 4.2b) (e.g. Guðmundsson et al., 1997; Corr & Vaughan, 2008).

Historical eruptions at more than 40 volcanoes worldwide have involved disruption of ice and snow (Major and Newhall, 1989), whereas numerous geological studies have enabled the recognition of interactions between volcanoes and ice or snow in ancient eruptions (e.g. Noe-Nygaard, 1940; Mathews, 1951; Gilbert et al., 1996; Smellie, 1999; Lescinsky & Fink, 2000; Mee et al., 2006). Volcanic deposits provide an invaluable record of palaeo-environmental change, such as fluctuations in ice thickness and extent (Smellie, 2008; Smellie et al., 2008; Tuffen et al., 2010), as well as the processes and hazards associated with various types of

Figure 4.1 (a) Explosive phreato-magmatic activity at Grímsvötn, Iceland on 2 November 2004. Photograph courtesy of Matthew Roberts, Icelandic Meteorological Office. (b) A small plume of ash and steam at the ice-covered summit of Mt Redoubt, Alaska in March 2009. (Photograph courtesy of Alaska Volcano Observatory.) (c) Lahar and flood deposits in the Drift River Valley following eruptions at Mt Redoubt in 2009. (Photograph courtesy of Game McGimsey, AVO/USGS.) (d) Aerial view of lahar deposits that destroyed the town of Armero in 1985 after the eruption of Nevado del Ruiz, Columbia. (Photograph courtesy of R. J. Janda, USGS.)

volcano–ice interaction (e.g. Smellie & Skilling, 1994; Smellie, 1999; Lescinsky & Fink, 2000; Tuffen & Castro, 2009; Carrivick, 2007).

Hazards for ice- and snow-covered volcanoes

The presence of ice and snow on volcanoes can greatly magnify hazards, principally because perturbation of ice and snow during eruptions can rapidly generate large volumes of meltwater which are released in destructive lahars and floods. Major and Newhall (1989) compiled a comprehensive global review of historical

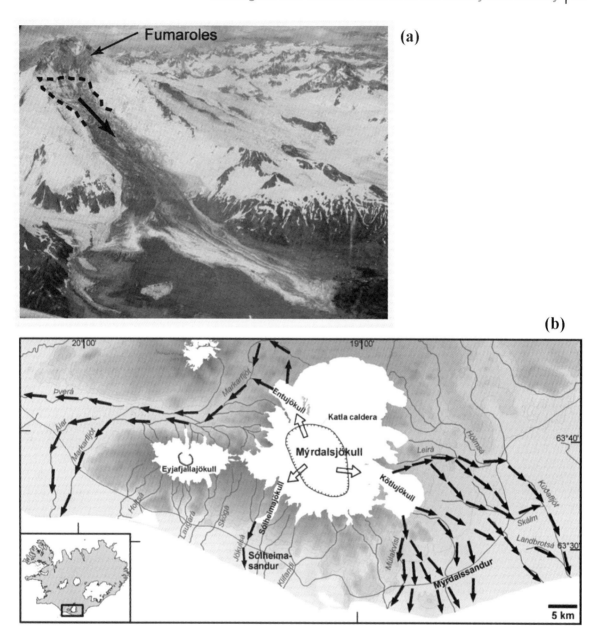

Figure 4.2 (a) Iliamna Volcano, Alaska, showing the path of an ice–rock avalanche that originated from a geothermally active zone high in the summit region. From Huggel, C (2009). Recent extreme slope failures in glacial environments: effects of thermal perturbation. Quat. Sci. Rev. 28, 1119–1130. Photograph courtesy of R.Wessels.) (b) Map of Myrdalsjökull ice cap, Iceland, showing potential drainage directions of jökulhlaups triggered by eruptions at the ice-covered Katla volcano. (From Eliasson J et al (2006). Probalistic model for eruptions and associated flood events in the Katia caldera, Iceland. Comp. Geosci. 10, 179–200 with permission.)

eruptions at more than 40 volcanoes during which ice and snow were perturbed and lahars or floods generated. Major loss of life occurred in several eruptions, including Nevados de Ruiz (Columbia, 1985), Villarrica (Chile, 1971), Tokachi-dake (Japan, 1926) and Cotopaxi (Ecuador, 1877). In 2010 an explosive eruption from Eyjafjallajökull, an ice-covered Icelandic volcano (Guðmundsson et al., 2010; Sigmundsson et al., 2010b), generated an ash plume that caused severe disruption to European aviation, as well as a *jökulhlaup* that caused local damage to infrastructure.

Pertubation of ice and snow by volcanic activity

Major and Newhall identified five distinct mechanisms that can cause perturbation of snow and ice or volcanoes: (1) mechanical erosion and melting by flowing pyroclastic debris or blasts of hot gases (e.g. Walder, 2000) (2) melting of the ice or snow surface by lava flows (e.g. Mee et al., 2006), (3) basal melting by subglacial eruptions or geothermal activity (e.g. Guðmundsson et al., 1997), (4) ejection of water by eruptions through a crater lake and (5) deposition of tephra onto ice and snow (e.g. Capra et al., 2004).

Subsequently, observations of volcanic activity in Columbia, Iceland, USA, New Zealand and Alaska (Waitt, 1989; Pierson et al., 1990; Guðmundsson et al., 1997, 2004, 2009; Carrivick et al. 2009a) have highlighted how rapidly meltwater may be generated during melting of the base of ice sheets and glaciers, and when pyroclastic debris move over ice and snow. Melting rates may exceed $0.5\,km^3$ per day during powerful subglacial eruptions (e.g. Guðmundsson et al., 2004). The hazards associated with meltwater production are exacerbated when transient accumulation occurs with craters or calderas, because this can lead to even higher release rates of meltwater when catastrophic drainage is triggered by dam Guðmundsson or floating of an ice barrier that allows rapid subglacial drainage (Pierson et al., 1990; Guðmundsson et al., 1997; Carrivick et al., 2004, 2009a).

The magnitude of meltwater floods (*jökulhlaups* and lahars) can exceed $40,000\,m^3/s$ (Major & Newhall, 1989; Pierson et al., 1990; Guðmundsson et al., 2004, 2009), creating significant hazards in river valleys and on outwash plains many tens of kilometres from the site of melting (see Figures 4.1c, 4.1d and 4.2b) (e.g. Pierson et al., 1990; Eliasson et al., 2006; Huggel et al., 2007a). The total volume of meltwater floods may be restricted by either the amount of pyroclastic material or lava available to cause melting, or the volume of ice and snow that can be melted.

Explosive eruptions

The hazards posed by explosive eruptions at ice- and snow-covered volcanoes are typical of those at other volcanoes, with the following important modifications:

- Interactions between magma and meltwater may trigger phreato-magmatic activity (see Figure 4.1a), even during basaltic eruptions that would not otherwise be explosive (e.g. Smellie & Skilling, 1994; Guðmundsson et al., 1997).
- When ice is thick the explosive phase of eruptions may partly or entirely take place beneath the ice surface (Tuffen, 2007), reducing the hazards associated with ashfall and pyroclastic debris.
- If explosive eruptions do occur then widespread perturbation of ice and snow by pyroclastic material may be important, both at the vent area and in more distal areas.

Edifice instability and collapse

Ice- and snow-covered volcanic edifices are especially prone to collapse, creating hazardous debris avalanches that may convert to lahars (e.g. Huggel et al., 2007a, 2007b) and reach many tens of kilometres from their source. Collapse is favoured by (1) constraint by ice, which may encourage the development of structurally unstable, over-steepened edifices, (2) melting of ice, which may create weak zones at ice–bedrock interfaces (Huggel, 2009), and (3) shallow hydrothermal alteration driven by snow and ice melt, which can greatly weaken volcanic edifices (e.g. Carrasco-Núñez et al., 1993; Huggel, 2009). Ice avalanches are a newly recognised phenomenon that may occur at ice-covered volcanoes (see Figure 4.2a) (Huggel et al. 2007b). Ice avalanches in the range $0.1–20 \times 10^{6}\,m^{3}$ in volume originate from steep areas near the summit of the Iliama volcano, Alaska, where the geothermal flux is high. These avalanches travel up to 10 km down the volcano flanks at speeds of 20–70 m/s (Huggel, 2009). The thermal perturbations that can trigger slope failure include volcanic/geothermal, glacier permafrost and climatically induced warming.

The distribution of hazards at active ice- and snow-covered volcanoes such as Citlaltepetl, Mexico (lahars; Hubbard et al., 2007), Nevado de Ruiz, Columbia (lahars, avalanches; Huggel et al. 2007a), Mt Rainier, Washington (lahars; Hoblitt et al., 1998), Ruapehu, New Zealand (lahars; Houghton et al., 1987), Iliama, Alaska (lahars, avalanches; Waythomas & Miller, 1999; Huggel et al. 2007b) and Katla, Iceland (*jökulhlaups*, Björnsson et al., 2000) reflect these different sources of volcanic hazard, principally meltwater floods, which potentially affect millions of people living close to these volcanoes. The 2010 Eyjafjallajökull eruption has also demonstrated that the impact of explosive eruptions from ice-covered volcanoes may be far wider reaching than previously anticipated.

How is ice thickness on volcanoes currently changing?

Rapid thinning and recession of ice have been noted on many active and potentially active volcanoes, including Popocatepetl and other Mexican volcanoes (Julio-Miranda et al., 2008), Columbian stratovolcanoes (Huggel et al. 2007a),

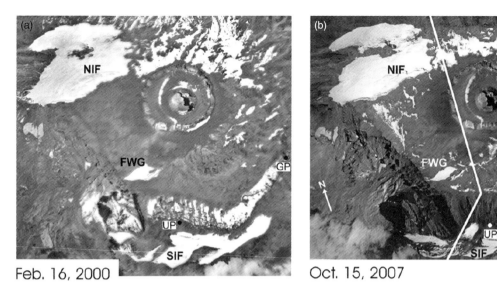

Figure 4.3 Dramatic loss of snow and ice from the summit of Kilimanjaro between 2000 and 2007. (From Thompson et al., (2009). Glacier loss on Kilimanjaro continues unabated. Proc. Nat. Acad. Sci. 106, 19770–19775 with permission.)

Villarrica and other Chilean volcanoes (Rivera et al., 2006), and Kilimanjaro, Tanzania (Figure 4.3) (Thompson et al., 2009). Ice sheets covering volcanic systems are also rapidly thinning, including Vatnajökull in Iceland (Björnsson & Pálsson, 2008) and parts of the West Antarctic Ice Sheet (Wingham et al., 2009). Selected measured or estimated rates of ice thinning and recession are provided in Table 4.1.

Whereas the changing mass balance of thick ice sheets is predominantly manifested in a reduction in ice surface elevation and therefore in ice thickness (e.g. Wingham et al., 2009), the surface area of smaller glaciers on many volcanoes is rapidly reducing, along with a rapid decrease in ice volume. Rates of thinning vary from 0.54 m/year on Kilimanjaro (Thompson et al., 2009) to 1.6 m/year (Pine Island Glacier, West Antarctic Ice Sheet; Wingham et al., 2009). Assuming that current rates of ice loss continue over the coming century, ice bodies on numerous volcanoes may therefore thin by approximately 50–150 m by 2100. Stratovolcanoes hosting thin glaciers, such as Kilimanjaro, may therefore become completely ice free in the coming century (Thompson et al., 2009). At some volcanoes this has already occurred, such as at Popocatepetl, Mexico where dramatic extinction of summit ice over the last 50 years reached completion in 2004 (Julio-Miranda et al., 2008).

In addition, there can be significant annual fluctuations in snow thickness at volcanoes with high annual accumulation and ablation, which are superimposed on any longer-timescale ice thickness fluctuations, e.g. the thickness of snow on Mýrdalsjökull ice cap, which overlies Katla volcano in Iceland, changes by up to 6 m between winter and summer (Albino et al., 2010).

Table 4.1 Current estimated rates of ice thinning and recession at selected volcanoes and ice sheets

Volcano	Last eruption	Area (A) or volume (V) of ice	Rate of thinning	Reference
Vatnajökull, Iceland	2004, 1998, 1996	$A = 8100\,km^2$ $V = 3100\,km^3$ (in 2000)	0.8 m/year average (1995–2008) Geothermal melting and eruptions melted 0.55 km³/year, annual surface ablation 13 km³/year	Pagli & Sigmundsson, 2008; Björnsson & Pálsson, 2008
Volcán Villarrica, Chile	2005, 2007, 2008	$A = 30.3\,km^2$	0.81 ± 0.45 m/year (1961–2004)	Rivera et al., 2006
Popocatepetl, Mexico	1994–2001	A was 0.729 km² in 1958; now 0 km²	1996 ~0.2 m/year 1999 ~4 m/year	Julio-Miranda et al., 2008
Nevado del Ruiz, Columbia	1991	$A = 19$–$25\,km^2$ (1985), 10.3 km² (2002–2003)	Not known	Ceballos et al., 2006; Huggel et al., 2007a
Cotopaxi, Ecuador	1940	$A = 19.2\,km^2$ (1976), 13.4 km² (1997)	3–4 m/year on snouts	Jordan et al., 2005
Kilimanjaro, Kenya/Tanzania	150–200 ka	$A = 2.5\,km^2$ (2000), 1.85 km² (2007)	0.54 m/year	Thompson et al., 2009
West Antarctic Ice Sheet	~2 ka?	$V = 2.2 \times 10^6\,km^3$	Pine Island Glacier ~1.6 m/year, accelerating over 1995–2006	Lythe & Vaughan, 2001; Shepherd et al., 2001; Corr & Vaughan, 2008; Wingham et al., 2009

Ice thinning due to climate change

Much of the current recession and thinning of glaciers and ice sheets covering volcanoes is attributed to the effects of global climate change, with increasing mean temperature and in some cases decreasing precipitation leading to negative glacier mass balance changes (e.g. Rivera et al., 2006; Bown & Rivera, 2007; Vuille et al., 2008). The equilibrium line altitude (ELA) is the altitude on a glacier where the annual accumulation and ablation rates are exactly balanced. The ELA of glaciers on Chilean stratovolcanoes such as Villarrica has migrated upwards by approximately 100 m between 1976 and 2004/2005 (Rivera et al., 2006), partly due to a mean temperature increase at 2000 m elevation of 0.023°C/year (Bown & Rivera, 2007). Ice thinning on Popocatepetl between 1958 and 1994 is likewise thought to be related to climatic change (Julio-Miranda et al., 2008), as is dramatic thinning of tropical mountain glaciers on Ecuadorian volcanoes such as Antizana and Cotopaxi (Vuille et al., 2008).

Thinning of the Vatnajökull ice sheet in Iceland is also pronounced, with an average thinning of 0.8 m/year between 1995 and 2008 (Björnsson & Pálsson,

2008; Pagli & Sigmundsson, 2008). Future prediction of mass balance changes at Vatnajökull in the coming century, which incorporate glacier dynamic models with predicted increases in mean temperature (2.8°C) and precipitation (6%), project a 25% volume loss by 2060 (Björnsson & Pálsson, 2008). The effects of climate change on the mass balance of glaciers and ice sheets on volcanoes is likely to be strongly location specific. This is because changes in temperature and precipitation are spatially heterogeneous and glacier dynamics highly variable. This local sensitivity is illustrated by the contrasting recession rates of glaciers on neighbouring Chilean volcanoes less than 50 km apart (e.g. Rivera et al., 2006; Bown & Rivera, 2007) and highlights the importance of studying individual ice-covered volcanoes, rather than applying a regional or global climate change model (Huybers & Langmuir, 2009) to predict local changes in ice thickness on specific volcanoes.

Ice thinning due to volcanic and geothermal activity

Volcanic and geothermal activity can strongly influence the mass balance of ice bodies on volcanoes during both eruptions and periods of quiescence. The 'background' ablation and accumulation rates determine the overall effects of volcanic and geothermal activity on glacier mass balance and dynamics (Magnusson et al., 2005; Guðmundsson et al., 2009). Mechanisms of ice loss include basal melting and ice disruption during and after subglacial eruptive activity (Figure 4.4a) (e.g. Guðmundsson et al., 1997; Jarosch & Gudmundsson, 2007), melting of the ice and snow surface by the heat of erupted debris (Julio-Miranda et al.,

Figure 4.4 (a) Disruption of ice at the site of the 1998 Grímsvötn eruption, Vatnajökull, Iceland. Photograph courtesy of Magnus Tumi Guðmundsson. (b) Tephra-covered blocks of ice were a last remnant of a now-extinct glacier on Popocatepetl in 2004. (From Julio-Miranda et al., (2008).Impact of the eruptive activity on glacier evolution at Popocatepeti Volacano (Mexico) during 1994–2004. J. Volcanol. Geothe. Res. 170, 86–98 with permission from Elsevier.)

2008), changes to surface albedo due to tephra cover (Figure 4.4b) (Rivera et al., 2006) and lubrication by sustained basal melting due to geothermal heat (e.g. Bell, 2008). Rapid melting, fracturing and mechanical erosion during eruptions can cause dramatic, localised thinning of ice above vents (Figure 4.4a) and meltwater drainage pathways (e.g. Guðmundsson et al., 1997), with removal of tens or hundreds of metres of ice in hours. Perturbations to the ice surface may be transient, however, because depressions formed may swiftly fill due to increased snow deposition and inward deformation of surrounding ice (Aðalgeirsdóttir et al., 2000).

There is strong evidence that volcanic and geothermal activity is hastening the demise of ice bodies on some volcanoes. Eruptive activity at Popocatepetl, Mexico from 1994 to 2001 led to the complete extinction of its small ($<1\,km^2$) summit glaciers (Figure 4.4b) (Julio-Miranda et al., 2008). This extinction reflects the negligible accumulation at a volcano located in an intertropical zone, which makes its glacier mass balance extremely sensitive to eruption-triggered ablation. It is speculated that the disappearance of ice on Popocatepetl was inevitable due to climate change, but greatly hastened by eruptive activity (Julio-Miranda et al., 2008). Recent changes in the mass balance of glaciers on Villarrica volcano, Chile reflect the effects of tephra cover on the ice surface (Rivera et al., 2006). At Villarrica mass balance is also strongly influenced by basal geothermal fluxes. Tephra cover drives enhanced melting when tephra is thin due to enhanced heat absorption, but thicker layers may insulate ice and snow and reduce melting (Rivera et al., 2006; Brock et al., 2007).

Melting and mechanical removal of ice during the 1985 eruption of Nevado del Ruiz, Columbia removed approximately 10% of the volume of ice on the volcano (Ceballos et al., 2006; Huggel et al. 2007a), which totalled $0.48\,km^3$ in 2003. This demonstrates that a single moderately large volcanic eruption (VEI 3) can have an appreciable impact on the mass balance of ice on Andean stratovolcanoes, due to the low ice accumulation rates on tropical glaciers (Ceballos et al., 2006).

By contrast, even considerable volcanically triggered melting probably has a negligible effect on the mass balance of Iceland's Vatnajökull ice sheet over a decadal timescale. This is because Icelandic glaciers and ice sheets are characterised by high annual accumulation and ablation due to temperate conditions and extremely high precipitation (Björnsson & Pálsson, 2008). At Vatnajökull on average $0.55\,km^3$/year was melted by volcanic eruptions during the period 1995–2008, but this amounted to only 4% of the total surface ablation from the ice sheet during this period ($13\,km^3$, Björnsson & Pálsson, 2008). However, the effects of geothermal heat fluxes may have significant effects on ice dynamics and mass balance over both long and short timescales: models of the volume of Vatnajökull at the last glacial maximum are highly sensitive to basal geothermal heat fluxes (Hubbard, 2006) and eruption-triggered jökulhlaups may also trigger surging, which affects glacier mass balance (Björnsson, 1998; Björnsson & Pálsson, 2008).

How has ice recession affected volcanic activity in the past?

Evidence for accelerated volcanism triggered by deglaciation

There is strengthening quantitative evidence linking periods of deglaciation with increased volcanic activity in many different volcanic settings. The best-established and most dramatic acceleration in activity occurred in Iceland, where vigorous volcanism is strongly affected by a temperate ice sheet that may almost completely cover the island during glacial periods and almost completely disappear during interglacials (Björnsson & Pálsson, 2008). Unloading of hundreds of metres to 2 km of ice during deglaciation in Iceland causes decompression which, according to current models, leads to a greater degree and depth range of mantle melting (Jull & McKenzie, 1996; Maclennan et al., 2002). This is reflected in a 30- to 50-fold increase in the rate of magma eruption on individual volcanic systems in the 1.5 ka after the deglaciation of each area, inferred from the volume of erupted deposits (Figure 4.5) (Maclennan et al., 2002). The short time delay between inferred ice unloading and enhanced volcanism shows that the 'extra' magma generated is rapidly transported from source to surface without prolonged storage in magma chambers, so that Icelandic volcanism responds swiftly to changes in ice thickness. In most other volcanic settings magma accumulation in chambers is the norm (e.g. volcanic arcs), in which case the mechanism for enhanced volcanism may differ. It may reflect the response of magma chambers to unloading, rather than the eruption of primitive melts directly to the surface.

Statistical analyses of eruption databases have shown quantitatively that patterns of volcanic activity elsewhere are also influenced by changes in ice thickness: globally (Huybers & Langmuir, 2009; Figure 4.6a), in eastern California (Jellinek et al., 2004) (Figure 4.6b) and in western Europe (Nowell et al., 2006). It is important to note that most statistical studies use a global climate proxy from marine $\delta^{18}O$ records as an indication of ice thickness changes, rather than local ice thickness changes on volcanoes themselves (which are poorly constrained). Further, only the number of eruptions is considered in analyses, rather than the volume of eruptions. Huybers and Langmuir (2009) used a database of global eruptions in the last 40 ka (Siebert & Simkin, 2002) to calculate the change in frequency of eruptions with VEI more than 2 before, during and after the last deglaciation. The increase in volcanic activity during deglaciation above modern values was found to be statistically highly significant ($p < 0.01$) and activity during deglaciation (18–7 ka) was significantly higher than glacial rates between 40 and 20 ka. Although there are doubts about the completeness of the eruption record, interesting trends emerge from the data. The timing of enhanced volcanism differs between localities (e.g. a global increase occurred at approximately 18 ka, but occurred later in Iceland, at about 12 ka). This may reflect differing regional deglaciation histories, although other factors such as the delay between deglaciation and magma reaching the surface may also differ and depend on the plumbing system of individual

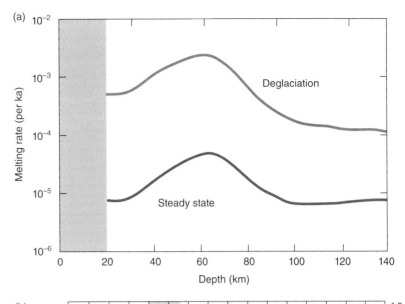

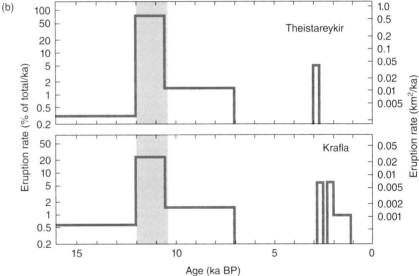

Figure 4.5 Modelled acceleration in melting of the Icelandic mantle during the last deglaciation (from Maclennan et al., 2002. The link between volcanism and deglaciation in Iceland. Geochem. Geophys. Geosyst., 3, 1062 with permission.) (a) Increased rate of melting versus depth in the mantle. The melting rate is the volume of melt produced from each unit volume of mantle per ka. (b) Modelled rate of melt production with a 'spike' between 12 and 11 ka.

volcanic complexes. There is currently no discussion in the literature about whether the magnitude of volcanic eruptions increases during deglaciation, or whether it is only the frequency of eruptions that is affected.

Qualitative evidence for accelerated volcanism at individual volcanic complexes during deglaciation includes studies at Mt Mazama, western USA (Bacon & Lanphere, 2006) and three Chilean volcanoes: Lascar, Puyehue and Nevados de Chillan (Gardeweg et al., 1998; Singer et al., 2008; Mee et al., 2009). However confidence about whether glacial–interglacial cycles truly influence eruptive

Figure 4.6 (a) The ratio of post-glacial (18–7 ka) to glacial (40–20 ka) activity at volcanoes worldwide plotted against a proxy for the amount of ice unloading from ice mass balance models (Huybers and Langmuir, (2009). Feedback between deglaciation, volcanism and atmospheric CO2. Earth Planet Sci. Lett. 286, 479–491 with permission from Elsevier.) Regions with a less negative ice volume balance are those that are most likely to have been glaciated, and thus have experienced significant unloading of ice during the last deglaciation. It is at these regions that the strongest acceleration in the rate of eruptions has occurred, suggesting a causal link between unloading of ice and enhanced volcanic activity. (b) Data from Jellinek et al. (2004) showing the SPECMAP $\delta^{18}O$ curve (a proxy for global ice volume) and the time series of eruptions in the Long Valley and Owens Valley volcanic fields, California. These data are used to show statistically significant correlation between ice unloading and accelerated volcanism. (Jellinek et al (2004). Did melting glaciers cause volcanic eruptions in Eastern California? Probing the mechanics of dike formation. J. Geophys. Res. 109.)

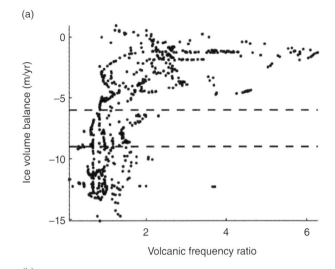

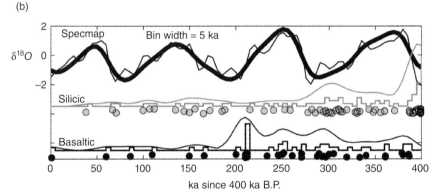

activity is generally low, because there are insufficient dated eruptions at individual volcanoes to adequately test statistical significances. In some cases there is no obvious increase in activity during the last deglaciation (e.g. Torfajökull, Iceland – McGarvie et al., 2006).

Recent studies have found that most of the largest-volume, caldera-forming silicic eruptions from ice-covered Kamchatkan volcanoes have occurred during thick ice conditions (Bindeman et al., 2010; Geyer & Bindeman, 2011). This suggests that the presence of thick ice favours the accumulation of large volumes of silicic magma in chambers, which is confirmed by two-dimensional models of the stress field beneath ice-covered volcanoes (Geyer & Bindeman et al., 2011). However, thick ice is expected to inhibit caldera fault formation and it is unclear whether abrupt ice thinning (e.g. interstadial events) is required for caldera-forming eruptions to occur. It is also unclear whether variations in ice thickness affected the total magma output from Kamchatkan volcanoes, or just the size of individual eruptions.

Edifice collapse triggered by ice recession

A mechanistic link between deglaciation and collapse of ice-covered strato-volcanoes has been proposed by Capra (2008), who noted the coincidence between major edifice collapses and periods of rapid ice recession in the last 30 ka for 24 volcanoes, predominantly located in Chile, Mexico and the USA. She proposed that abrupt climate change resulting in rapid ice melting may trigger edifice collapses through glacial debuttressing, and an increase in fluid circulation and humidity. However, more data are required to quantitatively test whether periods of rapid ice decline do indeed correlate with acceleration in the incidence of edifice collapse.

How does the rate and extent of current ice melting compare with past changes?

In order to assess whether the current changes in ice thickness and extent on many volcanoes are likely to trigger accelerated volcanic activity, the current rate of melting must be compared with inferred rates of melting during the last deglaciation. Precisely reconstructing rates of ice thinning during the last deglaciation is problematic, due to the limits of resolution provided by proxies for changing ice extent and thickness. Furthermore, the history of deglaciation was complex, with major stepwise advances and retreats including the Younger Dryas event at 11–10 ka and the Preboreal Oscillation at 9.9–9.7 ka (Geirsdóttir et al., 2000).

Quoted 'average' rates of deglaciation for Iceland, as used in mantle melting models, are 2 m/year (2 km in 1 ka; Jull & McKenzie, 1996; Pagli & Sigmundsson, 2008). Similarly, the mean rate of surface elevation change of the Laurentide ice sheet during Early Holocene deglaciation is estimated at 2.6 m/year (Carlson et al., 2008). However, it is inappropriate to assume a constant rate of ice unloading, because phases of dramatic warming, such as the end of the Younger Dryas event, are likely to have involved much more rapid recession over shorter time intervals. Indeed, there is geological evidence for bursts of considerably faster deglaciation during abrupt warming events (e.g. 100 m/year in Denmark between 18 and 17 ka, Humlum & Houmark-Nielsen, 1994). Rapid deglaciation in volcanically active areas could be further driven by positive feedback, with eruption-triggered *jökulhlaups* potentially playing an important role in glacier break-up (Geirsdóttir et al., 2000; Carrivick et al. 2009b).

Nevertheless, it is informative to compare data: the current rates are mostly about 20–40% of the mean estimated deglaciation rates for the Icelandic and Laurentide ice sheets (Figure 4.7). The extent of ice unloading is, however, very different, because rapid unloading has occurred only since the end of the Little Ice Age, e.g. Vatnajökull in Iceland has only shrunk since 1890 (Björnsson & Pálsson, 2008; Pagli & Sigmundsson, 2008). This means that total thinning of only about 60 m has occurred at Vatnajökull since 1890, compared with about

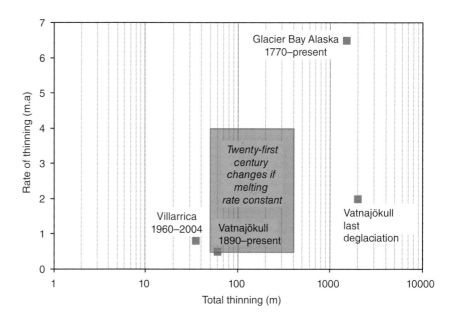

Figure 4.7 Some approximate rates and amounts of ice thinning since the Little Ice Age and during deglaciation, together with projections for the twenty-first century using current rates of ice melting. Note that total thinning may in many cases be limited by the complete extinction of ice e.g. Popocatepetl. (Based on data from Julio-Miranda et al., 2008).

2 km during the last deglaciation (Figure 4.7). Tropical glaciers in the Andes reached their maximum extents of the last millennium at between AD 1630 and AD 1730 (Jomelli et al., 2009), and have only rapidly retreated since the middle of the nineteenth century. Therefore current thinning has only been sustained over a 100- to 200-year period, which is considerably shorter than major deglaciation events. As a consequence the total reduction in ice thickness to date during current warming is probably less than 10% of that during major past deglaciation events.

However, there are marked local and regional discrepancies in how rapidly ice sheets and glaciers have receded during current (post-Little Ice Age) warming. Over 3030 km³ of ice has been lost from Glacier Bay, Alaska since 1770 (Larsen et al., 2005), with local thinning of up to 1.5 km at a mean rate of up to 6.5 m/year. This value is more comparable to changes during the main phases of deglaciation, but has not occurred in an active volcanic region.

How might hazards be affected by melting of ice and snow?

Ice unloading may encourage more explosive eruptions

The explosivity of eruptions beneath ice sheets is restrained by thick ice, as high glaciostatic pressures (>5 MPa) inhibit volatile ex-solution (Tuffen et al., 2010)

Figure 4.8 Results of modelling of rhyolitic eruptions under ice from a 1.5-km-long fissure. The evolution of subglacial cavities during melting and ice deformation is simulated and the combination of ice thickness and magma discharge rate likely to lead to explosive and intrusive eruptions is indicated. Explosive eruptions (above the line) are favoured by thin ice and high magma discharge rates. They are more hazardous than intrusion eruptions since meltwater is produced far more quickly (Guðmundsson, 2003) and eruptions may pierce the ice surface, producing tephra hazards. (Based on data from Tuffen et al., 2007.)

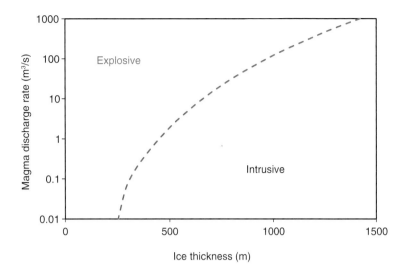

and rapid ice deformation can close cavities melted at the base of the ice, encouraging intrusive rather than explosive activity (Figure 4.8) (Tuffen et al., 2007). Thinning of ice covering a volcano may therefore encourage more explosive eruptions, which generate meltwater more rapidly than intrusive eruptions (Guðmundsson, 2003) and, if the ice surface were breached, would create hazards associated with tephra. Geyer and Bindeman (2011) have recently proposed that assimilation of water-enriched, hydrothermally altered wallrocks may increase the volatile content (and thus explosive potential) of magma residing in chambers within ice-covered edifices. They use two-dimensional stress models to show that periods of thick ice may favour magma accumulation within deep chambers, whereas rapid melting of ice favours initiation of caldera-bounding ring faults and thus large-volume explosive eruptions.

Where ice is thin (<150 m) there is generally comparatively little interaction between magma and meltwater, because thin ice fractures readily, offers little constraint to the force of eruptions and is inefficient at collecting meltwater around the vent (Smellie & Skilling, 1994; Smellie, 1999). Thinning of ice may therefore generally lead to more explosive eruptions at volcanoes that are currently covered by substantial thicknesses of ice (>300 m), especially those with deep ice-filled summit calderas such as Sollipulli, Chile (Gilbert et al., 1996) and Katla, Iceland (Björnsson et al., 2000). It is important to note, however, that there is currently no quantitative relationship between eruption explosivity and ice thickness. The models quoted only simulate a small part of the coupled volcano–ice system and thus are essentially qualitative; they do not incorporate feedbacks between the dynamics of magma storage, ascent and the response of the overlying ice.

Ice unloading and increased melting may trigger edifice stability

It has been hypothesised that melting and recession of ice on volcanic edifices may lead to instability and edifice collapse due to two independent mechanisms: first, debuttressing and the withdrawal of mechanical support from ice (Capra, 2008; Geyer & Bindeman, 2011) and, second, an increase in the pore fluid pressure within shallow hydrothermal systems, which may trigger movement on pre-existing weaknesses (Capra, 2008). However, this hypothesis currently remains unproven due to insufficient data. A significant proportion of glacier meltwater may enter the hydrothermal system of volcanoes (e.g. Antizana volcano, Ecuador; Favier et al., 2008). Seasonal seismicity at volcanoes such as Mt Hood (USA) is consistent with seismic triggering by an increase in meltwater input (Saar & Manga, 2003), illustrating that movement on pre-existing weaknesses is favoured by enhanced meltwater production.

Melting of ice and snow may decrease the likelihood and magnitude of meltwater floods

As the volume of ice and snow on a volcano decreases, the size of the reservoir of potential meltwater decreases. At volcanoes where a relatively small volume of ice and snow is present, the total volume of lahars may be restricted by the volume of ice and snow available for melting (Huggel et al. 2007a). This leads to the following qualitative prediction: as this volume decreases the total volume and magnitude of meltwater floods should decrease for a given size of eruption, thus reducing the associated hazards. Björnsson and Pálsson (2008) have shown that meltwater discharge from thinning Icelandic glaciers is likely to peak in 2040–2050 as ablation rates rise, but thereafter recede, reflecting the diminished volume of ice available for melting. Furthermore, as meltwater floods are triggered when tephra falls onto ice and snow (Major & Newhall, 1989; Walder, 2000; Julio-Miranda et al., 2008), a reduction in the area of ice and snow will reduce the probability that this will occur, therefore reducing the incidence of lahar generation. However, if the size of eruptions were to increase then in some cases a dwindling ice volume would not prevent an increase in the magnitude of meltwater floods, as recognised by Huggel et al. (2007a).

What are the likely effects of twenty-first century climate change on hazards at ice-covered volcanoes?

Unloading as ice and snow melt may trigger increased volcanic activity. Vexed questions include how quickly volcanic systems respond to ice thickness changes, which baselines for rates of volcanic activity are appropriate for the Holocene, and how to scale past accelerations in volcanic activity to changes in the twenty-first century.

Increased magma production and eruption in Iceland?

Melting of Icelandic ice sheets leads to increased mantle melting and eruption of magma to the surface (see Figure 4.5) (Jull & McKenzie, 1996; Maclennan et al., 2002; Pagli & Sigmundsson, 2008). It is estimated that melting of Vatnajökull between 1890 and 2003 (435 km^3 loss, with a thinning rate of about 0.5 m/year) led to a 1% increase in the rate of magma production (Pagli & Sigmundsson, 2008). If current melting rates continue throughout the twenty-first century a roughly similar additional rise in melt production may be anticipated. Any increase in the thinning rate would trigger a stronger acceleration in melt production. It is important to note, however, that the rate and amount of ice thinning are far lower than during the last deglaciation (see Figure 4.7), and the projected increases in the rate of melt production are far weaker (at most an increase of a few per cent, as opposed to a 30- to 50-fold increase). Although studies have shown that additional melt was transported to the surface at a rate of over 50 m/year (Maclennan et al., 2002), this only constrains the timescale of melt extraction to less than 2 ka.

There is incomplete evidence collected to date, but some preliminary data suggest that the timing of Icelandic volcanism during deglaciation may have coincided with rapid warming events, indicating a short delay between extra melting and eruption to the surface. The timing of large tuya-building eruptions in north Iceland appear to correspond with two most marked warming events during deglaciation – the Bolling warming and the end of the Younger Dryas (Licciardi et al., 2007).

We therefore have insufficient knowledge to predict whether the 'extra' melt generated would be erupted to the surface in the twenty-first century and whether any statistically significant increase in activity should be anticipated.

Increased magma production and eruption globally?

The pioneering study by Huybers and Langmuir (2009) attempts to relate changes in global volcanic activity during deglaciation to estimates of the rate of ice unloading. In it they use a simple glacier mass balance model to estimate modern changes in ice thickness at a number of glaciers. This model considers only relative accumulation versus ablation rates and ignores the ice dynamic processes (e.g. Bell, 2008; Wingham et al., 2009) and local variations in precipitation and temperature (e.g. Vuille et al., 2008) that strongly influence mass balance and ice sheet profiles (e.g. Hubbard, 2006). The results of eruption datasets are used to calculate glacial/deglacial and deglacial/Holocene eruption frequency ratios (see Figure 4.6a). Volcanoes with a current strong negative ice–volume balance are excluded from the analysis because they are assumed not to have been ice covered during the late Pleistocene, and therefore insignificant ice unloading is assumed to have occurred during deglaciation. Analysis of the eruption frequency of volcanoes considered to have been ice covered then produces an enhancement in the rate of volcanic activity by a factor of 2 and 6, between 12 and 7 ka. These figures were generated using a −6 m/year and a −9 m/year cut-off, respectively.

Estimates of the amount of increased melting and magma eruption to the surface are very approximate. Huybers and Langmuir assume that unloading 1 km of ice above a 60-km thick melting region triggers a 0.1% increase in the melt percentage, and that 10% of the melt then reaches the surface. They then estimate that 15% of the 1.8×10^6 km^3 of ice lost from mountain glaciers between the Last Glacial Maximum and today influences magma production. The validity of this percentage needs to be checked against the distribution of global ice loss from mountain glaciers, which is itself very difficult to constrain due to a lack of data and the complexity of local climatic variations (e.g. Vuille et al., 2008). The melting model also ignores diversity in melt zone depths and does not take into account crustal storage in magma chambers.

Elsewhere, Jellinek et al. (2004) examine statistical correlations between changes in ice thickness (assumed to be related to the time derivative of the SPECMAP $\delta^{18}O$ record) and the frequency (rather than magnitude) of documented volcanic eruptions in eastern California (see Figure 4.6b). They found a significant correlation, with increased frequency of volcanism after periods of inferred glacier unloading. Models indicated a delay between unloading and increased volcanism of 3.2 ± 4.2 ka and 11.2 ± 2.3 ka for silicic and basaltic eruptions, respectively. Although local ice thickness fluctuations are unlikely to relate consistently or linearly to the oxygen isotope record, this analysis does point to intriguing differences between the rate of response to unloading between different magma types and volcanic plumbing systems. Similarly, Nowell et al. (2006) found evidence for accelerated volcanism during deglaciation of western Europe.

These studies indicate that a statistically significant correlation exists between unloading of ice and increased volcanism. However, as is the case for Iceland, the timescale of the response of volcanic systems to ice unloading is not well constrained. Data from eastern California suggest that the volcanic response may be delayed by thousands of years. If this were the case, volcanism in the coming century could reflect changing ice thicknesses in the Mid- to Late Holocene, rather than melting of ice since the Little Ice Age. Scaling issues are also problematic. There is considerable uncertainty about how the magnitude of acceleration in melt production and magma eruption to the surface scale to the amount and rate of ice unloading. A simple linear relationship between melt production and unloading (e.g. Huybers & Langmuir, 2009) is not appropriate because the rate of melt production also depends on the previous loading and unloading history (Jull & McKenzie, 1996). Furthermore, magma residence in chambers may decouple the timing of melt production from that of magma eruption to the surface.

Potential effects on volcanic hazards

An increase in the rate of magma eruption to the surface would entail larger and/ or more frequent eruptions, thus increasing exposure to hazards. Indeed, analysis of tephra in the Greenland ice core (Zielinski et al., 1996) has shown that the greatest frequency of volcanic events in the last 110 ka occurred between 15 and 8 ka, closely corresponding to the timing of Northern Hemisphere deglaciation. The largest eruptions also occurred during a similar, overlapping interval, between 13 and 7 ka. To date most studies have focused solely on the frequency of eruptions (e.g. Jellinek et al., 2004; Nowell et al., 2006; Huybers & Langmuir, 2009). Increased eruptive frequency at a given volcano will increase risk exposure. The intensity and explosivity index (VEI) of eruptions also scale to their total volume (e.g. Newhall & Self, 1982; Pyle, 1999). There is currently insufficient evidence to determine whether the size or frequency of eruptions will increase in the twenty-first century.

The explosivity of eruptions beneath ice is expected to generally increase as the ice thins (Figure 4.8) (Tuffen et al., 2007). Therefore, where ice is over 150 m in thickness and thinning of more than 100 m occurs, the probability of more hazardous explosive eruptions will increase. This will be most relevant to volcanoes with deep ice-covered calderas such as Sollipulli, Chile (Gilbert et al., 1996). However, it is not currently possible to quantify the increased probability of explosive eruptions and whether it is significant.

There is stronger evidence that current ice recession may considerably increase hazards related to edifice instability. Capra (2008) has proposed that that the incidence of major volcano collapses is strongly affected by ice recession during deglaciation. Huggel et al. (2008) have noted an upturn in the rate of large-volume avalanches, which corresponds with and is attributed to recent climate change. Similar predictions are made for mountain instabilities due to recession of alpine glaciers (see Chapter 10). Melting and unloading of ice may have a much more rapid effect on edifice stability than on melt production and eruption. Modelling by Huggel (2009) shows that the thermal perturbations that may destabilize slopes are likely to occur over tens or hundreds of years (for conductive heat flow processes) and years to decades (for advective/convective heat flow processes). Perturbations triggered by volcanic activity may be effective over much shorter timescales.

Andean stratovolcanoes that host rapidly diminishing tropical glaciers are likely to be particularly sensitive to climate warming. Many glaciers are completely out of equilibrium with current climate and may completely disappear within decades (Vuille et al., 2008). Model projections of future climate change in the tropical Andes indicate a continued warming of the tropical troposphere throughout the twenty-first century, with a temperature increase that is enhanced at higher elevations. By the end of the twenty-first century, following the SRES A2 emission scenario, the tropical Andes may experience a massive warming on the order of 4.5–5°C (Vuille et al., 2008). This warming will drive edifice instability both by removing ice, increasing the amount of meltwater at high elevations on edifices, and by thawing ice-bedrock contacts, encouraging slippage.

Climate warming may in some incidences reduce lahar hazards, because the disappearance of small volumes of snow and ice from volcanoes such as Popocatepetl will reduce the volume of ice available for meltwater flood generation.

Dwindling areas of ice and snow will also reduce the probability of lahar generation. This reduction in lahar hazards may be notable only in volcanoes undergoing almost complete glacier extinction (Huggel et al. 2007a).

Was the 2010 eruption of Eyjafjallajökull triggered by climate change?

The April 2010 explosive eruption of Eyjafjallajökull (Figure 4.9) which had a huge economic impact, occurred from a summit caldera filled with ice up to approximately 200 m in thickness (Guðmundsson et al., 2010). The timing of the eruption coincided with the publication of a special issue of the *Philosophical Transactions of the Royal Society* addressing climate forcing of geological hazards, which was heavily featured in the mainstream media. Therefore one of the first questions that journalists asked was whether the explosive eruption in 2010 was triggered by climate change.

To adequately answer this question we require detailed knowledge of the timing and nature of past eruptions at Eyjafjallajökull (that started at approximately 1 Ma) and the thickness and extent of ice during its activity. Although mapping of the eruptive stratigraphy has been carried out (Loughlin, 2002; Óskarsson, 2009), with subglacial and subaerial units differentiated, there is little constraint on the dates of eruptions or on the thickness of overlying ice during those eruptions. There is therefore considerable uncertainty about the relationship between the style and volume of eruptions at Eyjafjallajökull and the thickness of the overlying ice. There is currently no detailed record of ice thickness changes at Eyjafjallajökull over a timescale of centuries to millennia and there have been few historical or dated eruptions (e.g. Guðmundsson et al., 2009), meaning that statistical relationships between ice thickness and patterns of eruption activity are unavailable.

In addition, Eyjafjallajökull is a stratovolcano with a complex plumbing system that involves a shallow, evolved magma chamber. Its eruptions may be triggered by intrusion of more basic magma from greater depths (Sigmundsson et al., 2010b). New models of stress distribution within ice-covered stratovolcanoes are beginning to be developed (Sigmundsson et al., 2010a; Geyer & Bindeman, 2011), but these models can consider only how storage and eruption of magma are controlled by the stress field around shallow chambers; they cannot account for intrusion triggering or other such complexities. We also have little knowledge of how the chemistry of erupted magma has changed throughout the eruptive history of Eyjafjallajökull, which would help to reveal whether magma storage, mixing and evolution are affected by the thickness and extent of overlying ice.

The 2010 eruption of Eyjafjallajökull created exceptionally fine-grained ash (see Figure 4.9), which, together with the north-westerly airflow, contributed to the severe impact on European aviation. The role of summit ice and meltwater in the explosive generation of this ash is, however, far from clear. Although it may seem

Figure 4.9 (a) Eyjafjallajökull volcano viewed from Þórólfsfell on 10 May 2010, showing the ash plume emerging from the summit crater. Ash blankets much of the ice and snow on the volcano. Visible patches of snow and ice indicate zones of disruption to the valley glacier Gígjökull, which was the pathway for the main jökulhlaup on 14 April 2010, the first day of the eruption, and for a subsequent lava flow. (Photograph courtesy of Martin Rietze.) (b) A close-up photograph of the summit crater taken on 10 May 2010 from Þórólfsfell, showing emission of incandescent lava bombs and a turbulent plume of fine-grained ash. The summit ice can only be locally seen in crevasses due to the thick ash cover. (Photograph courtesy of Martin Rietze.)

intuitive that explosive interactions between rising magma and glacial meltwater drove the fragmentation (Sigmundsson et al., 2010b), new evidence is suggesting that meltwater played a minor role, with fragmentation instead controlled by shallow crystallisation and degassing, which created a highly viscous magma plug (T. Thordarson, personal communication, 2011). Therefore the effect of the summit ice on the hazards and impacts of the eruption is unknown.

For these reasons we cannot currently answer the journalists' question. The following section outlines some of the research that needs to be carried out to help us understand links between climate change and hazards at Eyjafjallajökull and other ice-covered volcanoes that had have a vastly greater human impact, such as Nevados del Ruiz in Columbia.

Gaps in our knowledge and targets for future research

Important gaps in our knowledge of links between melting of ice and volcanic hazards remain, which include the following.

Uncertainty about the timescale of volcanic responses to ice unloading

We currently have only limited insight into the reasons for delayed volcanic responses (Maclennan et al., 2002) and the timescales involved (Jellinek et al., 2004); response times are likely to differ in different tectonic settings.

Poor constraint on how ice bodies on volcanoes will respond to twenty-first century climate change

The highly localised effects of topography, microclimates, and local geothermal and eruption-related processes on volcanoes conspire to create considerable diversity in the response of individual glaciers and ice sheets to climate change (e.g. Geirsdóttir et al., 2006; Rivera et al., 2006; Bown & Rivera, 2007; Brock et al., 2007).

The sensitivity of volcanoes to small changes in ice thickness or to recession of small glaciers on their flanks is unknown

Although there is strong evidence that wholesale ice removal during deglaciation can significantly accelerate volcanic activity, there is considerable uncertainty about how volcanic responses to unloading scale with the magnitude, rate and distribution of ice unloading. A simple linear relationship between the rates of ice melting and additional melt production is unlikely to be appropriate. The effects of recession of different scales of ice body need to be considered, from the largest ice sheets to the smallest summit glaciers. Recent work suggests that annual variations in snow thickness at the Mýrdalsjökull ice cap, Iceland may affect the timing of eruptions from the Katla volcano (Albino et al., 2010), highlighting the potential sensitivity of some volcanic systems to very small changes in load.

Lack of data on how past changes in ice thickness have affected the size and style of volcanic eruptions and associated hazards

Most statistical studies of the effects of changes in ice thickness on volcanism have focused exclusively on the frequency of eruptions, but some recent studies suggest that eruption sizes may also be affected (Geyer & Bindeman, 2011). More information about how the sizes of eruptions or the occurrence of large caldera-forming events relates to past changes in ice thickness is therefore urgently required.

It is not well known how localised ice withdrawal from stratovolcanoes will affect shallow crustal magma storage and eruption

Existing models for how loading by ice affects volcanism have focused on large (>50 km diameter), near-horizontal ice sheets and mantle melting (e.g. Jull & McKenzie, 1996; Pagli & Sigmundsson, 2008). Stratovolcanoes, which constitute the vast majority of ice- and snow-covered volcanoes worldwide, are entirely different systems, being characterised by smaller, thinner ice bodies and the existence of crustal magma chambers, and new stress models are beginning to explore the effects of changes in ice thickness (e.g. Albino et al., 2010; Sigmundsson et al. 2010a).

Broader feedbacks between volcanism and climate change remain poorly understood

A number of potential positive feedbacks during volcano–ice interactions exist, which could potentially greatly magnify the rate of ice recession and effects on volcanic activity. Feedbacks include the increased CO_2 emissions from accelerated volcanism during ice unloading, which may act to further warm the climate (Huybers & Langmuir, 2009). Enhanced basal melting may destabilise ice sheets, leading to more rapid ice recession (Bell, 2008). More locally, tephra covering the ice surface may affect the mass balance of glaciers (Rivera et al., 2006; Brock et al., 2007). Currently, little is known about the effects of these feedbacks and whether they will play an important role in the twenty-first century and beyond.

Future work required

In order to resolve these problems both new data and improved models are required. Existing databases of known volcanic eruptions need to be augmented by numerous detailed case studies of the Quaternary eruptive history of ice-covered volcanoes, especially in the Andes, to determine whether the frequency and size of their eruptions have been influenced by past changes in ice thickness. The volcanic response should be examined to both large-magnitude, long times-cale climatic changes such as glacial–interglacial cycles and smaller, briefer fluctuations in the last millennium such as the Little Ice Age. This will reveal the sensitivity and response time of volcanic systems to a range of forcing timescales and magnitudes.

The unique record of palaeo-ice thicknesses provided by subglacially erupted volcanic deposits (e.g. Mee et al., 2006; Licciardi et al., 2007; Smellie et al., 2008; Tuffen et al., 2010) must be exploited in order to precisely reconstruct fluctuating local ice thicknesses on volcanic edifices. In tandem, high-resolution dating techniques will be required, which stretch the limits of existing radiometric methods. Geochemical indicators of the residence time of magma in shallow magma chambers could reveal whether shallow magma storage is affected by variations in ice thickness.

Improved physical models are required to test how magma generation, storage and eruption at stratovolcanoes are affected by stress perturbations related to the waxing and waning of small-volume ice bodies on what is commonly steep topography. Finally, feedbacks between the mass balance of ice sheets and glaciers and volcanic activity need to be incorporated into future Earth System Models.

Acknowledgements

HT was supported by a NERC Research Fellowship (NE/E013740/1), a Royal Society University Research Fellowship and a NERC New Investigator Grant (NE/ G000654/1). Thanks to Seb Watts, David Pyle, John Maclennan, Jasper Knight, Freysteinn Sigmundsson, John Smellie and Jennie Gilbert for discussions and sharing materials. Thanks to Martin Rietze for generously sharing his photographs of the Eyjafjallajökull eruption. Chris Kilburn is thanked for efficient and insightful editorial handling.

References

Albino, F., Pinel, V. & Sigmundsson, F. (2010) Influence of surface load variations on eruption likelihood: application to two Icelandic subglacial volcanoes, Grímsvötn and Katla. *Geophysical Journal International* **181**, 1510–1524.

Aðalgeirsdóttir, G., Guðmundsson, M. T. & Björnsson, H. (2000) The response of a glacier to a surface disturbance: a case study on Vatnajökull ice cap, Iceland. *Annals of Glaciology* **31**, 104–110.

Bacon, C. & Lanphere, M. (2006) Eruptive history and geochronology of Mount Mazama and the Crater Lake region, Oregon. *Geological Society of American Bulletin* **118**, 1331–1359.

Bell, R. E. (2008) The role of subglacial water in ice-sheet mass balance. *Nature Geoscience* **1**, 297–304.

Bindeman, I. N., Leonov, V. L., Izbekov, P. E., et al. (2010) Large-volume silicic volcanismin Kamchatka: Ar–Ar and U–Pb ages, isotopic, and geochemical characteristics of major pre-Holocene caldera-forming eruptions. *Journal of Volcanology and Geothermal Research* **189**, 57–80.

Björnsson, H. (1998) Hydrological characteristics of the drainage system beneath a surging glacier. *Nature* **395**, 771–774.

Björnsson, H. & Pálsson, F. (2008) Icelandic glaciers. *Jökull* **58**, 365–386.

Björnsson, H., Pálsson, F. & Guðmundsson, M. T. (2000) Surface and bedrock topography of the Mýrdalsjökull ice cap, Iceland. *Jökull* **49**, 29–46.

Bown, F. & Rivera, A. (2007) Climate changes and recent glacier behaviour in the Chilean lake district. *Global Planetary Change* **49**, 79–86.

Brock, B., Rivera, A., Casassa, G., Bown, F. & Acuna, C. (2007) The surface energy balance of an active ice-covered volcano: Villarrica Volcano, Southern Chile. *Annals of Glaciology* **45**, 104–114.

Capra, L. (2008) Abrupt climatic changes as triggering mechanisms of massive volcanic collapses. *Journal of Volcanology and Geothermal Research* **155**, 329–333.

Capra, L., Poblete, M. A., Alvarado, R. (2004) The 1997 and 2001 lahars of Popocatepetl volcano (Central Mexico): textural and sedimentological constraints on their origin and hazards. *Journal of Volcanology and Geothermal Research* **131**, 351–369.

Carlson, A. E., Legrande, A. N., Oppo, D. W., et al. (2008) Rapid early Holocene deglaciation of the Laurentide ice sheet. *Nature Geoscience* **1**, 620–624.

Carrasco-Núñez, G., Vallance, J. W. & Rose, W. I. (1993) A voluminous avalanche-induced lahar from Citlaltépetl volcano, Mexico. Implications for hazard assessment. *Journal of Volcanology and Geothermal Research* **59**, 35–46.

Carrivick, J. L. (2007) Hydrodynamics and geomorphic work of jökulhlaups (glacial outburst floods) from Kverkfjöll volcano, Iceland. *Hydrology Proceedings* **21**, 725–740.

Carrivick, J. L., Russell, A. J., Tweed, F. S. & Twigg, D. (2004) Palaeohydrology and sedimentology of jökulhlaups from Kverkfjöll, Iceland. *Sedimentary Geology* **172**, 19–40.

Carrivick, J. L., Manville, V. & Cronin, S. (2009a) Modelling the March 2007 lahar from Mt Ruapehu. *Bulletin of Volcanology* **71**, 153–169.

Carrivick, J. L., Russell, A. J., Rushmer, E. L., et al. (2009b) Geomorphological evidence towards a deglacial control on volcanism. *Earth Surface Processes and Landforms* **34**, 1164–1178.

Ceballos, J. L. Euscátegui, C., Ramírez, J., et al. (2006) Fast shrinkage of tropical glaciers in Colombia. *Annals of Glaciology* **43**, 194–201.

Corr, H. F. J. & Vaughan, D. G. (2008) A recent volcanic eruption beneath the West Antarctic ice sheet. *Nature Geoscience* **1**, 122–125.

Eliasson, J., Larsen, G., Gudmundsson, M. T., Sigmundsson, F. (2006) Probabilistic model for eruptions and associated floods events in the Katla caldera, Iceland. *Comparative Geoscience* **10**, 179–200.

Favier, V., Coudrain, A., Cadier, E., et al. (2008) Evidence of groundwater flow on Antizana ice-covered volcano, Ecuador. *Hydrology and Science* **53**, 278–291.

Gardeweg, M. C., Sparks, R. S. J. & Matthews, S. J. (1998) Evolution of Lascar volcano, northern Chile. *Journal of the Geological Society of London* **155**, 89–104.

Geirsdóttir, Á., Hardardóttir, J. & Sveinbjörnsdóttir, Á. E. (2000) Glacial extent and catastrophic meltwater events during the deglaciation of Southern Iceland. *Quaternary Science Reviews* **19**, 1749–1761.

Geirsdóttir, Á., Johannesson, T., Björnsson, H., et al. (2006) Response of Hofsjökull and southern Vatnajökull, Iceland, to climate change. *Journal of Geophysical Research* **111**, F03001.

Geyer, A. & Bindeman, I. (2011) Glacial influence on caldera-forming eruptions. *Journal of Volcanology and Geothermal Research* **202**, 127–142.

Gilbert, J. S., Stasiuk, M. V., Lane, S. J., et al. (1996) Non- explosive, constructional evolution of the ice-filled caldera at Volcán Sollipulli, Chile. *Bulletin of Volcanology* **58**, 67–83

Guðmundsson, M. T. (2003) Melting of ice by magma-ice-water interactions during subglacial eruptions as an indicator of heat transfer in subaqueous eruptions. In: White, J. D. L., Smellie, J. L. & Clague, D. (eds), *Explosive Subaqueous Volcanism*. AGU Geophysical Monograph 140. Washington DC: American Geophysical Union, 61–72.

Guðmundsson, M. T., Sigmundsson, F. & Björnsson, H. (1997) Ice-volcano interaction of the 1996 Gjálp subglacial eruption, Vatnajökull, Iceland. *Nature* **389**, 954–957.

Guðmundsson, M. T., Sigmundsson, F., Björnsson, H. & Högnadóttir, Þ. (2004) The 1996 eruption at Gjálp, Vatnajökull ice cap, Iceland: Course of events, efficiency of heat transfer, ice deformation and subglacial water pressure. *Bulletin of Volcanology* **66**, 46–65.

Guðmundsson, S., Björnsson, H., Johannesson, T., et al. (2009) Similarities and differences in the response to climate warming of two ice caps in Iceland. *Hydrology Research* **40**, 495–502.

Guðmundsson, M. T., Pedersen, R., Vogfjörd, K., Thorbjarnardóttir, B., Jakobsdóttir, S. & Roberts, M. J. (2010) Eruptions of Eyjafjallajökull Volcano, Iceland. *Eos Transactions, American Geophysical Union, Jökull* **58**, 251–268.

Hoblitt, R. P., Walder, J. S., Driedger, C. L., Scott, K. M., Pringle, P. T. & Vallance, J. W. (1998) *Volcano Hazards from Mount Rainier, Washington*, Revised 1998. US Geological Survey Open-File Report 98-428. http://vulcan.wr.usgs.gov/Volcanoes/Rainier/Hazards/OFR98-428/framework.html (accessed Oct 2012).

Houghton, B. F., Latter, J. H. & Hackett, W. R. (1987) Volcanic hazard assessment for Ruapehu composite volcano, Taupo volcanic zone, New Zealand. *Bulletin of Volcanology* **49**, 737–751.

Hubbard, A. (2006) The validation and sensitivity of a model of the Icelandic ice sheet. *Quaternary Science Reviews* **25**, 2297–2313.

Hubbard, B. E., Sheridan, M. F., Carrasco-Nuñez, G., Díaz-Castellon, R. & Rodríguez, S. R. (2007) Comparative lahar hazard mapping at Volcan Citlaltépetl, Mexico using SRTM, ASTER and DTED-1 digital topographic data. *Journal of Volcanology and Geothermal Research* **160**, 99–124.

Huggel, C. (2009) Recent extreme slope failures in glacial environments: effects of thermal perturbation. *Quaternary Science Reviews* **28**, 1119–1130.

Huggel, C., Ceballos, J. L., Pulgarin, B., Ramirez, J. & Thouret, J-C. (2007a) Review and reassessment of hazards owing to volcano–glacier interactions in Colombia. *Annals of Glaciology* **45** 128–136 (2007).

Huggel, C., Caplan-Auerbach, J., Waythomas, C. F. & Wessels, R. F. (2007b) Monitoring and modeling ice-rock avalanches from ice-capped volcanoes: A case study of frequent large avalanches on Iliamna Volcano, Alaska. *Journal of Volcanology and Geothermal Research* **168**, 114–136.

Huggel, C., Caplan-Auerbach, J. & Wessels, R. (2008) Recent extreme avalanches: triggered by climate change? *Eos Transactions, American Geophysical Union* **89**, 469–470.

Humlum, O. & Houmark-Nielsen, M. (1994) High deglaciation rates in Denmark during the Late Weichselian – implications for the palaeoenvironment. *Geografisk Tidsskrift* **94**, 26–37.

Huybers, P. & Langmuir, C. (2009) Feedback between deglaciation, volcanism, and atmospheric CO_2. *Earth and Planetary Science Letters* **286**, 479–491.

Jarosch, A. H. & Gudmundsson, M. T. (2007) Numerical studies of ice flow over subglacial geothermal heat sources at Grímsvöt, Iceland, using the full Stokes equations. *Journal of Geophysical Research* **112**, F02008.

Jellinek, A. M., Manga, M., Saar, M. O. (2004) Did melting glaciers cause volcanic eruptions in eastern California? Probing the mechanics of dike formation. *Journal of Geophysical Research* **109**, B09206.

Jomelli, V., Favier, V., Rabatel, A., Brunstein, D., Hoffmann, G. & Francou, B. (2009) Fluctuations of glaciers in the tropical Andes over the last millennium and palaeoclimatic implications: A review. *Palaeogeography, Palaeoclimatology, Palaeoecology* **281**, 269–282.

Jordan, E., Ungerechts, L., Caceres, B., Penafiel, A. & Francou, B. (2005) Estimation by photogrammetry of the glacier recession on the Cotopaxi volcano (Ecuador) between 1956 and 1997. *Hydrological Sciences Journal* **50**, 949–961.

Julio-Miranda, P., Delgado-Granados, H., Huggel, C., et al. (2008) Impact of the eruptive activity on glacier evolution at Popocatepetl Volcano (Mexico) during 1994–2004. *Journal of Volcanology and Geothermal Research* **170**, 86–98.

Jull, M. & McKenzie, D. P. (1996) The effect of deglaciation on mantle melting beneath Iceland. *Journal of Geophysical Research* **101**, 21815–21828.

Larsen, C. F., Motyka, R. J., Freymueller, J. T., Echelmeyer, K. A. & Ivins, E. I. (2005) Rapid viscoelastic uplift in southeast Alaska caused by post-Little Ice Age glacial retreat. *Earth and Planetary Science Letters* **237**, 548–560.

Lescinsky, D. T. & Fink, J. H. (2000) Lava and ice interaction at stratovolcanoes: use of characteristic features to determine past glacial extents and future volcanic hazards. *Journal of Geophysical Research* **105**, 23711–23726.

Licciardi, J. M., Kurz, M. D. & Curtice, J. M. (2007) Glacial and volcanic history of Icelandic table mountains from cosmogenic ^{3}He exposure ages. *Quaternary Science Reviews* **26**, 1529–1546.

Loughlin, S. C. (2002) *Facies analysis of proximal subglacial and proglacial volcaniclastic successions at the Eyjafjallajökull central volcano, southern Iceland.* Special Publications 202. London: Geological Society, pp 149–178.

Lythe, M. B. & Vaughan, D. G. (2001) BEDMAP: A new ice thickness and subglacial topographic model of Antarctica. *Journal of Geophysical Research* **106**, 11335–11352.

McGarvie, D. W., Burgess, R., Tindle, A. J., Tuffen, H. & Stevenson, J. A. (2006) Pleistocene rhyolitic volcanism at the Torfajökull central volcano, Iceland: eruption ages, glaciovolcanism, and geochemical evolution. *Jökull* **56**, 57–75.

Maclennan, J., Jull, M., McKenzie, D. P., Slater, L. & Gronvold, K. (2002) The link between volcanism and deglaciation in Iceland. *Geochemistry, Geophysics, Geosystems* **3**, 1062.

Magnusson, E., Björnsson, H., Dall, J., et al. (2005) Volume changes of Vatnajökull ice cap, Iceland, due to surface mass balance, ice flow, and subglacial melting at geothermal areas. *Geophysical Research Letters* **32**, L05504.

Major, J. J. & Newhall, C. G. (1989) Snow and ice perturbation during historical volcanic eruptions and the formation of lahars and floods. *Bulletin of Volcanology* **52**, 1–27.

Mathews, W. H. (1951) The Table, a flat-topped volcano in southern British Columbia. *American Journal of Science* **249**, 830–841.

Mee, K., Tuffen, H. & Gilbert, J. S. (2006) Snow-contact volcanic facies at Nevados de Chillan volcano, Chile, and implications for reconstructing past eruptive environments. *Bulletin of Volcanology* **68**, 363–376

Mee, K., Gilbert, J. S., McGarvie, D. W., Naranjo, J. A. & Pringle, M. (2009) Palaeoenvironment reconstruction, volcanic evolution and geochronology of the Cerro Blanco subcomplex, Nevados de Chillán Volcanic Complex, central Chile. *Bulletin of Volcanology* **71**, 933–952.

Newhall, C. G., Self, S. (1982) The volcanic explosivity index (VEI): An estimate of explosive magnitude for historical volcanism. *Journal of Geophysical Research* **87**, 1231–1238.

Noe-Nygaard, A. (1940) Sub-glacial volcanic activity in ancient and recent times (studies in the palagonite system of Iceland, no. 1). *Folia Geographica Danica* **1**, 1–67.

Nowell, D., Jones, C. & Pyle, D. (2006) Episodic quaternary volcanism in France and Germany. *Journal of Quaternary Science* **21**, 645–675.

Óskarsson, B. V. (2009) The Skerin ridge on Eyjafjallajökull, south Iceland: Morphology and magma-ice interaction in an ice-confined silicic fissure eruption. Master's thesis, Faculty of Earth Sciences, University of Iceland.

Pagli, C. & Sigmundsson, F. (2008) Will present day glacier retreat increase volcanic activity? Stress induced by recent glacier retreat and its effect on magmatism at the Vatnajökull ice cap, Iceland. *Geophysical Research Letters* **35**, L09304.

Pierson, T. C., Janda, R. J., Thouret, J-C. & Borrero, C. A. (1990) Perturbation and melting of snow and ice by the 13 November 1985 eruption of Nevado del Ruiz, Colombia, and consequent mobilization, flow, and deposition of lahars. *Journal of Volcanology and Geothermal Research* **41**, 17–66.

Pyle, D. M. (1999) Sizes of volcanic eruptions. In: Sigurdsson, H. (ed.), *Encyclopaedia of Volcanoes*. San Diego, CA: Academic Press, pp 263–269.

Rivera, A., Bown, F., Mella, R., et al. (2006) Ice volumetric changes on active volcanoes in Southern Chile. *Annals of Glaciology* **43**, 111–122.

Saar, M. O. & Manga, M. (2003) Seismicity induced by seasonal groundwater recharge at Mt. Hood, Oregon. *Earth and Planetary Science Letters* **214**, 605–618.

Shepherd, A., Wingham, D. J., Mansley, J. A. D. & Corr, H. F. J. (2001) Inland thinning of Pine Island Glacier, West Antarctica. *Science* **291**, 862–864.

Siebert, L. & Simkin, T. (2002) Volcanoes of the world: an illustrated catalog of Holocene volcanoes and their eruptions. Smithsonian Institution, Global Volcanism Program, Digital Information Series, GVP-3.

Sigmundsson, F., Pinel, V., Lund, B., et al. (2010a) Climate effects on volcanism: influence on magmatic systems of loading and unloading from ice mass variations, with examples from Iceland. *Philosophical Transactions of the Royal Society of London A* **368**, 2519–2534.

Sigmundsson, F., Hreinsdottir, S., Hooper, A., et al. (2010b) Intrusion triggering of the 2010 Eyjafjallajökull explosive eruption. *Nature* **468**, 426–430.

Singer, B. S., Jicha, B. R., Harper, M. A., Naranjo, J. A., Lara, L. E. & Moreno, H. (2008) Eruptive history, geochronology, and magmatic evolution of the Puyehue-Cordón Caulle volcanic complex, Chile. *Geological Society of America Bulletin* **120**, 599–618.

Smellie, J. L. (1999) Subglacial eruptions. In: H. Sigurdsson (ed.), *Encyclopaedia of Volcanoes*. San Diego, CA: Academic Press, pp 403–418.

Smellie, J. L. (2008) Basaltic subglacial sheet-like sequences: evidence for two types with different implications for the inferred thickness of associated ice. *Earth Science Reviews* **88**, 60–88.

Smellie, J. L. & Skilling, I. P. (1994) Products of subglacial volcanic eruptions under different ice thicknesses – 2 examples from Antarctica. *Sedimentary Geology* **91**, 115–129.

Smellie, J. L., Johnson, J. S., McIntosh, W. C., et al. (2008) Six million years of glacial history recorded in volcanic lithofacies of the James Ross Island Volcanic Group, Antarctic Peninsula. *Palaeogeography, Palaeoclimatology, Palaeoecology* **260**, 122–148.

Thompson, L. G., Brecher, H. H., Mosley-Thompson, E., Hardy, D. R. & Mark, B. G. (2009) Glacier loss on Kilimanjaro continues unabated. *Proceedings of the National Academy of Sciences of the United States of America* **106**, 19770–19775.

Tuffen, H. (2007) Models of ice melting and edifice growth at the onset of subglacial basaltic eruptions. *Journal of Geophysical Research* **112**, B03203.

Tuffen, H. & Castro, J. M. (2009) An obsidian dyke erupted through thin ice: Hrafntinnuhryggur, Krafla, Iceland. *Journal of Volcanology and Geothermal Research* **185**, 352–366.

Tuffen, H., Gilbert, J. S. & McGarvie, D. W. (2007) Will subglacial rhyolite eruptions be explosive or intrusive? Some insights from analytical models. *Annals of Glaciology* **45**, 87–94.

Tuffen, H., Owen, J. & Denton, J. S. (2010) Magma degassing during subglacial eruptions and its use to reconstruct palaeo-ice thicknesses. *Earth Science Reviews* **99**, 1–18.

Vuille, M., Francou, B., Wagnon, P., et al. (2008) Climate change and tropical Andean glaciers -past, present and future. *Earth Science Reviews* **89**, 79–96.

Waitt, R. B. (1989) Swift snowmelt and floods (lahars) caused by great pyroclastic surge at Mount St Helens volcano, Washington, 18 May 1980. *Bulletin of Volcanology* **52**, 138–157.

Walder, J. S. (2000) Pyroclast/snow interactions and thermally driven slurry formation. Part 1: Theory for monodisperse grain beds. *Bulletin of Volcanology* **62**, 105–118.

Waythomas, C. F. & Miller, T. P. (1999) *Preliminary volcano-hazard assessment for Iliamna Volcano, Alaska*. US Geological Survey Open-File Report 99-373, 31 pp. http://www.avo.alaska.edu/downloads/classresults.php?citid=642 (accessed Oct 2012).

Wingham, D. J., Wallis, D. W. & Shepherd, A. (2009) Spatial and temporal evolution of Pine Island Glacier thinning 1995–2006. *Geophysical Research Letters* **36**, L17501.

Zielinski, G., Mayewski, P., Meeker, L., Whitlow, S. & Twickler, M. A. (1996) 110,000-yr record of explosive volcanism from the GISP2 (Greenland) ice core. *Quaternary Research* **45**, 109–118.

Multiple effects of ice load changes and associated stress change on magmatic systems

Freysteinn Sigmundsson[1], Fabien Albino[1], Peter Schmidt[2], Björn Lund[2], Virginie Pinel[3], Andrew Hooper[4] and Carolina Pagli[5]

[1] Nordic Volcanological Centre, Institute of Earth Sciences, University of Iceland, Iceland
[2] Department of Earth Sciences, Uppsala University, Sweden
[3] ISTerre, IRD R219, CNRS, Université de Savoie, Le Bourget du Lac, France
[4] Delft University of Technology, Delft, The Netherlands
[5] School of Earth and Environment, University of Leeds, UK

Summary

Ice retreat on volcanoes reduces pressure at the surface of the Earth and induces stress changes in magmatic systems. The consequences can include increased generation of magma at depth, increased magma capture in the crust and modification of failure conditions of magma chambers. We review the methodology to evaluate each of these effects, and consider the influence of ongoing ice retreat on volcanoes at the mid-Atlantic divergent plate boundary in Iceland. Evaluation of each of these effects requires a series of assumptions regarding the rheology of the crust and mantle, and the nature of magmatic systems, contributing to relatively large uncertainty in response of a magmatic system to climate warming and associated ice retreat. Pressure release melting caused by ice cap retreat in Iceland may at present times generate a similar amount of magma to plate tectonic processes, larger than previously realized. However, new modelling shows that part of this magma may be captured in the crust, rather than being erupted. Gradual retreat of ice caps steadily modify failure conditions at magma chambers, which is highly dependent on their geometry and depth, as well as the details of ice load variations. A model is presented where long-term ice retreat at the Katla volcano decreases the likelihood of eruption, because more magma is needed in the magma chamber to cause failure than in the absence of the ice retreat.

Climate Forcing of Geological Hazards, First Edition. Edited by Bill McGuire and Mark Maslin.
© 2013 The Royal Society and John Wiley & Sons, Ltd. Published 2013 by John Wiley & Sons, Ltd.

Introduction

The ongoing global warming drives retreat of ice caps and glaciers worldwide, many of which are located in volcanic regions. The reduced ice load on the surface of these volcanic areas modifies conditions in the subsurface by altering the stress field. The changes may include pressure decrease, possible rotation of the principal axes of the stress field and variation in differential stress. A review by Sigmundsson et al. (2010) is here complemented with additional considerations. The multiple effects of stress change at magmatic systems include: (1) effects on melting conditions, (2) influence on magma propagation and emplacement of dykes, and (3) influence on magma storage zones. In principle, a model of ice unloading can be applied to an Earth model to evaluate each of these effects. In practice, there are many assumptions to be made in order to carry out such modelling based on limited constraints on both Earth and ice retreat models. This leads to large uncertainty on the exact response of a particular volcano to stress change induced by global warming. The complexity of a real volcano interior needs to be simplified in these models. Various petrological, geochemical and geophysical evidence shows that the crust underlying volcanoes is often not homogeneous and may rather contain localized magma storage zones. Here these storage zones are modelled as fluid-filled cavities of simple geometries and referred to as magma chambers.

Effects of glacial unloading on deep magma generation

The amount of melt generated by glacial unloading depends on the pressure change and the relation between the pressure P and the degree of melting by weight, here denoted F (denoted by X by McKenzie, 1984 and Sigmundsson et al., 2010). We follow Jull and McKenzie (1996) by assuming isentropic decompressional melting, where the material derivatives of degree of melting and pressure are related as:

$$\frac{DF}{Dt} = \left(\frac{\partial F}{\partial P}\right)_S \left(\frac{\partial P}{\partial t} + \mathbf{v} \cdot \nabla P\right)$$

where $\mathbf{v}$ is the velocity of the solid and $(\partial F/\partial P)_S$ is the partial derivative of the degree of melting with respect to pressure at a constant entropy. This equation needs to be evaluated within a melting regime in the mantle. There are thus three important steps for evaluating the total amount of melt generated in the mantle by deglaciation in a particular volcanic area. These steps are: (1) evaluating the pressure change at depth, (2) evaluating what relation to use for pressure versus melting, and (3) estimating the geometry and extent of the melting regime, in order to know where to carry out integration of the deglaciation-induced melting process. All these steps are important when evaluating the deep magma generation caused by glacial unloading. We consider them below for the cause of the ongoing deglaciation of Iceland since 1890 as a result of climate warming.

An earlier study has evaluated mantle melting due to the ongoing retreat of only the largest ice cap in Iceland, Vatnajökull (Pagli & Sigmundsson, 2008). A study by Árnadóttir et al. (2009) revealed, however, that present geodetically measured uplift rates of 10–23 mm/year in central Iceland cannot be produced by deglaciation of Vatnajökull alone. Contributions from the smaller ice caps in Iceland are significant. In order to improve the estimate of deep melt generation under Iceland in response to the present deglaciation, we employed a refined version of the ice model by Árnadóttir et al. (2009), including the five largest ice caps at a spatial resolution of 2×2 km. The deglaciation rates in the model are based on the estimated 435 km^3 ice loss of Vatnajökull between 1890 and 2004, used by Pagli et al. (2007), and the mean annual mass balance between the years 1991–1992 and 2005–2006 (Björnsson & Pálsson, 2008). We linearly extend the ice model to cover the period 1890–2010 (120 model years). This refined ice retreat model yields a slight improvement in the fit of the predicted vertical uplift rates to the Global Positioning System (GPS) measurements reported by Árnadóttir et al. (2009). We used this ice model in a three-dimensional model of Glacial Isostatic Adjustment (GIA)-related decompressional melting, rather than in an axi-symmetrical model presented by Sigmundsson et al. (2010). For the Earth model, we use the preferred model of Árnadóttir et al. (2009) presented in Table 5.1. The uppermost layer in this model is a 10-km-thick elastic layer that we refer to as the elastic layer or upper lithosphere. Below this is a 30-km-thick viscoelastic layer of 10^{20} Pa s viscosity; the viscoelastic layer or lower lithosphere. Beneath is the mantle, modelled as a viscoelastic half-space of viscosity 10^{19} Pa s.

In order to estimate the total increase in melt production due to deglaciation, we use a similar triangular cross-section melting region as previously used by Jull and McKenzie (1996) and Pagli and Sigmundsson (2008). However, in our three-dimensional model the melting region traces the eastern volcanic zone of Iceland south of the Vatnajökull ice cap and the northern volcanic zone to the north of Vatnajökull (Figure 5.1). The length of the melting regime considered is approximately 340 km. Both Jull and McKenzie (1996) and Pagli and Sigmundsson (2008)

Table 5.1 Material parameters of the preferred three-dimensional Earth model by Árnadóttir et al. (2009), based on the fit to nation-wide GPS measurements of the present-day vertical uplift of Iceland

Layer	Depth (km)	Density (kg/m³)	Young's modulus (GPa)	Viscosity (Pa s)
Upper lithosphere	0–10	2800	40	∞
Lower lithosphere	10–40	3000	70	10^{20}
Mantle	40–∞	3200	130	10^{19}

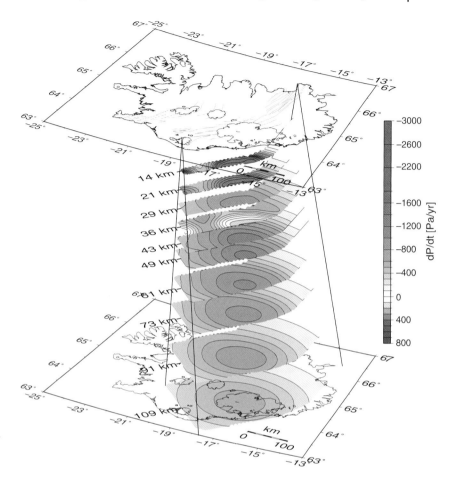

Figure 5.1 Rate of pressure change beneath Iceland due to ongoing glacial unloading (see text). Compression positive. Ice caps (closed outlines) are shown on the surface, as well as the fissure swarms of the volcanic zones. Vatnajökull is the largest ice cap.

assumed a triangular-shaped melting regime, with a ridge angle of 45°, in accordance with an assumption of passive upwelling and corner flow. However, the volcanic zones in Iceland are neither straight nor perfectly perpendicular to the direction of spreading, and the upwelling beneath Iceland in the presence of the mantle plume is unlikely to be governed by passive upwelling (Maclennan et al., 2002). Therefore, ridge angles of 45, 60 and 70° were considered here.

During the 120 years of the model run, the decompression rates vary about ±10%, but we consider the average values. Figure 5.1 shows the predicted GIA decompression at selected depths of the melting region. In the mantle, the decompression reaches 1390 Pa/year at 43 km depth beneath the centre of unloading, whereas at 109 km depth it is approximately 500 Pa/year. In the lowermost part of the lower viscoelastic lithosphere, pressure increases by up to 660 Pa/year beneath Vatnajökull.

The pressure change values form the basis for estimating the associated melt generation, but various forms of the relation between the two have been suggested. Here we have obtained $\partial F/\partial P$ by integrating Equation D7 of McKenzie (1984) and the melt parametrisation by McKenzie and Bickle (1988), from the solidus pressure. As we are interested in the melt volume due to ice removal we have to convert F, the degree of melting by weight, to ϕ, the degree of melting by volume McKenzie (1984) refers to ϕ as the porosity.

$$d\phi = \frac{\rho_l \rho_s}{\left(F\rho_s + (1-F)\rho_l \right)^2} dF$$

Here ρ_l is the density of the melt and ρ_s is the density of the solid. In contrast to the melt parametrisation given by Equation D8 of McKenzie (1984), which was used by Pagli and Sigmundsson (2008) in their study of glacially induced melting, the melt parametrisation by McKenzie and Bickle (1988) does not yield a constant magnitude of $\partial \phi/\partial P$. Assuming a potential temperature of 1500°C, which yields a solidus depth of approximately 112 km, $\partial \phi/\partial P$ attains its largest magnitude of about −0.132/GPa at a depth of approximately 100 km and decreases to a magnitude of about −0.048/GPa at the surface. The rate of degree of melting per volume is then given by the pressure change $dP/\partial t$ multiplied by $d\phi/\partial P$.

Integration from solidus depth to the base of the lithosphere (40 km depth) yields melt production rate estimates of 0.17, 0.11 and 0.07 km³/year for ridge angles of 45, 60 and 70°, respectively (Figure 5.2). A crude estimate of the steady-state melt production rate beneath Iceland can be obtained from the product of the mean thickness of the crust, the length of the spreading ridge and the full

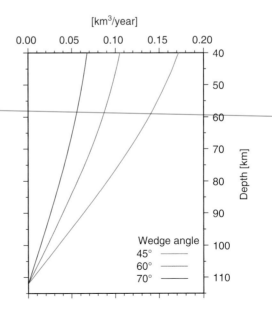

Figure 5.2 Melt production rates due to Glacial Isostatic Adjustment (GIA) decompression in Iceland integrated from the solidus depth, as a function of depth. Red line displays results for a ridge angle of 45°, blue line for a ridge angle of 60° and black line for a ridge angle of 70°.

spreading velocity. Assuming, as Pagli and Sigmundsson (2008), a mean thickness of 30 km, length of the ridge of 300 km and full spreading velocity of 19 mm/year, the melt production rate would be 0.17 km³/year. Note that this would be a minimum melt production rate because more melt could be produced in the mantle but not extracted to form crust. For a steady melt production rate of 0.17 km³/year, the inferred increase in melting due to present deglaciation would by 41–100%. This indicates that melt generation due to ongoing deglaciation of Iceland is presently of a similar magnitude to the plate tectonic melt production.

Influence on magma capture in the crust

Although a very significant volume of new magma is generated at depth in the mantle, it is uncertain if and when it reaches the surface of the Earth and is erupted. The rate of melt extraction from depth to the surface is uncertain. An average melt extraction velocity of >50 m/year would suggest a transport time of less than 1000 years from a depth of 50 km. Here we consider another effect, namely whether part of the magma generated by deglaciation will be captured by the crust and form intrusions, rather than feeding eruptions, as evaluated by Hooper et al. (2011).

The ascent of magma through the crust is driven primarily by buoyancy forces, which lead to opening of fractures above (e.g. Lister & Kerr, 1991). In an isotropic stress field, fractures below can contract as magma is squeezed upwards. However, if the principal components of the stress field within the crust are unequal, they are relaxed by non-reversible opening of the magma-filled fractures. Hence extensional stress environments favour intrusion over eruption (e.g. Segall et al., 2001). In general, the opening of a dyke scales with the excess dyke pressure, ΔP_{dyke}, e.g. a circular crack with uniform excess pressure has volume:

$$V = \frac{8a^3(1-v)}{3\mu} \Delta P_{dyke}$$

where a is the radius, v Poisson's ratio and μ the shear modulus (Segall, 2009). In the case where magma can flow until equilibrium pressure conditions are reached, the excess pressure is equal to the difference between the mean stress and the stress normal to the fracture. It follows that the magma volume that can be accommodated by the relaxation of the deviatoric stress in dykes is proportional to this excess pressure:

$$V \propto \Delta P_{dyke} = \frac{1}{3}(\sigma_1 + \sigma_2 + \sigma_3) - \sigma_n$$

where σ_1, σ_2 and σ_3 are the principal stresses in descending order of strength, and σ_n is the fracture-normal stress.

Figure 5.3 Contribution to express pressure in a dyke, ΔP_{dyke}, due to ice loss from all Icelandic ice caps between 1890 and 2003. (a) Location of Vatnajökull ice cap in Iceland. Black dots show relocated earthquake epicentres during 2007 in the Upptyppingar area in relation to a deep dyke intrusion (Based on data from Jakobsdóttir et al., 2008). Blue lines indicate pre-existing eruptive fissures parallel to the ice cap edge and yellow regions outline fissure swarms of volcanic systems. The black rectangle outlines the region shown in Figure 5.4. (b) A cross-sectional view of ΔP_{dyke} for dykes perpendicular to Δ_3 imparted by the ice loss. Short ticks show the orientation of Δ_3 projected on to the section and the long ticks indicate the dyke plane. (c) ΔP_{dyke} for dykes perpendicular to plate spreading. From Hooper et al. 2011.

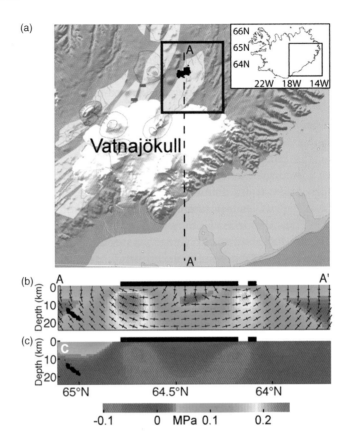

In Iceland, spreading across the rift zones of 19.7 mm/year (DeMets et al. 1994), is the primary contributor to ΔP in the crust, but changes in the ice load also affect it. Here we estimate the contribution to ΔP from the reduction in mass of Icelandic ice caps since glacial retreat began in about 1890 (Figure 5.3). We use the same Earth model from Árnadóttir et al. (2009), which consists of a 40-km-thick lithosphere overlying a Maxwell half-space with viscosity 10^{19} Pa s. The results are, however, not particularly sensitive to the Earth model used (Hooper et al., 2011). We calculate ΔP_{dyke} for two different scenarios, the first when a dyke is perpendicular to σ_3 induced by the melting ice cap and the second when the dyke is parallel with the rift zone (Figure 5.3). We see that whether intrusion is encouraged or discouraged depends on dyke orientation, although away from the edges of the ice cap, intrusive activity is always favoured. It is usually assumed that dykes propagate perpendicular to the direction of σ_3 and that σ_3 in this region is dictated by plate spreading, in which case the second scenario should apply.

This modelling has been put in context with an unusual dyke intrusion in the lower crust in Iceland in 2007–2008, in the Upptyppingar area north of the Vatna-

Figure 5.4 Deformation due to lower crustal intrusion at Upptyppingar, north of the Vatnajökull ice cap. Satellite radar interferogram from Envisat descending track 467 spanning 14 July 2007 to 28 June 2008. Each colour fringe represents 28 mm of displacement toward the satellite. Horizontal Global Positioning System (GPS) velocities, observed in black, with 95% confidence ellipses, and modelled in white. Also shown are the surface projections of the model patches of the inferred inclined dyke (white rectangles), and catalogue earthquake epicentres for the entire intrusion period from the Icelandic Meteorological Office (black circles). From Hooper et al. 2011.

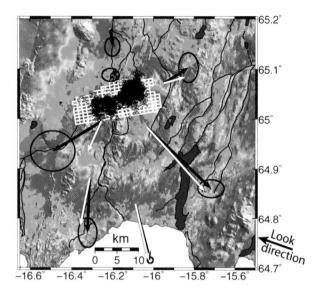

jökull ice cap. The dyke intrusion was captured well by satellite radar interferometric observations and GPS geodetic measurements (Figure 5.4). The geodetic study (Hooper et al., 2011) concludes that the strike and dip of the dyke are 81–84° and 42–43°, in good agreement with the seismic studies (Jakobsdóttir et al., 2008; White et al. 2011). The geodetic study allows the estimate of the magma volume involved: $42–47 \times 10^6$ m^3 at 95% confidence. The orientation of this dyke intruded in 2007–2008 into the lower crust near the largest ice cap in Iceland actually agrees well with it being perpendicular to σ_3 induced by the melting ice cap rather than being parallel with the rift zone. This raises the question of whether the stress field there is actually dominated by the melting ice cap. For the Upptyppingar intrusion, although the orientation of the dyke is perpendicular to σ_3 induced by the melting ice cap, the displacement across the dyke plane consists of both shearing and opening, indicating that actually the stress field is dominated by the plate tectonic process (Hooper et al., 2011).

Influence on shallow magma chambers

Pressure inside magma chambers, here denoted P_c, increases when magma flows into the chamber from deeper sources of melt, or by magma crystallization. When the pressure reaches a critical value, here denoted P_r, tensile rupture of the reservoir wall will occur and magma is transported towards the Earth's surface, eventually leading to an eruption (Albino et al., 2010). Any stress perturbations around a magma chamber cause a pressure change within it, ΔP_c, as well as a modification of the critical pressure value to initiate tensile fractures, ΔP_r. An unloading event at the surface of a volcano always induces a magma pressure decrease within the chamber. The amplitude of this pressure drop is largest for incompressible magmas and decreases for more compressible magmas (Pinel & Jaupart, 2005). However, as previously shown by Albino et al. (2010), the threshold pressure for dyke initiation P_r can either increase or decrease according to the geometry of the magma chamber and the surface unloading event. In all cases, the difference in the two terms, $\Delta P_r - \Delta P_c$, provides the relative evolution of a volcanic system between the initial and the final state and characterizes the effect of the surface perturbation on the ability of the system to erupt. A negative value signifies that the magma reservoir state moves closer to rupture conditions and eruption probability increases. Conversely, if the sign is positive, the magma reservoir evolves further away from its failure state and the likelihood of an eruption is reduced.

We have previously performed two dimensional numerical simulations to quantify the effect of short-term unloading events on idealized magma chambers of simple shapes (sphere and ellipsoids) filled with an inviscid fluid embedded in an elastic homogeneous crust (Albino et al., 2010; Sigmundsson et al., 2010). We showed, that either enhancement or reduction of eruption likelihood will occur as described above, depending on the magma chamber geometry, the magma compressibility as well as the spatial distribution of the surface load. A central unloading event occurring directly above a prolate reservoir will inhibit rupture initiation whereas the same event above a spherical or oblate reservoir will promote rupture. In the latter case, for the spherical shape, the triggering effect is maximal, and equal to the amplitude of the removed load.

This elastic model was applied to the Icelandic volcano, Katla, covered by the Mýrdalsjökull ice cap. This ice cap shows load variations at two different time-scales: (1) an annual load cycle in the centre part, with a difference up to 6 meters in snow thickness from winter to summer, and (2) a long-term ice thinning at the periphery due to global warming, with a rate around 4 m/year (Gudmundsson et al., 2007). For annual cycles, our elastic model predicts that, in the case of a spherical or horizontally elongated magma reservoir, eruptions at Katla are more likely when the seasonal snow cover at Mýrdalsjökull is smallest (Albino et al., 2010). This triggering effect is small, around a few kilopascals, but appears consistent with the fact that all the last nine known historical eruptions at Katla occurred during the warm season (e.g. Eliasson et al., 2006).

When considering surface load variation distributed over large areas or acting over a long periods as obviously the case for ice retreat, it becomes necessary to take into account the viscous response of the upper mantle and lower crust to fully evaluate the potential effect on magma reservoirs. This has not been considered before in our models. Based on this consideration, we performed a series of axi-symmetrical numerical models considering an upper elastic crust surrounding the magma reservoir and a lower medium characterized by a viscoelastic behaviour (Figure 5.5). The whole medium is submitted to gravity field and a buoyancy stress, σ_B, that acts at the bottom of the elastic medium when this dis-

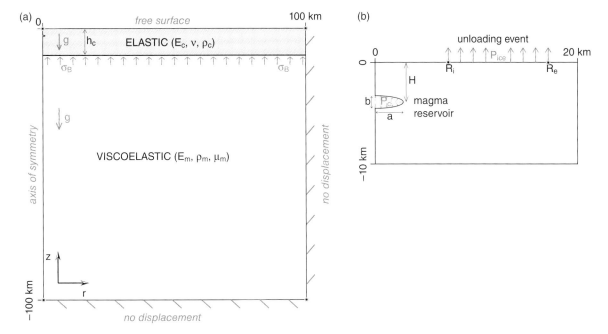

Figure 5.5 Model setup for evaluating surface unloading effects on a shallow magma chamber, applicable to Katla volcano Iceland. (a) The model is a $100 \times 100\,\text{km}$ box with two media: an elastic upper plate emplaced on the top of a lower viscoelastic part. Parameters taken into account are indicated (listed in Table 5.2). Stress conditions applied at boundaries are indicated in grey. (b) Zoom of the uppermost 10 km, which shows the characteristic of the unloading event as well as the geometry of the magma chamber.

continuity is displaced. Buoyancy stress is related to the vertical displacements of the interface between both media through the following relation:

$$\sigma_B = \rho_m g U_z$$

where ρ_m is the mantle density, and U_z the vertical displacement.

Solutions for the initial elastic response and the fully relaxed state, in absence of magma reservoir, were validated using analytical solutions provided by Pinel et al. (2007). As initial state, we take the fully relaxed state obtained after applying a constant magmatic pressure within the reservoir. We then apply a surface unloading, at a constant rate, over a given period of time. Failure of the magma reservoir will occur when the minimal compressive deviatoric stress, $\delta\sigma_3$, reaches the value of the rock tensile strength, T_s. For a whole range of magmatic pressure, we then calculate the temporal evolution of the parameter $\delta\sigma_3$ at each time step and compare to the tensile strength value. From these results, we are able to derive the threshold pressure P_r required for dyke initiation at each time step.

As all surface unloading events, a long term ice thinning does induce a magma pressure decrease, acting to move the system away from rupture conditions.

Table 5.2 Parameters of ice, Earth and magma chamber model used to quantify the effect of retreat of the Mýrdalsjökull ice cap on a shallow magma chamber at the Katla volcano

Ice retreat model	Symbol	Value	Crust (elastic)	Symbol	Value
Internal radius	R_i	7 km	Thickness	h_c	10 km
External radius	R_e	17 km	Young's modulus	E_c	30 GPa
Pressure decrease	P_{ice}	−35 kPa/year	Poisson's ratio	ν	0.25
			Density	ρ_c	2800 kg/m^3

Magma chamber	Symbol	Value	Mantle (viscoelastic)	Symbol	Value
Horizontal axis	a	2.5 km	Young's modulus	E_m	30 GPa
Vertical axis	b	0.5 km	Density	ρ_m	3100 kg/m^3
Centre depth	H	3 km	Viscosity	μ_m	3×10^{18} Pa s

However, such magma pressure evolution can easily be compensated by magmatic processes not considered in our model, as for example magma feeding from deep sources or even magma crystallisation. As for the elastic case, the pressure required to initiate intrusions from a magma chamber may either increase, indicating that the system is moving away from eruption conditions, or decrease, corresponding to an increase of eruption likelihood. The behaviour is always highly dependent on the geometry of the magma reservoir, the spatial distribution of the surface load and also its temporal evolution.

We applied our model to Katla Volcano in order to estimate the long term influence of Mýrdalsjökull's ice thinning using the set of parameters listed in Table 5.2. Viscosity of the lower medium is set at 3×10^{18} Pa s, consistent with volcanic systems being warmer and less viscous than their surroundings. Lithosphere thickness was here set at 10 km, same as the elastic upper lithosphere in the regional GIA study. For reservoir geometry and unloading distribution, we consider the same set of parameters as in the previous elastic study (Albino et al., 2010). Magma was considered incompressible. After a large advance during the Little Ice Age, Mýrdalsjökull, as all the glaciers in Iceland, started to retreat around 1890 (Björnsson, 1979). We thus model the Mýrdalsjökull retreat by applying a constant unloading rate of 35 kPa/year (corresponding to a 4 m loss of ice), at the periphery of the ice cap at distances between 7 and 17 km from the centre of the ice cap, as used by Albino et al. (2010). This unloading is applied over the last 120 years, from 1890 to present.

Figure 5.6a shows the resulting temporal evolution of the failure pressure change, ΔP_r. The threshold pressure increases gradually during the unloading period. The cumulative change reaches 0.34 MPa, which corresponds to around 7% of its initial failure pressure (the initial value in the model was 4.87 MPa and corresponds to the state before the unloading). This behaviour with an increase of failure pressure in the viscoelastic model is opposite to that compared to a fully

Figure 5.6 Model prediction of temporal evolution of failure conditions, of a shallow magma chamber at the Katla volcano (see text). (a) The failure pressure change, ΔP_r and (b) the magma pressure change, ΔP_c, of the Katla magma chamber due to the last 120 years of deglaciation at the Mýrdalsjökull ice cap. Ice retreat has a constant rate of −4 m/year at distances between 7 and 17 km from the centre of the ice cap, and starts at time = 0 year. The comparison between the viscoelastic case (black) and model for the same situation for a fully elastic Earth model (red) is shown. (c) Temporal evolution of $\Delta P_r - \Delta P_c$: this difference between both pressure changes is a key parameter for evaluating the evolution of failure conditions. All parameters used for the calculation are listed in Table 5.2.

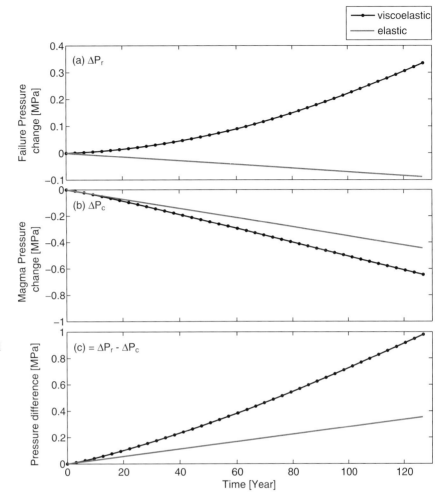

elastic model, where decrease is inferred (see Figure 5.6a). For magma pressure, both models, elastic and viscoelastic, predict a pressure decrease inside the Katla magma chamber (Figure 5.6b). By taking into account the viscous effect of lower crust, the rate of the magma pressure drop is larger than for the elastic model, with a difference about 0.1 MPa after 120 years. By evaluating the relative difference between both pressure changes, $\Delta P_r - \Delta P_c$, (Figure 5.6c), we are able to conclude if the probability of dyke injection from the Katla magma chamber will increase or decrease due to a century of icecap retreat at Mýrdalsjökull. The difference is positive, indicating that the cumulative change due to ice retreat may lead to a reduction of the eruption probability. The preventing effect on reservoir failure reaches an amplitude up to 1 MPa at the present time, after 120 years of icecap deglaciation. This may cause longer time between eruptions, and larger storage of magma in the chamber, available for eruptions. The rate of change in the failure conditions is more pronounced in the later part of the icecap evolution, suggesting that modulation of magmatic activity may be larger after few decades of consecutive retreat rather than when the retreat begins. The results presented here indicate the importance of taking into consideration viscous relaxation of the upper crust and the mantle for long-term unloading event, such as ice retreat, when considering their effects on shallow magma chambers. Indeed, results obtained are quite different if the Earth model is assumed to be fully elastic instead of viscoelastic, leading to a large underestimation of the effects of pressure variation in the case presented above.

Discussion and conclusions

The analyses presented here reveal some differences compared with an earlier summary of Sigmundsson et al. (2010) of the stress induced effects of unloading both on deep mantle melting and its influence on crustal magma chamber. The third effect presented above, on the increased magma capture in the crust was not considered in the previous overview. Regarding the influence on deep mantle melting due to present thinning of the Vatnajökull ice cap in Iceland, Pagli and Sigmundsson (2008) estimated that additional magma was generated at a rate of about 0.014 km³/year underneath Vatnajökull. The model calculations presented here suggest, however, a value of 0.07–0.17 km³/year for the whole of Iceland's rift zones, or about an order of magnitude more. There are four factors contributing to this difference, the first one being the area of study. The earlier study considered only the rift under Vatnajökull, but here we consider the full length of the rift zone across Iceland. This difference in the length of the rift considered contributes about a factor 3. A second contributor is the consideration of all the ice caps in Iceland rather than just the Vatnajökull ice cap. The smaller ice caps (see Figure 5.1) are also retreating and generally at a speed equal to the higher peripheral rates at Vatnajökull, adding about 50% to the weight loss compared to Vatnajökull alone. They also spread the pressure decrease over a wider geographic area, stimulating pressure decrease deep in the melting regime. An additional contributor to the high value of present day deglaciation induced mantle melting presented here is the use of the relation between pressure change and melting by McKenzie and Bickle (1988) rather than McKenzie (1984). Finally, the ridge angle and assumed depth of melting also influence the estimated melt volumes.

Despite the larger amount of melt generated under Iceland by ice unloading in present models, there is a large uncertainty in how much of this magma arrives at the surface of the Earth, and when it does so. Melt extraction rates from the mantle are finite and uncertain; and it may require decades or centuries for magma to travel from the melting regime to the surface. Furthermore, the excess melt generated by ice unloading will not arrive at the surface all at the same time, rather over a distributed time interval as comes from variable depth. On the way towards the surface, the additional magma may be captured by the crust as explained in Chapter 3, as the deglaciation can induce higher excess pressure in dykes, driving their formation. This allows more magma to be emplaced in the crust as dykes than in the absence of deglaciation. A deep dyke injection north of the Vatnajökull ice cap in Iceland 2007–2008 may have been influenced by the stress field deep in the crust, generated by ice retreat.

Finally, we have considered the influence of viscoelastic response of crust and mantle to long term ice retreat on stability of shallow magma chambers. The effects are very dependent on the geometry of magma chambers and ice unloading. The inferred response of the unloading on the magmatic systems can be significantly larger when viscoelastic effects of crust and mantle are evaluated, compared to that which elastic modelling would indicate. The effects will, however, not in all cases increase the probability of eruptions. For an oblate ellipsoidal model of the magma chamber at Katla volcano in Iceland, and an ice retreat model for the overlying Mýrdalsjökull ice cap consisting only of peripheral unloading at the ice cap edge as explained above, the effect of the reduction in the ice load will be to inhibit eruptions. Such effects could contribute to the present longer-than-average repose interval between major eruptions at this volcano breaking the ice cover. The most recent such eruption occurred in 1918, whereas prior to that the volcano has had major eruptions about one to three times per century since the twelfth century. In addition to the complications presented above in accurately evaluating the effects of climate driven stress change on magmatic systems, there are other ways ice unloading can affect magmatic systems. Ice load variation can influence the likelihood of ring fault formation at volcanoes (Geyer & Bindeman, 2011) and unloading can influence a delicate balance of dissolved volatiles in magma residing in shallow magma chambers. More modelling and evaluation of the effects of surface unloading on magmatic systems are therefore needed, as well as good monitoring and surveying of ice capped volcanoes experiencing ice retreat.

References

Albino, F., Pinel, V. & Sigmundsson, F. (2010) Influence of surface load variations on eruption likelihood: application to two Icelandic subglacial volcanoes, Grímsvötn and Katla. *Geophysical Journal International* **181**, 1510–1524.

Árnadóttir, T., Lund, B., Jiang, W., et al. (2009) Glacial rebound and plate spreading: Results from the first countrywide GPS observations in Iceland. *Geophysical Journal International* **177**, 691–716.

Björnsson, H. (1979) Glaciers in Iceland. *Jökull* **29**, 74–80.

Björnsson, H. & Pálsson, F. (2008) Icelandic glaciers. *Jökull* **58**, 365–368.

DeMets, C., Gordon, R. G., Argus, D. F. & Stein, S. (1994) Effect of recent revisions to the geomagnetic reversal time scale on estimates of current plate motions. *Geophysical Research Letters* **21**, 2191–2194.

Eliasson J., Larsen G., Gudmundsson M. T. & Sigmundsson, F. (2006) Probabilistic model for eruptions and associated flood events in the Katla caldera, Iceland. *Computational Geosciences* **10**, 179–200.

Geyer, A. & Bindeman, I. (2011) Glacial influence on caldera-forming eruptions. *Journal of Vocanology and Geothermal Research* **202**, 127–142.

Gudmundsson, M. T., Hognadottir, P., Kristinsson, A. B. & Guðbjornsson, S. (2007) Geothermal activity in the subglacial Katla caldera, Iceland, 1999–2005, studied with radar altimetry. *Annals of Glaciology* **45**, 66–72.

Hooper, A., Ofeigsson, B., Sigmundsson, F. et al. (2011) Increased capture of magma in the crust promoted by ice-cap retreat in Iceland. *Nature Geoscience* **4**, 783–786.

Jakobsdóttir, S. S., Roberts, M. J., Guðmundsson, G. B., Geirsson, H. & Slunga, R. (2008) Earthquake swarms at Upptyppingar, North-East Iceland: a sign of magma intrusion? *Studia Geophysica et Geodaetica* **52**, 513–528.

Jull, M. & McKenzie, D. (1996) The effect of deglaciation on mantle melting beneath Iceland. *Journal of Geophysical Research* **101**, 815–828.

Lister, J. R. & Kerr, R. C. (1991) Fluid-mechanical models of crack propagation and their application to magma transport in dykes. *Journal of Geophysical Research* **96**, 049–077.

Maclennan, J., Jull, M., McKenzie, D., Slater, L. & Grönvold, K. (2002) The link between volcanism and deglaciation in Iceland. *Geochemistry, Geophysics, Geosystems* **3**, 1062.

McKenzie, D. (1984) The generation and compaction of partially molten rock. *Journal of Petrology* **25**, 713–765.

McKenzie, D. & Bickle, M. J. (1988). The volume and composition of melt generated by extension of the lithosphere. *Journal of Petrology* **29**, 625–679.

Pagli, C. & Sigmundsson, F. (2008) Will present day glacier retreat increase volcanic activity? Stress induced by recent glacier retreat and its effect on magmatism at the Vatnajökull ice cap, Iceland. *Geophysical Research Letters* **35**, L09304.

Pagli, C., Sigmundsson, F., Lund, B., et al. (2007) Glacio-isostatic deformation around the Vatnajökull ice cap, Iceland, induced by recent climate warming: GPS observations and finite element modeling. *Journal of Geophysical Research* **112**, B08405.

Pinel, V. & Jaupart, C. (2005) Some consequences of volcanic edifice destruction for eruption conditions. *Journal of Vocanology and Geothermal Research* **145**, 68–80.

Pinel, V., F. Sigmundsson, F., Sturkell, E., et al. (2007) Discriminating volcano deformation due to magma movement and variable loads: Application to Katla subglacial volcano, Iceland. *Geophysical Journal International* **169**, 325–338.

Segall, P. (2009) *Earthquake and Volcano Deformation*. Princeton, NJ: Princeton University Press.

Segall, P., Cervelli, P., Owen, S., Lisowski, M. & Miklius, A. (2001) Constraints on dike propagation from continuous GPS measurements. *Journal of Geophysical Research* **106**, 301–318.

Sigmundsson, F., Pinel, V., Lund, B., et al. (2010). Climate effects on volcanism: influence on magmatic systems of loading and unloading from ice mass variations, with examples from Iceland. *Philosphical Transactions of the Royal Society A* **368**, 2519–2534.

White, R.S., Drew, J., Martens, H. R., Key, J., Soosalu, H. & Jakobsdóttir, S. S. (2011) Dynamics of dyke intrusion in the mid-crust of Iceland. *Earth and Planetary Science Letters* **304**, 300–312.

6 Response of faults to climate-driven changes in ice and water volumes at the surface of the Earth

Andrea Hampel[1], Ralf Hetzel[2] and Georgios Maniatis[1]

[1]Institut für Geologie, Leibniz Universität Hannover, Germany
[2]Institut für Geologie und Paläontologie, Universität Münster, Germany

Summary

Numerical models including one or more faults in a rheologically stratified lithosphere show that climate-induced variations in ice and water volumes on Earth's surface considerably affect the slip evolution of both thrust and normal faults. In general, the slip rate and hence the seismicity of a fault decreases during loading and increases during unloading. Here we present several case studies to show that a post-glacial slip rate increase occurred on faults worldwide in regions where ice caps and lakes decayed at the end of the last glaciation. It is noteworthy that the post-glacial amplification of seismicity was not restricted to the areas beneath the large Laurentide and Fennoscandian ice sheets but also occurred in regions affected by smaller ice caps or lakes, e.g. the Basin-and-Range Province. Our results do not only have important consequences for the interpretation of palaeo-seismological records from faults in these regions but also for the evaluation of the future seismicity in regions currently affected by deglaciation such as Greenland and Antarctica: shrinkage of the modern ice sheets due to global warming may ultimately lead to an increase in earthquake frequency in these regions.

Introduction

During the past two decades, much effort has been put into deciphering the interaction between climate changes and the deformation of the Earth's lithosphere using numerical modelling. The starting point of most models was that topography created by tectonic processes will ultimately be the target of erosion and sedimentation processes, which will lead to the re-distribution of mass at the Earth's surface. This mass re-distribution may, in turn, influence the localization and style of crustal deformation, as demonstrated by numerical modelling on different scales (e.g. Koons, 1989; Beaumont et al., 1992; Avouac & Burov, 1996; Batt & Braun, 1999). Erosion and sedimentation may even be able to accelerate fault slip, as shown by recent numerical models (Maniatis et al., 2009). Commonly, however, it has proven difficult to test the model predictions by field data, because cause and potential effect may be both temporally and spatially separated – in particular on the scale of an orogen – and hence not easily identified.

On timescales of 10^4–10^5 years, a clear link between the slip history of active faults and climate-driven mass fluctuations on Earth's surface was recently established by comparing predictions from numerical models with geological and palaeo-seismological data (Hetzel & Hampel, 2005; Hampel & Hetzel, 2006; Hampel et al., 2007, 2009, 2010; Turpeinen et al., 2008). Such mass fluctuations include variations in the volumes of ice caps, glaciers or lakes, and may considerably affect the slip behaviour of faults. The numerical models, which explicitly include one or several fault planes embedded in a rheologically layered lithosphere, showed that loading and flexure of the lithosphere by ice or water decreases the slip rates of nearby faults whereas unloading and rebound of the lithosphere accelerates the slip accumulation.

After a summary of the general set-up and results of these previous models, this chapter presents an overview of prominent examples from Scandinavia and the Basin-and-Range Province, for which numerical models showed that the post-glacial increase in seismicity documented by field data can be explained by rapid changes in the volume of nearby ice caps or lakes. Finally, we describe regions where similar processes could have played a role and discuss the implications for regions that currently experience ice loss or lake regression.

General model set-up and results

The principal set-up of both the two- and three-dimensional numerical models includes a lithosphere that is divided into an elastic upper crust, a viscoelastic lower crust and a viscoelastic lithospheric mantle (Figure 6.1). In the upper crust, one or more faults may be embedded, which may have variable dip and – in three-dimensional models – also variable strike. Fault slip is governed by a Mohr–Coulomb criterion. To simulate a contractional or extensional tectonic setting and to initiate slip on the fault, the model is shortened or extended by applying a velocity boundary condition (typically a few millimetres per year) at the model sides. If the models are applied to specific case studies, the shortening/extension rate is adjusted to velocities derived from space-geodetic measurements. The load, i.e. the ice or water volume, is represented by a pressure, which is applied to the top of the model and may vary in both space and time. Detailed information on

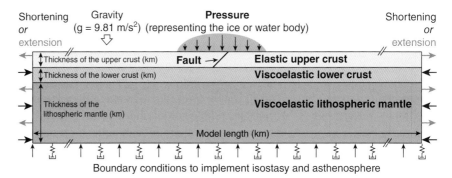

the set-up and the rheological parameters can be found in Hampel and Hetzel (2006) and Turpeinen et al. (2008).

The models are calculated using the commercial finite-element software ABAQUS. Each experiment starts with the establishment of isostatic equilibrium and an initial phase, during which the model is shortened or extended for usually several hundred thousand years. During this phase, the fault reaches a steady-state slip rate (Figure 6.2a). Afterwards the cycle of loading and subsequent unloading begins, with the shortening/extension rate unchanged. In most experiments of our sensitivity analyses (Hampel & Hetzel, 2006; Turpeinen et al., 2008; Karow, 2009), the fault responds to loading and flexure of the lithosphere by a decreasing slip rate, whereas subsequent unloading and rebound accelerates fault slip (Figure 6.2a). Parameters that have a major influence on the slip rate variations include the characteristics of the load (volume and spatial distribution), the viscosity of the lithospheric layers and the shortening/extension rate, which is applied as a boundary condition. In particular, the slip rate variations are considerably amplified with increasing magnitude of the load. Parameters of minor influence include the thickness of the lithosphere and the viscosity of the asthenosphere.

For both thrust and normal faults, the slip rate variations are caused by temporal changes in the differential stress, i.e. the difference between the maximum principal stress σ_1 and the minimum principal stress σ_3. For a fault located beneath the load, the differential stress $(\sigma_1 - \sigma_3)$ decreases during loading and flexure of the lithosphere and increases during unloading and rebound (Figure 6.2b). As a consequence the fault slip rate first decreases and subsequently increases. Remarkably, both thrust and normal faults show a similar slip pattern during the loading–unloading cycle even though the orientation of the principal stresses is different for contractional and extensional tectonic regimes (i.e. σ_1 is horizontal in contractional settings, whereas σ_1 is vertical in extensional regimes). The underlying reason is that – for both settings – the horizontal stress changes, which are caused by flexure and rebound of the lithosphere, exceed the vertical stress changes, which are caused by addition and removal of the load. Hence the differential stress shows a similar temporal evolution for both thrust and normal faults.

Figure 6.2 (a) Typical slip evolution of a model fault. After a period of shortening or extension, slip is initiated and the fault reaches a steady-state slip rate. During the subsequent loading–unloading cycle, the slip rate of the fault first decreases during loading and then increases during removal of the load. Depending on the magnitude of the load, the slip rate during unloading may increase by up to a factor of 10 compared with the steady-state rate. (b) Typical temporal evolution of the changes in the vertically oriented principal stress, the horizontally oriented principal stress and the resulting differential stress during a loading–unloading cycle, as obtained from points in the centre of the upper crust near the model fault. Note that, although both principal stresses increase during loading and decrease during unloading, the net effect is a decrease and subsequent increase in the differential stress, which explains the slip rate variations of the fault. The reason is that the change in the horizontally oriented stress caused by flexure and rebound of the lithosphere is more pronounced than the change in the vertically oriented principal stress caused by the changing volume of the load.

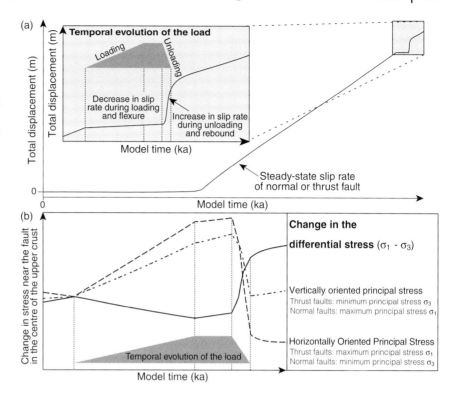

Case studies

Scandinavia

One region where a close spatiotemporal correlation between melting of a large ice sheet and the occurrence of palaeo-earthquakes was already recognised in the 1970s is the Lapland Fault Province in northern Scandinavia (Figure 6.3a) (Lundquist & Lagerbäck, 1976; Mörner, 1978; Lagerbäck, 1979). In this region – commonly considered as a tectonically quiet continental shield – up to 15-m-high scarps along faults up to 150 km long document the occurrence of palaeo-earthquakes with magnitudes as large as M_w (moment of magnitude) of approximately 8 (e.g. Arvidsson, 1996; Dehls et al., 2000; Lagerbäck & Sundh, 2008). Most of the faults have a reverse sense of slip and strike north-north-east to south-south-west. Prominent examples include the Pärvie fault in northern Sweden (Figure 6.3b,c) and the Stuoragurra fault in Norway. As the fault scarps offset deposits from the last glacial period, they must be of post-glacial age. Trenching and dating of earthquake-induced soil liquefaction features and landslides revealed that the palaeo-earthquakes in the Lapland Fault Province clustered approximately 9000 years before the present (BP) (Lagerbäck, 1992; Mörner,

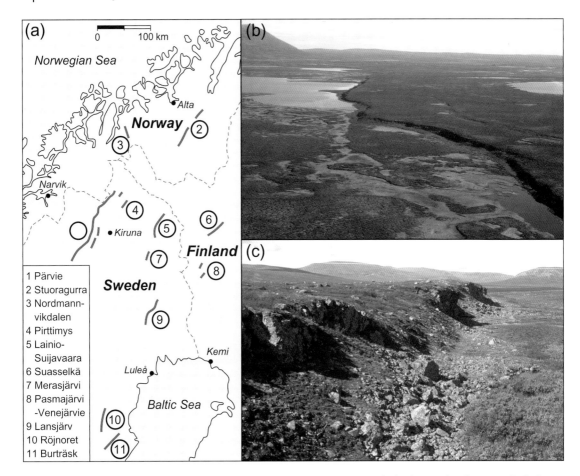

Figure 6.3 (a) Map showing the late and post-glacial faults in the Lapland Fault Province, northern Scandinavia. Fault traces were adopted from Stewart et al. (2000) and Lagerbäck and Sundh (2008). (b) Aerial photograph of the most prominent fault, the 150-km-long Pärvie fault. (c) Photograph showing the up to 15-m-high Pärvie fault scarp. (Photographs courtesy of Robert Lagerbäck.)

2005), which coincides in space and time with the deglaciation of region (e.g. Lundquist, 1986).

To investigate whether the palaeo-earthquakes in Scandinavia were triggered by the melting of the more than 2-km-thick Fennoscandian ice sheet, we used a two-dimensional model, in which the width of the ice sheet, its temporal evolution as well as the tectonic shortening rate were adjusted to northern Scandinavia (for details see Turpeinen et al., 2008). The model results showed that faulting is suppressed during the entire lifetime of the Fennoscandian ice sheet, regardless of temporal variations in its thickness between 700 m and 3400 m (Figure 6.4). Melting of the ice sheet triggers a strong slip rate increase: between about 10.5 ka and 8 ka, the fault accumulates about 15 m of slip. This is in excellent agreement with the timing and displacement of the palaeo-earthquakes in the Lapland Fault Province. The model results further reveal that most of the slip is accumulated on

Figure 6.4 Fault slip history resulting from a two-dimensional thrust fault model, in which the magnitude and width of the load were adjusted to the part of the Fennoscandian ice sheet that covered northern Scandinavia during the last glaciation. The temporal evolution of the ice thickness was simplified from Talbot (1999). A shortening rate of 1 mm/a was applied as a velocity boundary condition to the 1000-km-long model (compare Milne et al., 2001). Note the abrupt accumulation of approximately 15 m of slip between 8 and 10 ka when the deglaciation of northern Scandinavia was almost completed.

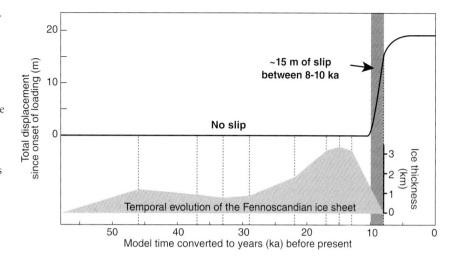

the fault before deglaciation is completed, which is in accordance with the palaeo-seismological record. Note that the model does not predict the number of earthquakes by which the total displacement is achieved. In the Lapland Fault Province, some large faults similar to the Pärvie and Lansjärv faults probably ruptured in a single event (Arvidsson, 1996; Dehls et al., 2000). In summary, our models support the view that deglaciation caused the increased seismicity in northern Scandinavia. The same may be true in central and southern Scandinavia, where palaeo-earthquakes also coincided in space and time with the deglaciation. In central Sweden an increase in seismicity is documented at approximately 10,500 varve years BP (Mörner, 2005), whereas palaeo-seismological investigations in southern Sweden revealed a cluster of earthquakes between approximately 12,000 and 12,400 varve years BP. Hence, it appears that deglaciation-induced seismicity migrated northward together with the margin of the melting Fennoscandian ice sheet.

Basin-and-Range Province

In contrast to Scandinavia, the Basin-and-Range Province in the western USA is characterised by active east–west extension at a rate of several millimetres per year (e.g. Wernicke et al., 2000), which led to the formation of alternating normal-fault-bounded ranges and sedimentary basins. For several of the many normal faults in the Basin-and-Range Province, detailed geological and palaeo-seismological records constrain the slip histories on timescales of up to 10^5 years. Interestingly, some of these records revealed considerable variations in the fault slip rates on a 10^4-year timescale (e.g. Wallace, 1987; Byrd et al., 1994; Friedrich et al., 2003; Wesnousky & Willoughby, 2003; Wesnousky et al., 2005). Such highly variable faulting rates are difficult to reconcile with quasi-periodic earthquake recurrence models and led to a new recurrence model that combines earthquake clustering on short timescales with uniform long-term strain accumulation (Wallace, 1987). It is intriguing, however, that several faults, including the prominent Wasatch and Teton faults, show an increase in their slip rate at the end of the last glacial period (Byrd et al., 1994; McCalpin, 2002; Ruleman & Lageson, 2002). Although the Basin-and-Range Province was not covered by a large ice sheet similar to Scandinavia, it experienced considerable loading by two large lakes – Lake Bonneville and Lake Lahontan – and an ice cap centred on the Yellowstone Plateau during the last glacial period. However, the potential relationship between the glacial loads and accelerated fault slip is not as obvious as in Scandinavia owing to the active crustal extension. In the following, we present two case studies, for which well-documented fault slip variations could be linked to changing ice and water volumes (Hetzel & Hampel, 2005; Hampel et al., 2007; Karow & Hampel, 2010).

The first case study is dedicated to the approximately 350-km-long Wasatch fault in the eastern Basin-and-Range Province and its response to the regression of Lake Bonneville (Figure 6.5) (Hetzel & Hampel, 2005; Karow & Hampel, 2010). This predecessor of the Great Salt Lake occupied an area of approximately 52,000 km^2 and reached a maximum depth of 350 m. Well-preserved shorelines on islands and the flanks of mountain ranges (Figure 6.5b,c) reveal three highstands (Gilbert, 1890) at 17,500 calibrated ^{14}C years BP: 17.5 ka (Bonneville highstand), 16.7 ka (Provo highstand) and 11.4 ka (Gilbert highstand), respectively (Oviatt et al., 1992). As first noted by Gilbert (1890), the shorelines in the former centre of Lake Bonneville occur at an elevation 60–70 m higher than the shorelines along the lake periphery, which documents that the lake regression caused pronounced isostatic uplift (Bills et al., 1994). In addition to Lake Bonneville, the area east of the Wasatch fault was loaded by glaciers in the Uinta and Wasatch Mountains (Figure 6.5d,e), which locally reached a thickness of several hundred metres (Atwood, 1909; Madsen & Currey, 1979).

A wealth of palaeo-seismological data exists for the different segments of the Wasatch fault, and the datasets from the Salt Lake City and Brigham City segments even extend beyond the Holocene (e.g. Personius, 1991; McCalpin et al., 1994; McCalpin & Nishenko, 1996; McCalpin, 2002; McCalpin & Forman, 2002). The Salt Lake City segment experienced a phase of seismic quiescence lasting from at least 26 ka to the end of the Bonneville highstand, which was followed by seven earthquakes. The first of these earthquakes occurred at approximately 17 ka, i.e. shortly after the rapid regression of the lake; the six other seismic events followed between about 9 ka and present. Similar to the Salt Lake City segment, the Brigham City segment was ruptured by an earthquake that occurred after the Bonneville flood but before the end of the Provo highstand. Another palaeo-earthquake probably occurred between 9 and 13 ka BP, followed by a sequence of five earthquakes during the Holocene. From the cumulative post-Bonneville displacements on the Salt Lake City and Brigham City segments an average slip rate of approximately 1 mm/year since regression of Lake Bonneville was inferred (e.g. Friedrich et al., 2003). In contrast, the average slip rate of the central Wasatch fault during

Figure 6.5 (a) Map showing Late Pleistocene Lake Bonneville and glaciers in the Uinta Mountains. Additional valley glaciers existed in the footwall of the Wasatch fault, the Wasatch Mountains. (b) Photograph showing the palaeo-shorelines of Lake Bonneville on Antelope Island, Utah. Shorelines from the Bonneville and Provo highstands are indicated by white and light blue arrows, respectively. (c) Shorelines cut into the flanks of the Wasatch Mountains near Brigham City (white arrows: Bonneville highstand; light blue: Provo highstand). (d) Photograph showing the Little Cottonwood Canyon south of Salt Lake City. Note the fault scarp of one splay of the Wasatch fault in the foreground (red arrows). (e) Moraines of the last glaciation (Pinedale), offset by splays of the Wasatch normal fault near Salt Lake City.

the last 130 ka was only 0.1–0.6 mm/year, as derived from tectonically offset Bull Lake glacial deposits and pre-Bonneville lacustrine deposits. This implies that the post-Bonneville slip rate is about twice as high as the average slip rate on a 10^5-year timescale. Slip accumulation on the Wasatch fault varied also along strike because the cumulative displacement since the regression of Lake Bonneville is 18–20 m on the Brigham City segment but only 5–10 m on the southern Nephi segment.

Using two-dimensional models including a normal fault, Hetzel and Hampel (2005) showed that the rebound induced by regression of the lake and melting of glaciers increases the slip rate of the model fault by temporally altering the stress field of the crust. To investigate potential along-strike slip variations that may result from the spatial distribution of Lake Bonneville and the valley glaciers in the Uinta and Wasatch mountains, Karow and Hampel (2010) recently created a three-dimensional numerical model, which takes the spatial distribution of the lake and glaciers into account. This model shows a maximum rebound-induced uplift of 69 m, which agrees well with the uplift derived from deformed lake shorelines, and reveals substantial along-strike variations in fault slip (Figure 6.6). During rise of the lake level, from 35 ka to 17.5 ka, most slip is accumulated in the southern part of the model fault. From 17.5 ka to the present, the northern and central segments experience the most slip (Figure 6.6). In particular, the model predicts a pronounced slip rate increase on the northern and central segments shortly after the Bonneville flood (Figure 6.6b), which agrees well with the occurrence of the first post-Bonneville earthquakes on the Brigham City and Salt Lake City segments approximately 17 ka ago (McCalpin, 2002; McCalpin & Forman, 2002). In accordance with the geological and palaeo-seismological data (McCalpin & Nishenko, 1996; McCalpin, 2002; Friedrich et al., 2003), the slip rate of the model fault since 17.5 ka is higher than its long-term slip rate. Finally, the model shows that the central Brigham City and Salt Lake City segments accumulate about three times more displacement than the southern Nephi segment, which agrees with the observed along-strike variations in fault slip.

The second case study from the Basin-and-Range Province sheds light on the influence of the Yellowstone ice cap on the Teton normal fault (Figure 6.7). The Teton Range in the fault footwall is famous for its spectacular glacial geomorphology and high relief. Its highest peak, the Grand Teton, stands about 2400 m above Jackson Hole, the hanging wall of the fault (Figure 6.7b,c). Both elevation and relief along the range decrease along strike toward the north and south (Figure 6.7b). Maximum estimates for the slip rate of the Teton fault on a 10^6-year timescale range of between 0.5 and 1.2 mm/year (Love, 1977; Smith et al., 1993; Byrd et al., 1994). Fault scarps, which offset deposits from the last glacial period (Pinedale), provide evidence for its seismic activity during the Late Pleistocene and Holocene (Figure 6.7d). Vertical offsets derived from these scarps reach approximately 30 m in the central part of the fault and diminish to the north and south, thus mimicking the along-strike variations in elevation and relief of the range. Assuming that the scarps formed since about 15 ka, a vertical slip rate of about 2 mm/year was derived for the central Teton fault (Machette et al., 2001).

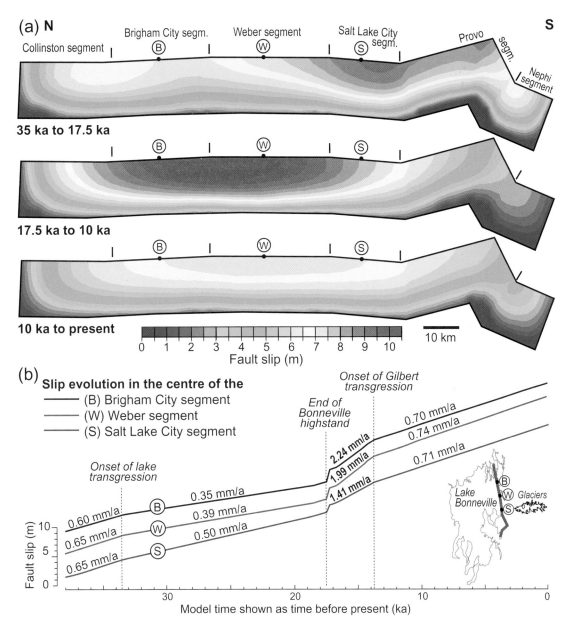

Figure 6.6 Model results for the slip evolution on the Wasatch fault. The temporal evolution of the load is adjusted to the lake level curve of Lake Bonneville (Oviatt et al., 1992) and the glaciation history of the Uinta and Wasatch mountains (Atwood, 1909; Madsen & Currey, 1979). (a) Contour plots of cumulative slip projected on the fault plane for the periods 35–17.5 ka, 17.5–10 ka and 10 ka to the present. Black circles mark the points for which slip histories are shown in (b). (b) Temporal evolution of slip at points coinciding with the centres of the Brigham City, Weber and Salt Lake City segments of the Wasatch fault as predicted by the model. Points for which the slip curves are shown are marked by black circles in the inset map and in (a).

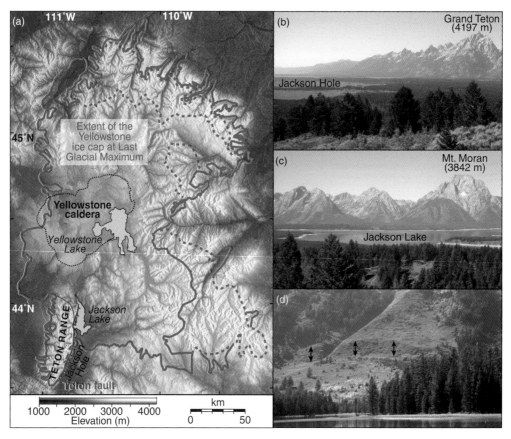

Figure 6.7 (a) Map showing the location of the Teton Range, the Teton normal fault and the extent of the Yellowstone ice cap and the glaciers in the Teton Range during the Pinedale glaciation (thick blue line; after Love et al., 2003). Dashed blue lines indicate areas that were presumably covered by Pinedale ice but are partly unmapped. (b) Photograph of Teton Range and Jackson Hole, located on the footwall and hanging wall of the Teton fault, respectively (view to the south). (c) Photograph showing U-shaped valleys of the Teton Range (view to the west) and Jackson Lake, which is bounded by forested Pinedale moraines to the south. (d) Post-glacial scarp in the central part of the Teton fault near String Lake (black arrows; view to the west).

Trenching across the southern Teton fault revealed two Holocene earthquakes at approximately 8 ka and approximately 5 ka with a total vertical offset of approximately 4 m (Smith et al., 1993; Byrd et al., 1994). As the surface offset near the trench site is approximately 15 m, approximately 11 m of vertical displacement must have been produced between 8 ka and the onset of deglaciation (Lageson et al., 1999). Hence about 70% of the post-glacial slip on the Teton fault occurred before 8 ka but after the end of the Pinedale glaciation (16–14 ka; Licciardi et al., 2001, Licciardi and Pierce, 2008). In the valleys adjacent to the Grand Teton and Mount Moran (Figure 6.7b,c), the Pinedale glaciers attained a maximum thickness of approximately 1000 m (Love et al., 2003). Additional loading of

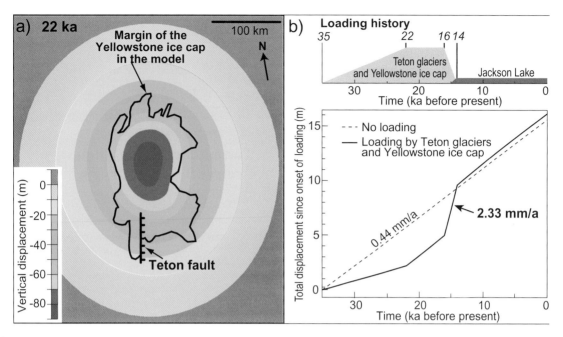

Figure 6.8 (a) Contour lines illustrating the subsidence at 22 ka resulting from loading by Yellowstone ice cap and Teton glaciers. (b) Slip histories at the fault centre at the model surface from experiments without loading (dashed line) and with loading by Teton glaciers and Yellowstone ice cap (solid line).

the Teton region was caused by the Yellowstone ice cap, which covered an area of approximately 16,500 km^2 and reached a surface altitude of approximately 3400 m (Love et al., 2003). Terminal moraines indicate that its southernmost lobes reached into northern Jackson Hole, where melting of the ice formed Jackson Lake (Figure 6.7c).

To evaluate whether the melting of the Yellowstone ice cap and the Teton valley glaciers could trigger the post-glacial slip rate increase on the Teton fault, a three-dimensional numerical model is required because the ice cap is mostly located north of the Teton fault (Figure 6.7a). The results of our three-dimensional model (Hampel et al., 2007) show that the ice load caused more than 80 m of subsidence beneath the ice cap centre 22 ka ago (Figure 6.8a). Growth and subsequent melting of the ice causes considerable slip rate variations on the model fault. The slip rate at the fault centre on the model surface decreases markedly during growth of the ice cap. During deglaciation, the slip rate increases by a factor of about five with respect to the steady-state rate and reaches a value of 2.33 mm/year (Figure 6.8b). The model further reveals pronounced variations in the post-glacial slip along strike of the fault, because the influence of the Yellowstone ice cap on fault slip increases from south to north (compare Hampel et al., 2007). To compare our results with the palaeo-seismological data, we calculated the ratio between the slip accumulated from 16 ka to 8 ka and the slip accumulated from 8 ka to the present. For the model without loading, the slip accumulated in both time intervals is equal. In contrast, loading by the Teton valley glaciers and the Yellowstone ice cap leads to a slip distribution with two-thirds of the slip occurring between 16 and 8 ka (Figure 6.8b), which agrees with the palaeo-seismological data. Our model further predicts that a considerable amount of slip occurred between 16 and 14 ka. Unfortunately, the existing palaeo-seismological data do not allow a detailed comparison with this model prediction because the timing of palaeo-earthquakes between 16 and 8 ka is still unknown.

Implications for other formerly glaciated mountain ranges and for regions currently experiencing ice-mass loss

Spatial and temporal slip rate variations are often taken as indicative of processes during fault growth, interaction among faults and fault linkage (Anders & Schlische, 1994; Cowie, 1998; Roberts & Michetti, 2004). In particular, postseismic relaxation and reloading may affect the future seismic activity of fault networks (Meade & Hager, 2004; Kenner & Simons, 2005). Temporal slip rate variations may further arise from changes in the friction coefficient of the fault, which may in turn be caused by climate-controlled changes in pore pressure (Chéry & Vernant, 2006). In some cases, an apparent increase in seismicity during the Holocene may simply be the result of a sampling bias toward faults with higher Holocene slip rates because these faults have a clearer morphological expression than faults with low slip rates (Nicol et al., 2009). However, if several faults show a synchronous increase in their slip rate with a timing correlated with major climate transitions, an external driving force from changing loads on the Earth's surface should be considered. During the Last Glacial Maximum (LGM), numerous mountain ranges worldwide were covered by ice caps or valley glaciers, e.g. the New Zealand Alps, the North American Cordillera, the southern Andes and the European Alps (e.g. Ehlers & Gibbard, 2004). It can thus be expected that more faults than the examples presented above were affected by post-glacial unloading and rebound. For the European Alps, which were covered by up to 2 km of ice during the LGM, an integrated field and modelling study shed light on the formation of unusual uphill-facing fault scarps occurring on hillslopes in the central Swiss Alps (Ustaszewski et al., 2008). These up to 10-m-high scarps run parallel to the axes of formerly glaciated valleys and locally offset deposits from the last glaciation. Finite-element experiments showed that their formation is linked to the deglaciation of the valleys, which caused uplift of the valley floors with respect to the ridge crests. A recent three-dimensional numerical models suggests that deglaciation of the European Alps after the LGM caused more than 130 m of rebound, which still continues today at a rate of approximately 0.3 mm/year (Norton & Hampel, 2010). Based on these model results, we speculate that the increase in seismicity after the LGM inferred from palaeo-seismological records in the French and Swiss Alps (Beck et al., 1996; Becker et al., 2005 and references therein) may be triggered by post-glacial unloading and rebound. Another example from this region is a normal fault in the southern Upper Rhine Graben, which was related to the Basel earthquake in 1356. Palaeoseismological investigations suggest that five earthquakes with a cumulative slip of approximately 3 m occurred on this fault during the last 13 ka, which is equivalent to a vertical displacement rate of 0.27 mm/year (Ferry et al., 2005). This rate is by a factor of three higher than the average faulting rate of 0.11 mm/year over the last 0.7 Ma. The increased postglacial slip accumulation of the fault may be related to both the deglaciation of the Alps and of glaciers in the southern Black Forest, Germany, and the southern Vosges, France (compare Ehlers & Gibbard, 2004 and references therein).

With respect to currently glaciated regions, our models indicate that the currently low level of seismicity in Greenland and Antarctica is caused by the presence of the large ice sheets. They further imply that the frequency of earthquakes should increase in the future if the ice continues to melt. This effect may be important even on timescales of 10–100 years, as shown by studies in south–central Alaska where the level of seismicity during the last decades was modulated by glacial mass fluctuations (e.g. Sauber & Ruppert, 2008). In particular, the number and size of earthquakes increased during a large-scale redistribution of ice, which occurred between 1993 and 1995 and unloaded an area near the Bering glacier in the eastern Chugach Mountains.

Conclusions

Numerical models show that mass fluctuations on the Earth's surface affect the slip evolution of both normal and reverse faults and provide quantitative predictions on this type of interaction between tectonics and climate. In general, a low level of seismicity can be expected in the presence of a load whereas slip is accelerated during unloading and rebound. Hence glacial loading and subsequent postglacial unloading provide a feasible mechanism to explain the synchronous increase in seismicity reported from formerly glaciated regions worldwide. Parameters exerting a major control on fault behaviour include the characteristics of the load (volume, spatial distribution), the viscosity of the lower crust and the lithospheric mantle, and the rate of shortening or extension. Based on the model results, it is expected that the decay of the modern ice sheets in Greenland and Antarctica will ultimately increase the level of seismicity in these regions.

Acknowledgements

Over the last years, a number of scientific partners have contributed to the topic presented in this overview chapter, by either contributions to previous publications or inspiring discussions. In particular, we thank A. Densmore, M. Ustaszewski, A. Pfiffner, G. Roberts, J. Sauber and I. Stewart as well as our former MSc student H. Turpeinen and doctoral student T. Karow. R. Lagerbäck kindly provided the two pictures from the Pärvie fault. Funding by the German Research Foundation (DFG) within the framework of an Emmy-Noether fellowship to A. Hampel (grant HA 3473/2-1) is gratefully acknowledged.

References

Anders, M. H. & Schlische, R. W. (1994) Overlapping faults, intrabasin highs, and the growth of normal faults. *Journal of Geology* **102**, 165–180.

Arvidsson, R. (1996) Fennoscandian earthquakes: Whole crustal rupturing related to postglacial rebound. *Science* **274**, 744–746.

Atwood, W. W. (1909) *Glaciation of the Uinta and Wasatch Mountains.* USGS Professional Paper 61.

Avouac, J-P. & Burov, E. B. (1996) Erosion as driving mechanism of intracontinental mountain growth. *Journal of Geophysical Research* **101**, 17747–17769.

Batt, G. E. & Braun, J. (1999) The tectonic evolution of the Southern Alps, New Zealand: insights from fully thermally coupled dynamical modelling. *Geophysical Journal International* **136**, 403–420.

Beaumont, C., Fullsack, P. & Hamilton, J. (1992) Erosional control of active compressional orogens. In: McClay, K. R. (ed.), *Thrust Tectonics.* New York: Chapman & Hall, pp 1–18.

Beck, C., Manalt, F., Chapron, E., Rensbergen, P. V. & Batist, M. D. (1996) Enhanced seismicity in the early post-glacial period: Evidence from the post-Würm sediments of Lake Annecy, northwestern *Alps Journal of Geodynamics* **22**, 155–171.

Becker, A., Ferry, M., Monecke, K., Schnellmann, M. & Giardini, D. (2005) Multi-archive paleoseismic record of late Pleistocene and Holocene strong earthquakes in Switzerland. *Tectonophysics* **400**, 153–77.

Bills, B. G., Currey, D. R. & Marshall, G. A. (1994) Viscosity estimates for the crust and upper mantle from patterns of lacustrine shoreline deformation in the Eastern Great Basin. *Journal of Geophysical Research* **99**, 22059–22086.

Byrd, J. O. D., Smith, R. B. & Geissman J. W. (1994) The Teton fault, Wyoming: Topographic signature, neotectonics, and mechanisms of deformation. *Journal of Geophysical Research* **99**, 20095–22122.

Chéry, J. & Vernant, P. (2006) Lithospheric elasticity promotes episodic fault activity. *Earth and Planetary Science Letters* **243**, 211–217.

Cowie, P. A. (1998) A healing-reloading feedback control on the growth rate of seismogenic faults. *Journal of Structural Geology* **20**, 1075–1087.

Dehls, J. F., Olesen, O., Olsen, L. & Harald-Blikra, L. (2000) Neotectonic faulting in northern Norway; the Stuoragurra and Nordmannvikdalen postglacial faults. *Quaternary Science Review* **19**, 1447–1460.

Ehlers, J. & Gibbard, P. (2004) *Quaternary Glaciations – Extent and Chronology.* Amsterdam: Elsevier.

Ferry, M., Meghraoui, M., Delouis, B. & Giardini, D. (2005) Evidence for Holocene palaeoseismicity along the Basal-Reinach active normal fault (Switzerland): a seismic source for the 1356 earthquake in the Upper Rhine graben. *Geophysical Journal International* **160**, 554–572.

Friedrich, A., Wernicke, B. P., Niemi, N. A., Bennett, R. A. & Davis, J. L. (2003) Comparison of geodetic and geologic data from the Wasatch region, Utah, and implications for the spectral character of Earth deformation at periods of 10 to 10 million years. *Journal of Geophysical Research* **108**, 2199.

Gilbert, G. K. (1890) *Lake Bonneville.* USGS Monograph 1.

Hampel, A. & Hetzel, R. (2006) Response of normal faults to glacial-interglacial fluctuations of ice and water masses on Earth's surface. *Journal of Geophysical Research* **111**, B046406.

Hampel, A., Hetzel, R. & Densmore, A. L. (2007) Postglacial slip rate increase on the Teton normal fault, northern Basin and Range Province, caused by melting of the Yellowstone ice cap and deglaciation of the Teton Range? *Geology* **35**, 1107–1110.

Hampel, A., Hetzel, R., Maniatis, G. & Karow, T. (2009) Slip rate variations on faults during glacial loading and postglacial unloading: Implications for the viscosity structure of the lithosphere. *Journal of Geophysical Research* **114**, B08406.

Hampel, A., Hetzel, R. & Maniatis, G. (2010) Response of faults to climate-driven changes in ice and water volumes on Earth's surface. In: W. McGuire (ed.) *Climate forcing of geological and geomorphological hazards. Philosophical Transactions of the Royal Society of London A* 368, 2501–2517.

Hetzel, R. & Hampel, A. (2005) Slip rate variations on normal faults during glacial-interglacial changes in surface loads. *Nature* **435**, 81–84.

Karow, T. (2009) *Three-dimensional finite-element modeling of slip rate variations on faults caused by glacial-interglacial changes in ice and water volumes: Parameter study and application to nature.* Doctoral thesis, Ruhr-University.

Karow, T. & Hampel, A. (2010) Slip rate variations on faults in the Basin-and-Range Province caused by regression of Late Pleistocene Lake Bonneville and Lake Lahontan. *International Journal of Earth Sciences* **99**, 1941–1953.

Kenner, S. J. & Simons, M. (2005) Temporal clustering of major earthquakes along individual faults due to postseismic reloading. *Geophysical Journal International* **160**, 179–194.

Koons, P. O. (1989) The topographic evolution of collisional mountain belts: A numerical look at the Southern Alps, New Zealand. *American Journal of Sciences* **289**, 1041–1069.

Lagerbäck, R. (1979) Neotectonic structures in northern Sweden. *Geologiska Foreningens I Stockholm Förhandlinger* **100**, 271–278.

Lagerbäck, R. (1992) Dating of Late Quaternary faulting in northern Sweden. *Journal of the Geological Society of London* **149**, 285–291.

Lagerbäck, R. & Sundh, M. (2008) *Early Holocene faulting and paleoseismicity in northern Sweden.* Research Paper C 836, Geological Survey Sweden.

Lageson, D. R., Adams, D. C., Morgan, L., Pierce, K. L. & Smith, R. B. (1999) Neogene-Quaternary tectonics and volcanism of southern Jackson Hole, Wyoming and south-eastern Idaho. In Hughes, S. S. & Thackray, G. D. (eds), *Guidebook to the Geology of Eastern Idaho*. Pocatello: Idaho Museum of Natural History, pp 115–130.

Licciardi, J. M., Clark, P. U., Brook, E. J., et al. (2001), Cosmogenic ^{3}He and ^{10}Be chronologies of the late Pinedale northern Yellowstone ice cap, Montana, USA. *Geology* **29**, 1095–1098.

Licciardi, J. M. & Pierce, K. L. (2008) Cosmogenic exposure-age chronologies of Pinedale and Bull Lake glaciations in greater Yellowstone and the Teton Range, USA. *Quaternary Science Review* **27**, 814–831.

Love, J. D. (1977) *Summary of the upper Cretaceous and Cenozoic stratigraphy, and of tectonic and glacial events in Jackson Hole, northwest Wyoming*. Wyoming Geological Association Guidebook, 29th Annual Field Conference, pp 585–593.

Love, J. D., Reed, J. C. Jr & Pierce K. L. (2003) *Creation of the Teton landscape*. Grand Teton Natural History Association, Moose, Wyoming, USA.

Lundquist, J. (1986) The Weichselian glaciation and deglaciation in Scandinavia. *Quaternary Science Review* **5**, 269–292.

Lundquist, J. & Lagerbäck, R. (1976) The Pärve fault: A lateglacial fault in the Precambrian of Swedish Lapland. *Geologiska Föreningens i Stockholm Förhandlingar* **98**, 45–51.

McCalpin, J. P. (2002) *Post-Bonneville Paleoearthquake Chronology of the Salt Lake City Segment, Wasatch Fault Zone, from the 1999 'Megatrench' Site*. Utah Geological Survey Miscellaneous Publication 02-7.

McCalpin, J. P. & Forman, S. L. (2002) *Post-Provo Paleoearthquake Chronology of the Brigham City Segment, Wasatch Fault Zone, Utah*. Utah Geological Survey Miscellaneous Publication 02-9.

McCalpin, J. P. & Nishenko, S. P. (1996) Holocene palaeoseismicity, temporal clustering, and probabilities of future large (M >7) earthquakes on the Wasatch fault zone, Utah. *Journal of Geophysical Research* **101**, 6233–6353.

McCalpin, J. P., Forman, S. L. & Lowe, M. (1994) Reevaluation of Holocene faulting at the Kaysville site, Weber segment, of the Wasatch fault zone Utah., *Tectonics* **13**, 1–16.

Machette, M. N., Pierce, K. L., McCalpin, J. P., Haller, K. M. & Dart, R. L. (2001) *Map and data for Quaternary faults and folds in Wyoming. USGS Open-File Report 01-461*.

Madsen, D. B. & Currey, D. R. (1979) Late Quaternary glacial and vegetation changes, Little Cottonwood Canyon, Wasatch Mountains, Utah. *Quaternary Research* **12**, 254–270.

Maniatis, G., Kurfeß, D., Hampel, A. & Heidbach, O. (2009) Slip acceleration on normal faults due to erosion and sedimentation – results from a new three-dimensional numerical model coupling tectonics and landscape evolution. *Earth and Planetary Science Letters* **284**, 570–582.

Meade, B. J. & Hager, B. H (2004) Viscoelastic deformation for a clustered earthquake cycle. *Geophysical Research Letters* **31**, L10610.

Milne, G. A., Davis, J. L., Mitrovica, J. X., et al. (2001) Space-geodetic constraints on glacial isostatic adjustment in Fennoscandia. *Science* **291**, 2381–2385.

Mörner, N-A. (1978) Faulting, fracturing and seismic activity as a function of glacial-isostasy in Fennoscandia. *Geology* **6**, 41–45.

Mörner, N.-A. (2005) An interpretation and catalogue of paleoseismicity in Sweden. *Tectonophysics* **408**, 265–307.

Nicol, A., Walsh, J., Mouslopoulou, V. & Villamor, P. (2009) Earthquake histories and Holocene acceleration of fault displacement rates. *Geology* **37**, 911–914.

Norton, K.P. & Hampel, A. (2010) Postglacial rebound promotes glacial advances – a case study from the European Alps. *Terra Nova* **22**, 297–302.

Oviatt, C. G., Currey, D. R. & Sack, D. (1992) Radiocarbon chronology of Lake Bonneville, Eastern Great Basin, USA. *Palaeogeography, Palaeoclimatology, Palaeoecology* **99**, 225–241.

Personius, S. F. (1991) *Paleoseismic analysis of the Wasatch fault zone at the Brigham City trench site, Brigham City, Utah and Pole Patch trench site, Pleasant View, Utah*. Utah Geological Survey Special Study 76.

Roberts, G. P. & Michetti, A. M. (2004) Spatial and temporal variations in growth rates along active normal fault systems: an example from the Lazio-Abruzzo Apennines, central Italy. *Journal of Structural Geology* **26**, 339–376.

Ruleman, C. A. & Lageson, D. R. (2002) Late Quaternary tectonic activity along the Madison fault zone, southwest Montana. Geological Society of American Rocky Mountains. 54th Annual Meeting, Cedar City, Utah.

Sauber, J. & Ruppert N. A. (2008) Rapid ice mass loss: does it have an influence on earthquake occurrence in southern Alaska? In: Freymüller, J. T., Haeussler, P. J., Wesson, R. L. & Ekström, G. (eds), *Active Tectonics and Seismic Potential of Alaska*. Geophysical Monograph Series 179. Washington DC: American Geophysical Union, pp 369–384.

Smith, R. B., Byrd, J. O. D. & Susong, D. D. (1993) The Teton fault, Wyoming: seismotectonics, Quatery history, and earthquake hazards. In: Snoke, A. W., Steidtmann, J. R. & Roberts, S. M. (eds), *Geology of Wyoming*. Geological Survey of Wyoming Mem. 5, 628–67.

Stewart, I. S., Sauber, J. & Rose, J. (2000) Glacio-seismotectonics: ice sheets, crustal deformation and seismicity. *Quaternary Science Review* **19**, 1367–1389.

Talbot, C. J. (1999) Ice ages and nuclear waste isolation. *Engineering Geology* **52**, 177–192.

Turpeinen, H., Hampel, A., Karow, T. & Maniatis, G. (2008) Effect of ice sheet growth and melting on the slip evolution of thrust faults. *Earth and Planetary Science Letters* **269**, 230–241.

Ustaszewski, M., Hampel, A. & Pfiffner, O. A. (2008) Composite faults in the Swiss Alps formed by the interplay of tectonics, gravitation and postglacial rebound:

an integrated field and modelling study. *Swiss Journal of Geoscience* **101**, 223–235.

Wallace, R. E. (1987) Grouping and migration of surface faulting and variation in slip rates on faults in the Great Basin province. *Bulletin of the Seismological Society of America* **77**, 868–877.

Wernicke, B., Friedrich, A. M., Niemi, N. A., Bennett, R. A. & Davis, J. L. (2000) Dynamics of Plate Boundary Fault Systems from Basin and Range Geodetic Network (BARGEN) and Geologic Data. *GSA Today* **10**, 1–7.

Wesnousky, S. G. & Willoughby, C. H. (2003) Neotectonic note: The Ruby-East Humboldt Range, northeastern Nevada. *Bulletin of the Seismological Society of America* **93**, 1345–1354.

Wesnousky, S. G., Barron, S. D., Briggs, R. W., Caskey, S. J., Kumar, S. & Owen, L. (2005) Paleoseismic transect across the northern Great Basin. *Journal of Geophysical Research* **110**, B05408.

7

Does the El-Niño – Southern Oscillation influence earthquake activity in the eastern tropical Pacific?

Serge Guillas[1], Simon Day[2] and Bill McGuire[2]

[1]Department of Statistical Science & Aon Benfield UCL Hazard Centre, University College London, UK
[2]Aon Benfield UCL Hazard Centre, Department of Earth Sciences, University College London, UK

Summary

Statistical evidence is presented for a temporal link between variations in the El Niño – Southern Oscillation (ENSO) and the occurrence of earthquakes on the East Pacific Rise (EPR). We adopt a zero-inflated Poisson regression model to represent the relationship between the number of earthquakes in the Easter Microplate on the EPR and ENSO (expressed using the Southern Oscillation Index [SOI] for East Pacific Sea-level Pressure anomalies) from February 1973 to February 2009. We also examine the relationship between the numbers of earthquakes and sea level rises, as retrieved by Topex–Poseidon from October 1992 to July 2002. We observe a significant (95% confidence level) positive influence of SOI on seismicity: positive SOI values trigger more earthquakes over the following 2–6 months than negative SOI values. There is a significant negative influence of absolute sea levels on seismicity (at 6 months' lag). We propose that increased seismicity is associated with ENSO-driven sea-surface gradients (rising from east to west) in the equatorial Pacific, leading to a reduction in ocean-bottom pressure over the EPR by a few kilopascals. This relationship is the opposite of reservoir-triggered seismicity and suggests that EPR fault activity may be triggered by plate flexure associated with the reduced pressure.

Introduction

It is becoming increasingly apparent that small changes in environmental conditions provide a means whereby physical phenomena involving the atmosphere and hydrosphere can elicit a response from the Earth's crust. McNutt and Beavan (1987) and McNutt (1999) propose, for example, that eruptions of the Pavlof

(Alaska) volcano, from the early 1970s to the late 1990s, were modulated by ocean loading involving yearly, non-tidal variations in local sea level as small as 20 cm. On a broader scale, Mason et al. (2004) present evidence in support of a seasonal signal in global volcanic activity, which they attribute to surface deformation accompanying the movement of surface water mass during the annual hydrological cycle. In relation to active faults, Rubinstein et al. (2008) have been able to correlate episodes of slow slip and accompanying seismic tremor at subduction zones in Cascadia (Pacific North West) and Japan with the rise and fall of ocean tides. Liu et al. (2009) show that such slow-slip earthquakes occurring beneath eastern Taiwan can also be triggered by reduced atmospheric pressure associated with passing typhoons. For the active Slumgullion landslide in south-west Colorado, Schulz et al. (2009) demonstrate a correlation between daily slip and diurnal tidal variations in atmospheric pressure.

Improving our understanding of such relationships between the atmosphere and hydrosphere, on the one hand, and the geosphere, on the other, are important for a number of reasons: (1) where a system, such as a fault, volcano or unstable rock mass, is critically poised, a small change in external environmental conditions may trigger a potentially hazardous response in the form, respectively, of earthquake, eruption or landslide; (2) major, short-period, climatic signals, such as ENSO (El Niño-Southern Oscillation) and NAO (North Atlantic Oscillation), which involve significant variations in sea level and atmospheric pressure, e.g. Cane (1983), Rasmusson and Wallace (1983) and Woolf et al. (2003), may also elicit a discernable and geologically significant reaction from the crust; and (3) the driving mechanisms, such as ocean loading and atmospheric pressure change, have the potential to promote an enhanced geospheric response in a future warmer world characterised by accelerated sea levels and more intense storms associated with lower central pressures and higher surges. Here we address point (2) by means of a statistical analysis of a putative relationship (Walker 1988, 1995, 1999) between the ENSO and episodic seismicity along the East Pacific Rise.

ENSO and the seismicity of the East Pacific Rise

The El Niño Southern Oscillation is the largest climate signal on Earth after the seasons and has widespread ramifications for global weather patterns. In the atmosphere, the prime manifestation of ENSO is a seesaw in sea-level pressure (the Southern Oscillation) between the south-east Pacific subtropical high and a region of low pressure that stretches across the Indian Ocean, from Africa to northern Australia. This difference is normally expressed in the form of the Southern Oscillation Index (SOI), determined from the normalised sea-level pressure difference between Tahiti and Darwin. In the ocean, ENSO is characterised by episodic warming (El Niño) or cooling (La Niña) of surface waters in the central and eastern Pacific, with an oscillation period that typically ranges from 2 years to 7 years (Guilyardi et al. 2009). The meteorological influence of ENSO is global, with its El Niño phase, in particular, resulting in potentially hazardous weather conditions in regions both adjacent to and remote from its source. These include drought in Australia, southern Africa, and south and south-east Asia, increased heavy precipitation in north-western South America, and more winter storms and floods across the US Gulf Coast (Cane, 1983). A detailed explanation of ENSO and its drivers can be found in Cane (2005), but for the purposes of the present study the key effect of ENSO is on the level of the ocean at low latitudes in the Pacific. Broadly speaking, normal conditions in the equatorial Pacific involve the easterly trade winds pushing warm surface water westwards, resulting in upwelling of deeper, colder water in the eastern Pacific. During a La Niña event (when SOI is high), this cooling in the east is enhanced. In contrast, El Niño events (when SOI is exceptionally low) are associated with a faltering, or even reversal, of the trade winds, leading to warm surface waters 'sloshing' back eastwards (Tsonis et al. 2005). Thus, high SOI (La Niña) is associated with low sea levels in the Eastern Pacific and low SOI (El Niño) is associated with high sea levels in the Eastern Pacific (Figure 7.1).

A link between the El Niño phase of ENSO and crustal seismicity was first proposed by Walker (1988), who recognised a coincidence of extreme lows in the SOI and elevated seismic activity in the vicinity of the East Pacific Rise (EPR). The relationship was subsequently (Walker 1995, 1999) elaborated and updated by the same author, focusing on the segment of the EPR between 20° S and 40° S. Walker (1999) proposes that all El Niño events since 1964 (up to and including 1998) have been preceded by anomalous seismicity along parts of the EPR, where 'anomalous' is defined as months having eight or more reported earthquakes (compared with an apparent long-term average of less than two), or months with seismic energy values of 0.9×10^{20} ergs or more. According to Walker (1999), the lead time between onset of anomalous seismicity and extreme lows in the SOI ranges from 5 months to 15 months. This the author presents as supporting a link between the elevated levels of seismicity and a trigger for El Niño emergence. Walker (1995) notes that the EPR between 20° S and 40° S contains one of the most rapidly spreading ridge segments on the planet, characterised by high levels of hydrothermal activity and submarine volcanic activity. This behaviour, the author suggests, could underpin the apparent temporal link between El Niño conditions, reflected in extreme lows in the SOI, and elevated levels of seismicity along this part of the EPR. Although admitting the absence of direct evidence, Walker (1995, 1999) goes on to consider that the latter may be indicative of massive episodes of submarine eruptive activity. As a corollary, he speculates that the heating effect of such volcanism could result in thermal expansion of deep water, sufficient to produce the approximately 2-millibar reduction in atmospheric pressure at sea level required to promote the development of El Niño conditions. Walker (1999) presents other possibilities whereby submarine volcanism could warm surface waters sufficiently to reduce atmospheric pressure in the region and produce SOI lows that flag El Niño conditions. These include 'massive plumes' of hot water that penetrate the thermocline, and 'massive episodic heating' of the ridge system triggering large earthquakes,

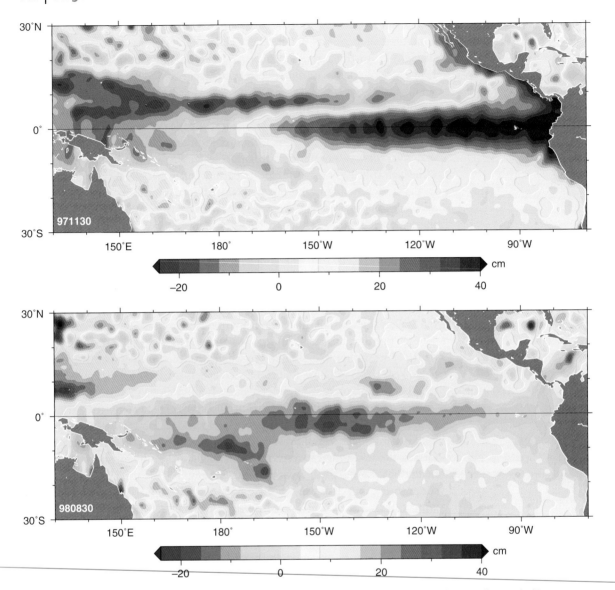

Figure 7.1 Sea level anomalies for 30 November 1997 (971130, top) and 30 August 1998 (980830, bottom). (Source: ESR/Envisat courtesy of Remko Scharroo)

increased numbers of earthquakes, sea-floor volcanism, hydrothermal venting and sea-floor spreading.

In a reply to Walker (1999), Hunt (2000) suggests that, rather than the observed seismicity reflecting a causative volcanic trigger for El Niño, it may be a consequence of changing environmental conditions precursory to El Niño emergence, specifically, systematic variations in sea level with the potential to induce small, additional crustal stresses. Hunt (2000) notes that every El Niño since 1960 (possibly with the exception of the one in 1982–1983) was preceded by the development of anomalously large upward sea-surface slopes from east to west, driven by the trade winds. This resulted, in turn, in the development of anomalously low sea levels in the Eastern Pacific before the onset of El Niño conditions (Wyrtki 1975). If higher sea levels associated with El Niño conditions prove to be capable of inducing increased seismicity in the EPR, then lowered sea levels, speculates Hunt (2000), might be capable of eliciting a similar response. Notwithstanding these suggestions, Hunt goes on to note that sea-level variations of tens of centimetres associated with the wholesale transfer of water mass during ENSO cycles would result in stress variations on the crust of less than 0.1 bar, which he regards as being insufficient to affect earthquake frequencies. Hunt (2000) also emphasises the fact that the role of atmospheric winds in triggering El Niño conditions is well established, making it difficult to accept the Walker model for triggering due to volcanic heating of the deep ocean. Hunt concludes that the apparent correlation between El Niño emergence and EPR seismicity is coincidental.

It is noteworthy that the speculations of both Walker (1988, 1995, 1999) and Hunt (2000) have been undertaken in the absence of robust statistical analysis of the putative correlation between 'anomalous' seismicity and changing environmental conditions associated with the ENSO cycle and, in particular, with El Niño onset. In order to address this omission we present here the results of a statistical study designed to shed light on the relationship between earthquake activity across the area of the EPR (20–40° S and 100–120° W) considered by Walker (1988) and (1) ENSO, as measured by the SOI, or (2) variations in sea-surface elevation. The period covered is from 1 February 1973 to 28 February 2009. Data sources are as follows: earthquake data National Earthquake Information Center (NEIC: http://neic.usgs.gov/neis/epic – accessed 23 March 2009); SOI data (version Equatorial Eastern Pacific SLP, Standardized Anomalies) from the National Oceanic and Atmospheric Administration (NOAA) – Climate Prediction Center (CPC: www.cpc.noaa.gov/data), not the more standard SOI obtained as the standardised difference in sea-level pressure between Tahiti and Darwin, because we want to focus on the Eastern Pacific; sea-surface elevation data are from the joint NASA/CNES satellite altimeter, Topex/Poseidon (http://ibis.grdl.noaa.gov/SAT/hist/tp_products/topex.html); these are sea level deviations averaged in cells with dimensions of $4 \times 1°$ (longitude × latitude), heights are relative to the 3-year mean (1993–1995), deviations are then averaged over the 20–40° S and 100–120° W region, and the resulting monthly time series spans the period October 1992 to July 2002.

A number of authors have addressed the impact of ENSO on sea-level variations (Wyrtki, 1975; Cane, 1983; Rasmusson and Wallace, 1983; Woolf et al., 2003). Over the region 20–40° S and 100–120° W and the period October 1992 to July 2002, we corroborate the previously established correlation between ENSO and sea level (Figure 7.2). We computed estimates and confidence intervals of these correlations, with a lag ranging from 0 to 12 months after which sea level is compared with the aforementioned SOI (Table 7.1).

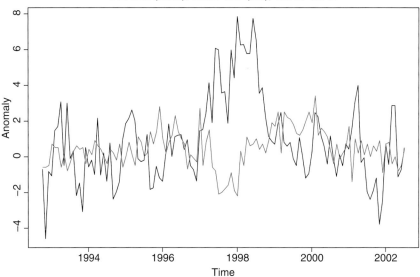

Figure 7.2 Monthly Southern Oscillation Index or SOI (sea-level pressure anomalies), October 1992 to July 2002 (black line); monthly indices of sea level deviations from the Topex–Poseidon satellite instrument, averaged over the region 20–28° S, 110–120° W, October 1992 to July 2002 (red line). We can see that absolute sea levels and SOI vary in opposite directions.

Table 7.1 Correlations estimates, with 95% confidence intervals, between El Niño – Southern Oscillation (ENSO) and sea levels lagged by 0–12 months

Lag	Lower	Estimate	Higher
0	−0.38	−0.22	−0.04
1	−0.36	−0.2	−0.02
2	−0.37	−0.2	−0.02
3	−0.35	−0.18	0
4	−0.53	−0.39	−0.22
5	−0.47	−0.32	−0.14
6	−0.56	−0.42	−0.26
7	−0.54	−0.39	−0.22
8	−0.51	−0.35	−0.18
9	−0.41	−0.24	−0.05
10	−0.34	−0.16	0.03
11	−0.33	−0.15	0.04
12	−0.24	−0.05	0.14

Origins and distribution of seismic activity on the East Pacific Rise

Seismic activity on the EPR in the area of interest for this study, from the equator to 30° S latitude, originates from three types of structure associated with this plate boundary, perhaps the type example of a fast-spreading mid-oceanic ridge system

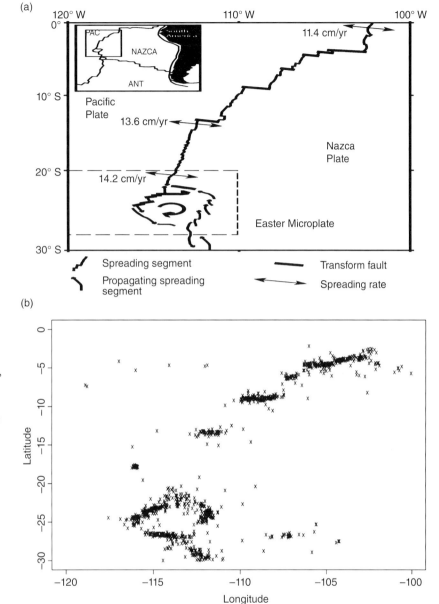

Figure 7.3 (a) Simplified tectonic map of the 0–30° S sector of the East Pacific Rise, modified after Searle et al. (1993). Dashed line encloses area of Easter Microplate, the seismicity of which we consider in detail (Figure 7.4). (b) Map of earthquake epicentres in the area of (a) taken from the NEIC earthquake catalogue for the period 1 February 1973 to 28 February 2009. Almost all earthquake foci were placed within 33 km of the surface by the NEIC earthquake location procedure.

(Figure 7.3). Recent reviews of the structure of the EPR in our area of interest, emphasising its subdivision and the development of microplates along the EPR, are provided by Searle et al. (1993) and Hey (2004).

Three different sources of seismic activity on the EPR

It is important to distinguish these three different seismic source types because they may respond in different ways to environmental forcing by ENSO. Furthermore,

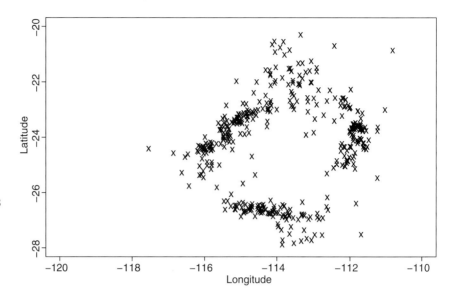

Figure 7.4 Map of earthquake epicentres in the region 20–28°S, 110–120°W taken from the NEIC earthquake catalogue for the period 1 February 1973 to 28 February 2009. Almost all earthquake foci were placed within 33 km of the surface by the NEIC earthquake location procedure.

the levels of activity differ markedly across the three, particularly in respect of the numbers of earthquakes large enough to be recorded as tele-seismic events on global seismic networks and thus appear in our primary data source, the NEIC earthquake catalogue for the time period 1 February 1973 to 28 February 2009.

Transform faults: zones of strike-slip faulting linking offset spreading centres

These structures are especially common in the northern part of the areas of interest and are the sources of numerous large earthquakes. However, an analysis of the time–space distribution of these earthquakes along these and similar transform faults, on the parts of the EPR to the north of the equator by McGuire (2008), indicates that seismic cycles of around 5 years are present: once a large (moment of magnitude, $M_w \sim 6$) earthquake occurs on any one part of a transform fault, it is about 5 years before another such earthquake occurs in the same location. McGuire (2008) attributes this short seismic cycle time to the moderate size of the individual earthquakes and the high rates of deformation and seismic loading that result from the rapid spreading of the EPR. For our purposes, however, the similarity between the proposed seismic cycle times and the interval between ENSO events makes statistical analysis of the occurrence of transform fault earthquakes problematic, because of the potential for aliasing of the time distribution signal between the two sources of time variation. Future analysis of the variation in time interval between large earthquake pairs in transform faults as a function of ENSO may help to elucidate interaction between the seismic cycles and ENSO, but at present the numbers of such earthquake pairs (16 in the McGuire [2008] dataset) are too small to permit such an analysis.

Spreading centres: the main plate boundary structures

Owing to the high rates of magma supply to the EPR, plate boundary spreading in the area of interest appears to be almost entirely accommodated by dyke intrusion. There are very few large earthquakes in the NEIC dataset located on or near these spreading centres and therefore interactions between the spreading centres and ENSO will not be revealed by analysis of that dataset. However, distributions of micro-seismicity recorded by temporary deployments of ocean bottom seismometers (OBSs) as part of research into the behaviour of the EPR spreading centres (Hey, 2004; Stroup et al., 2009) indicate significant modulation of the micro-seismicity on diurnal and semi-diurnal, tidal timescales. This is attributed to variations in fluid flow into the highly permeable crust at the spreading centre, along the dyke swarms and out of the crust again at the celebrated 'black smoker' high temperature hot springs, located at intervals along the EPR spreading centres. We speculate that longer-term OBS deployments on the EPR may reveal ENSO-related time variations in the level of micro-seismicity on these spreading segments and perhaps also links between ENSO and the larger less frequent events that occur on the spreading centres, such as dyke intrusions and seabed eruptions, but at present data are insufficient to investigate this point. For the present purpose, perhaps the most significant result to be derived from the studies of spreading centre micro-seismicity is the remarkably high permeability of the spreading centre crust ($\sim 10\text{--}13$ to $10\text{--}12 \, \text{m}^2$) derived by Stroup et al. (2009), which implies that pressure variations associated with ENSO may be transmitted into the pore fluids within the uppermost ocean crust on short timescales. We consider this point further below.

Microplates along the East Pacific Rise

A number of small microplates, some tens of kilometres across, are located along the EPR. These are thought to have been produced by unsteady propagation or jumping of spreading centre segments across transform faults, which result in isolation of small pieces of young oceanic crust between overlapping spreading centres connected by two rather than one broadly strike-slip fault segment. In plate tectonic terms, this geometry is unstable and as a result the microplates experience rapid rotation and high levels of internal deformation (Searle et al. 1993; Cogne et al. 1995; Hey, 2004). They are therefore characterised by high levels of seismic activity, typically occurring in swarms lasting periods of the order of days, suggesting some component of magma intrusion as part of the deformation, rather than in mainshock–aftershock sequences. One such microplate, the Easter Microplate (Searle et al. 1993), lies within our area of interest and despite its small size produces about 40% of all the earthquakes in the NEIC catalogue for the 0–30° S segment of the EPR. As seen in Figure 7.5, over the Easter Microplate, the relative position of the largest earthquake in groups of at least three earthquakes is roughly uniform; it cannot correspond to mainshock–aftershock sequences in which the largest earthquakes occur at the start of the sequences.

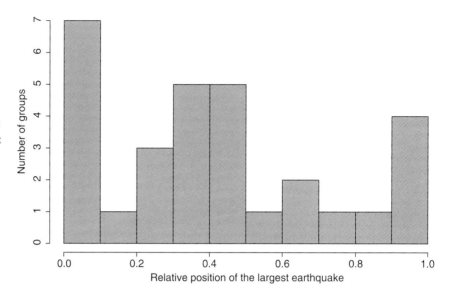

Figure 7.5 Histogram of the relative position of the largest earthquake in groups of at least three earthquakes. A group is defined as a sequence of earthquakes following each other by 2 days or less. The relative position is defined as the ration (position − 1) / (size of the group −1).

Once the transform faults have been eliminated from the analysis for reasons discussed above, earthquakes within and around the microplates dominate the EPR earthquakes in the NEIC catalogue. Much of our statistical analysis therefore concentrates on the Easter Microplate area (20–28° S, 110–120° W). Conversely, the lack of large earthquakes on the spreading segments means that large-scale eruptive activity on them will only be indirectly related, through stress transfer, to levels of seismic activity at the main earthquake sites, the transform faults and the microplates. This presents some difficulties for the Walker (1999) model because the seismic response to stress transfer after the postulated 'massive' eruptions will not be immediate, thereby weakening the postulated time sequence of earthquakes followed by thermal effects in the ocean. Under these circumstances, we consider two alternative hypotheses for how ENSO and seismic activity on the EPR might be linked, both focusing on how ENSO-driven changes in sea level over the EPR might affect this very active plate boundary:

1. Modulation of pore-fluid pressure within the crust by sea-level variations. Pore-fluid pressure variations in fault zones can influence seismicity by changing the frictional resistance to fault slip. If permeability of the crust down to the seismogenic zone is sufficiently high, short-term variations in ocean bed water pressure may propagate down into the crust and change fluid pressure on the fault zones where the earthquakes occur.

2. Flexure of the crust and unclamping of fault zones by sea level variations. As the EPR is a plate boundary where the lithosphere on either side is much stronger than the boundary itself, the two plates on either side may flex in response to sea-level changes, and so produce significant changes in the stresses transmitted across the plate boundary, especially into the microplates located within the EPR such as the Easter Microplate.

Pore fluids and permeability in mid-ocean ridge systems compared with continental crust: implications for possible mechanisms of linkage between ENSO and EPR earthquakes

A link between fluid pressure variations and seismic activity has long been recognised in the case of earthquakes in continental crust triggered by filling of surface reservoirs (see Talwani [1997] for a review). Perhaps the most notorious example is the sequence of damaging earthquakes triggered by the filling of the Koyna reservoir in the Deccan region of India (Simpson et al. 1988). Notably, although the reservoir was first filled in 1962–1963, the seismic activity was initially limited but became more intense and damaging from late 1967, with the largest events of magnitude 5.5 and 6.2 (comparable to the largest EPR earthquakes) causing a large amount of damage and a number of deaths. These delayed earthquakes at Koyna and other comparable examples are commonly relatively distant and deep (~10 km). In other cases the onset and peak of seismic activity are more immediate and directly linked to the reservoir filling, with the earthquake foci at shallow depths and close to the reservoir; in these cases the earthquakes are generally small. Simpson et al. (1988) attributed the two contrasted types of seismic activity to, on the one hand, poroelastic deformation in effectively sealed but compressible fluid reservoirs where loading by the water in the reservoir compresses the pores and raises pore fluid pressure immediately, and, on the other, to the delayed effect of fluid diffusion into reservoirs at greater depths and distances which raises pore pressure over timescales of the order of years. Furthermore, they showed that the values of large-scale diffusivity implied by the timescale of delayed seismicity, even in the relatively rigid plate interiors, were relatively high, of the order of $100 \, m^2/s$. As noted by Simpson et al. (1988), these values are much higher than those obtained by laboratory measurements on intact rock samples, implying that the large-scale diffusivity is dominated by fluid flow through fracture systems.

The high levels of seismicity and the dense fault systems in the EPR microplates, such as the Easter Microplate, imply high rates of ongoing brittle deformation and therefore the presence of abundant newly formed fractures. The well-known occurrences of hot brine springs ('black smokers') on the EPR also imply high crustal permeabilities. Although high thermal gradients and the chemically reactive nature of oceanic crust mean that such fractures will tend to seal over time, they are nevertheless likely to lead to high crustal permeability in the microplate and the vicinity of the plate boundary as a whole, and therefore significant diffusion of fluid pressure variations from the seabed to depth within the crust. As noted by Jupp and Schultz (2004), and demonstrated for the EPR by Stroup et al. (2009), the depths to which such seabed pressure variations will be felt depends on the timescale of the variations (expressed in terms of a 'skin thickness' within which diffusion is effective). At constant diffusivity (mainly influenced in fractured rocks by permeability variations – Lister, 1974) the skin thickness will increase as the square root of the timescale of diffusion. Therefore, if diurnal tidal variations are effective only within approximately the topmost 3 km of crust (Stroup et al., 2009), the months' timescales of ENSO-related sea-level variations would be expected to produce diffusion-controlled pore pressure variations in approximately the topmost 30 km of crust, comparable to or significantly greater than the depth range of the observed earthquakes in oceanic crust according to its age. In the young, hot oceanic crust of the EPR microplates it is therefore likely that the depth range of pore-fluid pressure variations will be limited by a marked downward decrease in fracture abundance through the seismogenic zone into hot crust and mantle which deforms by ductile mechanisms and is therefore impermeable on the large scale. We conclude that the whole of the seismogenic zone of the EPR microplates could therefore be susceptible to the effects of ENSO-modulated pore pressure variations: it is reasonable to expect that even the largest microplate earthquakes could be susceptible to ENSO. Whether or not this is actually the case will depend critically on the permeability structure of the seismogenic layer of the crust, so in principle the nature of the statistical relationship between ENSO and seismic activity will provide insights into the permeability structure of the Easter Microplate.

Although reservoir-related seismicity is usually associated with initial reservoir filling, in some cases the seismicity has continued for decades and is associated with fluctuations in reservoir level (Talwani, 1997). In these cases pore-fluid pressure fluctuations are thought to be associated with inflow and outflow of pore water as the reservoir level changes: thus deformation and the resulting seismicity are linked to the pressure changes.

The comparison with reservoir-induced seismicity therefore suggests two alternative hypotheses for permeability-controlled linkages between ENSO and seismicity around the Easter Microplate. We refer to these as P1 and P2.

Hypothesis P1

A falling sea level reduces seabed pressure, and hence overall load in the rocks beneath; but finite and relatively low permeability means that pore pressure at depth decreases more slowly, so the pore pressure is a higher fraction of total load. This may then lead to increased fault movement. This is opposite to the normal reservoir-related seismicity relationship but may be appropriate to the EPR because of the crustal permeability structure noted above.

Hypothesis P2

An elevated sea-level increases seabed pressure, and then absolute pore pressure as inflow into the sub-seabed rocks occurs. Again, the elevated pore fluid pressure may lead to increased fault movement. This hypothesis is similar to the normal relationship between reservoir levels and reservoir-induced seismicity. The effectiveness of this mechanism may be increased in the hot rocks near the EPR because the inflowing fluids are cooler and denser than the ambient fluids, so fluid pressure gradients with depth are increased.

Either hypothesis can be framed in two ways, in terms of absolute sea level or rate of sea level change: hypothesis P1 implies that increased seismicity will be associated with falling sea level or low absolute sea level, whereas hypothesis P2 implies that increased seismicity will be associated with rising sea level or high absolute sea level. However, we investigated the relationship between monthly sea level changes (i.e. differences of two consecutive months) and seismicity and, unlike the correlations we report in the next section, we found no significant influence of sea-level changes on seismicity (probabilities of a non-significant effect being in the range 28–54%). Hence we focus on the effects of absolute sea level and not rate of sea-level change.

Sea levels, plate flexure and EPR seismicity

The alternative hypothesis for the interrelationship of ENSO, sea levels and EPR seismicity is similar to the models for modulation of fault activity at plate boundaries. As noted above, Rubinstein et al. (2008) have been able to correlate episodes of slow-slip and accompanying seismic tremor at subduction zones in Cascadia (Pacific North West) and Japan, with the rise and fall of ocean tides, whereas Liu et al. (2009) show that such slow-slip earthquakes occurring beneath eastern Taiwan can also be triggered by reduced atmospheric pressure associated with typhoons. In these cases the concept is that reduction in pressure normal to low-angle faults unclamps them and enables fault slip through a reduction in frictional resistance to slip. In the case of the faults on the EPR, because these are high-angle normal and strike-slip faults the relationship may be more complex. Reduction in ocean bed pressure over the plates to either side of the EPR will tend to allow them to flex upwards, especially in the relatively shallow water to either side of the EPR itself. This will tend to induce additional extension across the plate boundary, in particular freeing up the microplates to rotate and so triggering seismicity within them. Conversely, increased sea level will cause the main plates to flex down and towards each other, so locking the microplates. This hypothesis (F) therefore predicts that increased seismicity will be associated with high SOI and La Niña conditions and, most notably, with the anomalously reduced sea levels that are a precursor to El Niño emergence, similar to the prediction of hypothesis P1 for the pore pressure–seismicity link but opposite to the prediction of hypothesis P2.

Statistical modelling of sea level and ENSO influence on earthquakes

Zero-inflated Poisson regression model

Poisson regression models are used to model the relationship between counts and other variables, and are a particular case of generalised linear models (GLMs) (McCullagh & Nelder, 1989). In GLMs, the response variable is not related directly

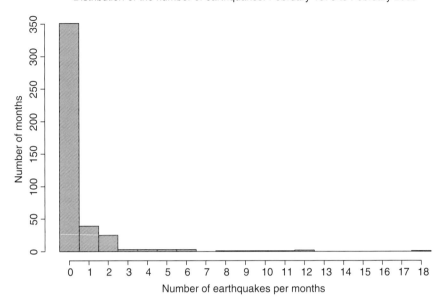

Figure 7.6 Frequencies of monthly number of earthquakes over 1973–2009 across the region 20–28° S, 110–120° W.

to the explanatory variables, a function – called link function – of the parameter of the distribution from which the response variable is drawn is assumed to be a linear combination of the explanatory variables. In Poisson regression models, the link function is the logarithm. We model the logarithm of the parameter of the Poisson distribution, which is also the mean and variance, as a linear combination of the explanatory variables.

The histogram of number of earthquakes per months over February 1973 to February 2009 can be seen in Figure 7.6. Many months have no earthquake. After further examination of the NEIC record, earthquake occurrence times (Figure 7.7) are clustered over periods of up to a few months, suggesting two regimes: one non-susceptible and one susceptible, when some conditions are satisfied for seismicity to take place. The largest earthquakes generally do not occur at the start of each cluster, but are distributed through the clusters (see Figure 7.5). Therefore, these clusters of earthquakes are not mainshock–aftershock sequences but represent groups of events that may be related to magmatic activity in the Easter Microplate, or form sequences of fault ruptures propagating around the microplate boundaries.

When two regimes are present, a zero-inflated Poisson regression model (ZIP) can represent better the relationship between counts and explanatory variables than a standard Poisson regression model (Lambert, 1992). Indeed, in a ZIP, a probability of being in the non-susceptible mode is estimated, and only in the susceptible mode is the standard Poisson regression model fitted. We use the R package 'ZIGP' to fit the ZIP model (Czado et al., 2007). We considered various timescales, over which we explained the number of earthquakes occurring over

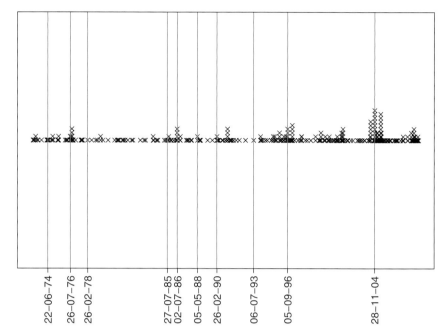

Figure 7.7 Daily occurrence of earthquakes from February 1973 to February 2009 on the East Pacific Rise (20–28° S, 110–120° W) (stacked when occurring on the same day); times of earthquakes with magnitude greater than 6 indicated by red lines.

that period, according to sea level or SOI variations: 2, 3 and 6 months. Indeed, the high permeability of the crust, as discussed in the previous section, is likely to enable the seismicity to be triggered by sea level or ENSO over short timeframes of a few months maximum. A period of 1 month only, however, was deemed to be too short because earthquakes occurring at the start of the month after a high value of ENSO or sea level ought to be counted.

Let us denote Y_i the number of quakes over the months i to $i + (m - 1)$, where m is the length in months of the considered period ($m = 2, 3, 6$). In the ZIP, the distribution of Y_i is assumed to be:

$Y_i \sim 0$, with probability p_i
$Y_i \sim \text{Poisson } (\mu_i)$, with probability $1 - p_i$

where the mean μ satisfies:

$$\log(\mu) = \beta_0 + \beta_1 X \tag{1}$$

and X is the explanatory variable (here ENSO or sea level index at month i). There is a so-called 'logit' relationship between the probability of being in the non-susceptible mode p_i and the potential linear combination of variables leading to the non-susceptibility state of the system, as in GLMs with a binary response variable (McCullagh & Nelder, 1989). However, here we do not have access to information that could give some insight about the susceptibility of the crust to changes in sea level or ENSO, so we assume a constant probability of susceptibility:

$$\text{logit } (p) = \log(p/(1 - p)) = \omega \tag{2}$$

ENSO results

Using East Pacific sea-level pressure anomalies, we retrieve the monthly SOI over February 1973 to February 2009. We show below the results of fitting the ZIP to explain the variation in the number of earthquakes in the region 20–28° S, 110–120° W by the variation of ENSO.

In all cases, we find significantly positive coefficients β_1 in the regression (1) (Table 7.2). For 2-, 3- and 6-month periods, the probability of such coefficient being statistically insignificant is less than 0.012; indeed the z value (or ratio between the estimated and its estimated standard deviation) is large. Therefore, for the 2-, 3- and 6-month periods, we find that higher values of the SOI are indeed significantly triggering a larger number of earthquakes with 95% confidence (our scientific hypothesis P1 in the previous section).

The 6-month analysis does not suffer from a zero-inflated phenomenon because the number of earthquakes is not often zero when we consider 6-month periods. As a result, the probability of being in a non-susceptible state p is estimated to be only 0.03. This probability was estimated to be 0.30 and 0.18 respectively for the 2- and 3-month periods.

The average impact of ENSO on the number of earthquakes can be computed as a result of the analysis. A one-unit increase in the SOI triggers 0.095, 0.087 or

Table 7.2 ZIP regression for 2-, 3- and 6-month periods, μ ranges of [1.96, 3.66], [2.57, 4.55], [3.60, 8.97] and p ranges of [0.30, 0.30], [0.18, 0.18], [0.03, 0.03] (logit relationship 2), respectively

			Estimate	Standard error	z value	Pr (> |z|)
2-month period		μ regression				
	β_0	Intercept	0.975	0.041	23.908	$< 2 \times 10^{-16}$
	β_1	ENSO	0.095	0.038	2.524	0.012
		ω regression				
	ω	Intercept	−0.834	0.119	7.002	$< 2.5 \times 10^{-12}$
3-month period		μ regression				
	β_0	Intercept	1.220	0.032	38.508	$< 2 \times 10^{-16}$
	β_1	ENSO	0.087	0.030	2.931	0.003
		ω regression				
	ω	Intercept	−1.500	0.137	−10.975	$< 2 \times 10^{-16}$
6-month period		μ regression				
	β_0	Intercept	1.723	0.022	79.623	$< 2 \times 10^{-16}$
	β_1	ENSO	0.138	0.020	6.839	$< 8.0 \times 10^{-12}$
		ω regression				
	ω	Intercept	−3.401	0.288	−11.826	$< 2 \times 10^{-16}$

0.138 more earthquakes in the log scale over 2, 3 and 6 months, respectively (with uncertainties, as measured by the standard errors, of respectively 0.038, 0.03 and 0.02), e.g. in the case of the 2-month period, and in the susceptible state, comparing the case where the SOI is 2 vs 0 with the estimated intercept of 0.975 (neglecting the standard error of 0.041), we obtain a mean (as well as the variance) number of earthquakes over 2 months of 3.206 vs 2.651. Adding the effect of the increase variance, these two distributions of earthquakes can yield dramatically different numbers of earthquakes. For 6 months, the numbers are respectively 7.38 and 5.60, also showing a large difference. Similar calculations for a SOI −2 vs 0 also show that lower SOI would reduce the number of earthquakes: 2.192 vs 2.651 on average for the following 3 months in the susceptible state.

Sea-level results

As the physical mechanism explaining the variation of earthquake numbers is likely to be the absolute sea level, we focus on the estimation of the effect of absolute sea level, as measured by the Topex–Poseidon satellite instrument, on the number of earthquakes. The sea-level data are averaged deviations over the region 20–28° S, 110–120° W, over October 1992 to July 2002. The results are not as significant as in the ENSO regressions, but show that lower sea levels trigger more earthquakes, and higher sea levels produce fewer earthquakes, because the coefficients β_1 for sea levels influence on earthquakes, in the log scale, according to Equation (1), are always negative: −0.048, −0.045 and −0.063, respectively for the 2, 3 and 6 months following periods (Table 7.3).

One should expect more significant results for sea levels than for ENSO, because sea level is more closely related to the amount of pressure exerted on the seabed than SOI variations. However, the existing short period of sea level data only permits us to conclude at the 95% confidence level that lower sea levels are linked to more earthquakes (higher sea levels are linked to fewer earthquakes) over the following 6 months (for a 3-month period, we only have a 90% confidence level). Calculations similar to the ones in the last section can quantify these differences.

Evaluation of alternative hypotheses

Our results have implications for the three alternative hypotheses discussed above. We argue against hypothesis H2 because its prediction of increased seismicity associated with raised sea level and low SOI (El Niño conditions) is contrary to the results of our analysis. We cannot distinguish between hypotheses H1 and F because their predictions are broadly similar, but both are generally consistent with the results of the analysis. We consider how further work might allow us to distinguish between them below.

Table 7.3 ZIP regression for 2-, 3- and 6-month periods, μ ranges of [1.83, 3.34], [2.48, 4.37], [3.96, 8.66] respectively, and p ranges of [0.23, 0.23], [0.12, 0.12], [0.03, 0.03] (logit relationship 2), respectively

		Estimate	Standard error	z value	Pr (> |z|)
Two-month period	μ regression				
β_0 Intercept		0.983	0.076	12.995	$< 2 \times 10^{-16}$
β_1 sea level		−0.048	0.031	−1.551	0.121
ω regression					
ω Intercept		−1.231	0.268	−4.595	$< 4.3 \times 10^{-6}$
Three-month period	μ regression				
β_0 Intercept		1.266	0.059	21.626	$< 2 \times 10^{-16}$
β_1 sea level		−0.045	0.024	−1.854	0.064
ω regression					
ω Intercept		−1.590	0.321	−6.083	$< 1.2 \times 10^{-9}$
Six-month period	μ regression				
β_0 Intercept		1.870	0.039	47.855	$< 2 \times 10^{-16}$
β_1 sea level		−0.063	0.017	−3.731	0.000
ω regression					
ω Intercept		−3.389	0.541	−6.265	$< 3.7 \times 10^{-10}$

Discussion

Having a better understanding of the physical mechanisms leading to a susceptible regime of seismicity would enable us to pin down some potentially observable variables that may be a proxy for these physical mechanisms. As a result, we could fit much better the ZIP by explaining the variability in the susceptibility through such a variable, and hence provide a finer assessment of the impact that sea level has on submarine earthquakes. The presence of gaps in the seismicity over timescales of 1 or 2 months but not on timescales of order 6 months, indicated by the marked decrease in the probability term p_i in the zero-inflated distribution model, indicates frequent switching between earthquake-susceptible and non-earthquake-susceptible states on timescales of a few months. Some possible mechanisms for the switching between susceptible and non-susceptible states include:

1. One potential mechanism is that seismicity in the Easter Microplate is driven on short timescales by dyke emplacement in the adjacent spreading segments to north and south, with seismicity more likely soon after dyke emplacement episodes have loaded the adjacent crust. However, the switching from susceptible to non-susceptible states on a timescale of months is probably too short for this mechanism to be viable, because increments of deformation on these timescales will be

only a fraction of the annual spreading rates of approximately 14 cm/year (see Figure 7.3) and therefore small compared with the width of typical dykes (of the order of 1–3 m).

2. Another potential mechanism for the switching is short-term variation in the permeability of the crust of the Easter Microplate, such that at times the permeability is high and ENSO-linked pore pressure changes can propagate down into the seismogenic region of the crust on timescales of the order of months, producing the statistical relationship that we observe, while at other times the permeability is lower. Rapid permeability changes seem plausible because of the high temperature gradient and highly reactive nature of the oceanic crust, which will lead to high rates of mineral deposition in, and clogging of, fracture permeability, coupled with the rapid creation of fracture permeability by the intense brittle deformation associated with rotation of the microplate.

3. Finally, the seismicity may be self-modulating, in the sense that a fault may be brought into the susceptible regime through stress loading by recent earthquakes on adjacent faults, particularly if there is an increase in compression and consequent poroelastic fluid pressure increases, and/or flow of pore fluid out of the faulted region. A particularly interesting point is that such stress loading may amplify the effect on seismicity of the small initial effect of the ENSO-induced fluid pressure changes, with an increased level of seismicity itself producing more seismicity in a positive feedback. Testing this model, however, requires a more detailed examination of the spatial distribution of successive earthquakes within the fault zones in and around the Easter Microplate that is beyond the scope of the present study.

A formal study of the amount of data collection necessary to detect a significant effect, under various assumptions, is beyond the scope of the chapter. Nevertheless, it is of interest to know how long a satellite series such as Topex–Poseidon needs to be kept in space to observe such a relationship with high confidence. This type of study has already been undertaken for other regression models (Guillas et al., 2004).

The negative relationship between sea level and numbers of earthquakes that we have identified has implications for the permeability structure of the crust of the Easter Microplate. As the rejected hypothesis H2 requires a high-permeability crust whereas hypothesis H1 implies a lower permeability in the seismogenic zone where the large earthquakes are generated, and hypothesis F is independent of permeability, our results point to lower large-scale permeability in the relevant regions of the Easter Microplate. Further study of the permeability structure of the microplate may allow a distinction to be made between these two hypotheses. However, another alternative approach would be to determine whether there is in fact significant flexure of the plates as required by hypothesis F, e.g. by direct measurement of variations in ocean floor pressure at suitable points on and distant from the EPR using the DART buoy technologies developed for the Pacific Tsunami Warning System (Bernard et al., 2006), which could then be compared with measurements of sea surface elevations.

Quantification of such plate flexure would enable a better understanding of the stress changes within the lithosphere at the plate boundary produced by ocean bottom pressure changes. El Niño emergence is preceded by negative sea elevation anomalies in the Eastern Pacific and accompanied by positive anomalies, the combined change in elevation approaching 0.5 m in the strongest events, corresponding to an ocean-bottom load pressure change of approximately 5 kPa. Sea-level reduction precursory to El Niño emergence ranges up to about 20 cm, resulting in a load pressure change of 1–2 kPa. Notwithstanding the scepticism of Hunt (2000), stress changes of these orders are increasingly becoming recognised as being sufficient to trigger a seismic response. Heki (2003), for example, demonstrates that snow load seasonally influences the seismicity of Japan through increasing compression on active faults and reducing the Coulomb failure stress by a few kilopascals. For neighbouring Taiwan, Liu et al. (2009) show that slow earthquakes in eastern Taiwan are triggered by stress changes of approximately 2 kPa on faults at depth, associated with atmospheric pressure falls caused by passing tropical cyclones. In the ocean basins, Wilcock (2001) provides convincing evidence for micro-earthquakes on the Endeavour segment of the Juan de Fuca Ridge (North East Pacific) being triggered by the loading effect of ocean tides, which result in vertical stress variations of 30–40 kPa. We emphasise, further, that plate flexure in response to these loads in the case of the Easter Microplate at the EPR could produce larger stress concentrations at particular levels in the crust at the plate boundaries of the microplate itself. Furthermore, given particular permeability variations in the crust, the effects of the stress variations could be amplified by the imbalances in pore fluid pressures invoked in our hypothesis P1.

Conclusion

As speculated by Hunt (2000), we conclude that increased seismicity in the region of the EPR is associated with the development of a strong sea–surface gradient rising from east to west (Wyrtki 1975), leading to lower sea level across the EPR and a reduction in ocean-bottom pressure of a few kilopascals. The development of the gradient is driven by strong south-east Trade Winds and strengthening of the South Equatorial Current, leading to a build-up of water in the western equatorial Pacific. The length of sea-level data permits us to conclude only at the 95% confidence level that lower sea levels are linked to more earthquakes over the next 6 months. Using a longer time period, we are able to establish that higher values of SOI triggers more earthquakes over the following 2, 3 and 6 months (and lower values of SOI-less earthquakes). As proposed by Walker (1999), we suggest that monitoring of seismicity in the EPR may have utility as an independent (non-meteorological) means of predicting future El Niño events. In a broader context, robust statistical correlation between ENSO and seismicity in the EPR provides a further example of how variations in the atmosphere and hydrosphere can drive very small changes in environmental conditions, which can in turn elicit responses from the Earth's crust. Although there are no hazard implications in this case, in a future warmer world similar small (a few kilopascals to a few tens of kilopascals) pressure changes associated with ocean loading due to rising sea levels, or atmospheric unloading as a consequence of more intense cyclones, may have the potential to trigger significant earthquakes at major marine or coastal fault systems that are in a critical state.

Acknowledgements

The authors gratefully acknowledge Seymour Laxon for his help in accessing the Topex–Poseidon data and Remko Scharroo for providing figures of Pacific sea-level anomalies from ERS/Envisat altimetry.

References

Bernard, E. N., Mofjeld, H. O., Titov, V., Synolakis, C. E. & Gonzalez, F. I. (2006) Tsunami: scientific frontiers mitigation, forecasting and policy implications. *Philosophical Transactions of the Royal Society of London A* **364**, 1989–2006.

Cane, M. (1983) Oceanographic events during El Niño. *Science* **222**, 1189–1195.

Cane, M. (2005) The evolution of El Niño, past and future. *Earth and Planetary Science Letters* **230**, 227–240.

Cogne, J., Francheteau, J., Courtillot, V., et al. (1995) Large rotation of the Easter Microplate as evidenced by oriented paleomagnetic samples from the ocean floor. *Earth and Planetary Science Letters* **136**, 213–222.

Czado, C., Erhardt, V., Min, A. & Wagner, S. (2007) Zero-inflated generalized poisson models with regression effects on the mean, dispersion and zero-inflation level applied to patent outsourcing rates. *Statistical Modelling* **7**, 125–153.

Guillas, S., Stein, M. L., Wuebbles, D. J. & Xia, J. (2004) Using chemistry transport modeling in statistical analysis of stratospheric ozone trends from observations. *Journal of Geophysical Research* **109**, D22303.

Guilyardi, E., Wittenberg, A., Fedorov, A., et al. (2009) Understanding El Niño in ocean-atmosphere general circulation models: progress and challenges. *Bulletin of the American Meterological Society* **90**, 325.

Heki, K. (2003) Snow load and seasonal variation of earthquake occurrence in Japan. *Earth and Planetary Science Letters* **207**, 159–164.

Hey, R. (2004) Propagating rifts and microplates at mid-ocean ridges. In: Selley, R. C. R.C. & Plimer, I. (eds), *Encyclopedia of Geology*. London: Academic Press, pp 396–405.

Hunt, A. G. (2000) Comment on an atmospheric stochastic trigger for El Niño. *Eos Transactions AGU* **81**(24), 266.

Jupp, T. & Schultz, A. (2004) A poroelastic model for the tidal modulation of sea floor hydrothermal systems. *Journal of Geophysical Research* **109**, B03105.

Lambert, D. (1992) Zero-inflated poisson regression, with an application to defects in manufacturing. *Technometrics* **34**, 1–14.

Lister, C. (1974) Penetration of water into hot rock. *Geophysical Journal of the Royal Astronomical Society* **39**, 465–509.

Liu, C., Linde, A. T. & Sacks, I. S. (2009) Slow earthquakes triggered by typhoons. *Nature* **459**, 833–836.

McCullagh, P. & Nelder, J. A. (1989) *Generalized Linear Models.* New York: Chapman & Hall.

McGuire, J. (2008) Seismic cycles and earthquake predictability on East Pacific Rise transform faults. *Bulletin of the Seismological Society of America* **98**, 1067–1084.

McNutt, S. (1999) Eruptions of Pavlof Volcano, Alaska, and their possible modulation by ocean load and tectonic stresses: re-evaluation of the hypothesis based on new data from 1984–1998. *Pure and Applied Geophysics* **155**, 701–712.

McNutt, S. & Beavan, R. (1987) Eruptions of pavlof volcano and their possible modulation by ocean load and tectonic stresses. *Journal of Geophysical Research* **92**, 11509–11523.

Mason, B., Pyle, D., Dade, W. & Jupp, T. (2004) Seasonality of volcanic eruptions. *Journal of Geophysical Research* **109**, B04206.

Rasmusson, E. & Wallace, J. (1983) Meteorological aspects of the El-Niño Southern oscillation. *Science* **222**, 1195–1202.

Rubinstein, J. L., La Rocca, M., Vidale, J. E., Creager, K. C. & Wech, A. G. (2008) Tidal modulation of nonvolcanic tremor. *Science* **319**, 186–189.

Schulz, W. H., Kean, J. W. & Wang, G. (2009) Landslide movement in southwest Colorado triggered by atmospheric tides. *Nature Geoscience* **2**, 863–866.

Searle, R., Bird, R., Rusby, R. & Naar, D. (1993) The development of 2 oceanic microplates – Easter and Juan-Fernandez microplates, East Pacific Rise. *Journal of the Geological Society* **150**(Part 5), 965–976.

Simpson, D., Leith, W. & Scholz, C. (1988) Two types of reservoir-induced seismicity. *Bulletin of the Seismological Society of America* **78**, 2025–2040.

Stroup, D. F., Tolstoy, M., Crone, T. J., Malinverno, A., Bohnenstiehl, D. R. & Waldhauser, F. (2009) Systematic along-axis tidal triggering of microearthquakes observed at 9 degrees 50′ N East Pacific Rise. *Geophysical Research Letters* **36**, L18302.

Talwani, P. (1997) On the nature of reservoir-induced seismicity. *Pure and Applied Geophysics* **150**, 473–492.

Tsonis, A., Elsner, J., Hunt, A. & Jagger, T. (2005) Unfolding the relation between global temperature and ENSO. *Geophysical Research Letters* **32**.

Walker, D. (1988) Seismicity of the East Pacific Rise: Correlations with the southern oscillation index? *Eos Transactions AGU* **69**, 857.

Walker, D. (1995) More evidence indicates link between El Niños and Seismicity. *Eos Transactions AGU* **76**(4), 33–36.

Walker, D. (1999) Seismic predictors of El Niño revisited. *Eos Transactions AGU* **80**, 281.

Wilcock, W. (2001) Tidal triggering of micro earthquakes on the Juan de Fuca Ridge. *Geophysical Research Letters* **28**, 3999–4002.

Woolf, D., Shaw, A. G. P. & Tsimplis, M. N. (2003) The influence of the North Atlantic Oscillation on sea-level variability in the North Atlantic region. *Journal of the Atmospheric Sciences* **9**, 145–167.

Wyrtki, K. (1975) El Niño – dynamic-response of equatorial pacific ocean to atmospheric forcing. *Journal of Physical Oceanography* **5**, 572–584.

Submarine mass failures as tsunami sources – their climate control

David R. Tappin

British Geological Survey, Nottingham, UK

Summary

Recent research on submarine mass failures (SMFs) shows that they are a source of hazardous tsunamis, with the tsunami magnitude mainly dependent on water depth of failure, SMF volume and failure mechanism, cohesive slump or fragmental landslide. A major control on the mechanism of SMF is the sediment type, together with its post-depositional alteration. The type of sediment, fine or coarse grained, its rate of deposition, together with post-depositional processes, may all be influenced by climate. Post-depositional processes, termed sediment 'preconditioning' is known to promote instability and failure. Climate may also control the triggering of SMFs, e.g. through earthquake loading or cyclic loading from storm waves or tides. Instantaneous triggering by other mechanisms such as fluid over-pressuring and hydrate instability is controversial, but is here considered unlikely. However, these mechanisms are known to promote sediment instability. SMFs occur in numerous environments, including the open continental shelf, submarine canyon/fan systems, fjords, active river deltas and convergent margins. In all these environments there is a latitudinal variation in the scale of SMF. The database is limited, but the greatest climate influence appears to be in high latitudes where glacial/interglacial cyclicity has considerable control on sedimentation, preconditioning and triggering. Consideration of the different types of SMF in the context of their climate controls provides additional insight into their potential hazard in sourcing tsunamis, e.g. in the Atlantic, where SMFs are common, the tsunami hazard under the present-day climate may not be as great as their common occurrence suggests.

Climate Forcing of Geological Hazards, First Edition. Edited by Bill McGuire and Mark Maslin.
© 2013 The Royal Society and John Wiley & Sons, Ltd. Published 2013 by John Wiley & Sons, Ltd.

Introduction

Tsunamis, especially destructive ones, are mainly (~70–80%) caused by earthquakes. However, they can also be sourced by failure of sediment and rock both on land and at the seabed. Most of these sediment/rock failures are in submarine sediments or from volcanic lateral collapse. Historical records of destructive tsunamis caused by lateral collapse include eighteenth century examples from Japan, such as Oshima-Oshima in 1741 (Satake and Kato, 2001; Satake, 2007) and Unzen in 1792 (Siebert et al., 1987), as well as the AD 1888 collapse of Ritter Island in Papua New Guinea (Johnson, 1987; Ward and Day, 2003). Such historical ocean-entering landslides are recognised from eyewitness accounts. In contrast, prehistoric lateral collapses, such as those in Hawaii and the Canary Islands, are identified from geological evidence (e.g. Moore et al., 1994; Urgeles et al., 1997). This difference in identification is significant in terms of hazard recognition. Submarine seabed failures, termed here 'submarine mass failures', are less easily recognised than those on land, and there are few historical accounts of these events. Thus SMFs may have been underestimated as a tsunami source. In fact, until recently, they were discounted as a cause of destructive tsunamis (e.g. Jiang and LeBlond, 1994; LeBlond and Jones, 1995). Tsunamis such as the Grand Banks in 1929 (e.g. Heezen et al., 1954; Piper and Asku, 1987) and those associated with the Good Friday 1964 earthquake in Alaska, at Seward and Valdez (e.g. Lee et al., 2003) might have flagged the hazard but they did not. It was not until 1998, when a submarine slump caused the devastating tsunami at Sissano Lagoon in Papua New Guinea, in which 2200 people died, that the threat from submarine landslides was fully realised (Tappin et al., 1999, 2001, 2008).

Since 1998 there have been major advances in understanding SMFs as sources of tsunamis. Applying new mapping methodologies, such as multibeam bathymetry, the numerous architectures of SMFs have been identified. In addition, there has been an improved understanding of the mechanisms of SMF, including their formation and triggering. Based on the new knowledge, advanced parametric mathematical models of SMFs have been developed that form the basis of more realistic tsunami wave propagation and run-up (Tappin et al., 2008). Thus SMFs and their potential to generate hazardous tsunamis are now much better understood. An aspect of this improved understanding is the influence of climate on both their formation and their triggering. It is, therefore, the objective of this chapter to present an overview of the tsunami hazard from SMFs in the context of their climate control(s) – specifically to consider (1) the temporal evidence on how climate change may relate to SMF, (2) how climate change may influence the stability of submarine sediments leading to mass failure, (3) whether climate can control the triggering of SMFs and (4) how climate, and associated sea level change, may influence the preservation potential of tsunami sediments derived from SMFs and thereby our potential to recognise in the geological record SMFs as tsunami sources. The focus of the chapter is on non-volcanic SMFs, although it is recognised that volcanic mass failure may also pose a significant tsunami hazard.

Submarine mass failures

In the context of tsunami generation SMFs may be considered to range from cohesive slumps to translational landslides (Hampton et al., 1996), although there are other mechanisms of failure (Figure 8.1). Study of SMFs has advanced our understanding of (1) their role in the transport of sediment from the land to the ocean, (2) their potential as deep-water hydrocarbon reservoirs and (3) their hazard to seabed infrastructure, particularly in relation to the hydrocarbon industry. The recent realisation that SMFs may cause hazardous, if not devastating, tsunamis has also led to new research in this context. Controls on SMF are many and varied and include their sediment properties resulting from initial deposition and post-depositional alteration, termed 'preconditioning', together with 'triggers' that are instantaneous events (such as earthquakes) causing sediment failure. Deposition and post-depositional alterations are the result of uplift and erosion, and basin subsidence. Tectonics undoubtedly influences sedimentation by creating depositional, 'accommodation' space with uplift of sediment source areas leading to erosion and deposition. There has been considerable debate about how the environment of sediment deposition may contribute to sediment failure by 'preconditioning' sediment physical properties (e.g. Biscontin et al., 2004; Masson et al., 2009). Earthquakes are recognised as the main SMF trigger (e.g. Bugge, 1983; Laberg et al., 2000), but other mechanisms include salt movement, storm wave loading and low tides (Prior et al., 1982a, 1982b; Twichell et al., 2009). Triggers, such as increases in sediment pore pressures and hydrate destabilisation, are more controversial (see discussion in Dugan and Stigall, 2009; Grozic, 2009).

Another first-order control on sedimentation of SMFs is global climate change (Figure 8.2). During recent Earth history, over the past hundreds of thousands of years, global climate has fluctuated with remarkable cyclicity. A major result of these fluctuations is the variation in the expansion and contraction of continental ice sheets which have resulted in eustatic sea-level changes of up to 120 m. A consequence of these glacial–interglacial cycles is a variation in the rate of sediment delivery to the ocean. The combined effects of climate and sea-level change result in changes in the location of SMFs along the ocean margins, e.g. at high latitudes the large volumes of sediment delivered to the ocean margins during

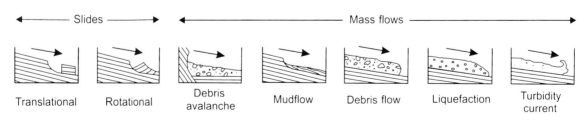

Figure 8.1 General landslide classification. (After Lee et al., 2007.)

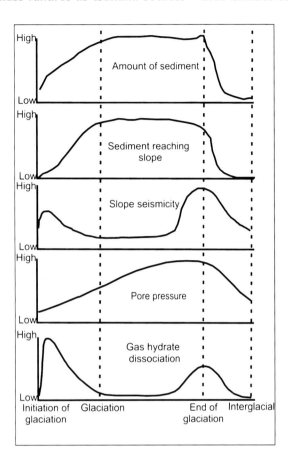

Figure 8.2 Approximate impact of time on several factors that influence the stability of submarine slopes (Modified from Lee et al, 2009. Timing of occurence of large submarine landslides on the Atlantic ocean margin. Marine Geology, Vol 264, 53–64.)

glacial periods are destabilised by earthquakes that result from glacio-isostatic uplift as the ice sheets melt and contract (Bryn et al., 2005). At mid and lower latitudes the glacial influence is less evident and the database less substantial, but here also there is evidence of a climate control on SMFs (e.g. Gee et al., 1999; Henrich et al., 2009).

Landslide territories

SMFs take place in many different environments and Hampton et al. (1996) introduced the term 'landslide territory' for areas where they are more common than elsewhere (Figure 8.3). These environments include the open continental shelf, submarine canyon/fan systems, fjords, active river deltas and volcanic islands. In addition, Lee (2005) identified convergent margins as an environment where submarine landslides also take place. Recent research in the South Pacific

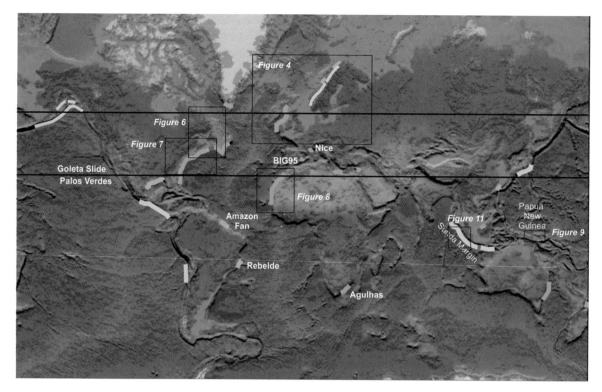

Figure 8.3 Global distribution of mapped submarine mass failures (SMFs). Green lines: SMFs on continental shelves and fan systems – no identified tsunami. Yellow lines: SMFs located along convergent margins, no identified tsunami. Red lines: locations of SMF-sourced tsunamis, or where there may be a SMF contribution. Pale-blue lines: active river systems – no tsunami identified,. Orange lines: SMFs along strike-slip margins. Pale-yellow lines: SMFs along fjord margins. Black latitudinal lines are the divisions of Weaver et al. (2000), marking the boundaries of glacial and non-glacial SMF influence – see text for discussion. Boxes are locations of Figures 8.4, 8.6, 8.7, 8.8, 8.9 and 8.11.

on reef-front environments suggest that these too may prove to be a location where SMFs can cause hazardous tsunamis, as in the Suva, Fiji event of 1953 (Rahiman et al., 2007).

The locations of SMFs are identified by unique combinations of sedimentology and physiography, with common factors including thick sedimentary deposits, sloping seafloor and high environmental stresses. Several dominant controls on slope stability are recognised which include (1) the rate, volume and type of sediment delivery to the continental margins, (2) sediment thickness, (3) changes in seafloor conditions, which can influence hydrate stability and the possible generation of free gas, (4) variations in seismicity and (5) changes in groundwater flow. With regard to tsunami generation, there is a recognisable close and genetic relationship to most landslide territories. On open continental slopes and canyons, fjords and convergent margins, there is evidence that SMFs can result in hazardous, if not devastating, tsunamis. Examples include the Grand Banks event of 1929, Alaska in 1964, Papua New Guinea in 1998 and Storegga at 8200 years before the present (BP). However, on active river deltas, there is no evidence of tsunamis sourced from SMFs, e.g. submarine landslides are common on the Mississippi Delta (Prior and Coleman, 1982) yet no associated tsunami has been recognised.

Open continental slope and eise

Along continental margins major sediment transfer takes place from land to sea, with slope canyons resulting from erosion and fan/delta systems from deposition. During changes in sea level, sediment in these regions has the potential to become unstable, in the process triggering tsunami. The North Atlantic is one of the best-studied regions (McAdoo et al., 2000; Hühnerbach et al., 2004; Lee, 2009), and here there is an extensive database of SMFs, which includes their depositional character, age and proposed triggering mechanisms (Figures 8.4–8.7 and see Figure 8.3). Off Norway, the discovery of the second largest gas field in Europe, located beneath the largest submarine landslide in the North Atlantic, resulted in one of the most intensive investigations into offshore slope stability ever undertaken (Bryn et al., 2005; Solheim et al., 2005).

SMFs at all scales are common along the Atlantic continental margins. Mechanisms of sediment mass movement include debris flows, landslides, turbidite flows and slumps. Generally, they are more abundant in the western North Atlantic (off Canada and the USA) than in the eastern North Atlantic (off Europe and Africa) (see Figure 8.3). In the west, SMFs are smaller than those in the east. On both sides of the Atlantic, most SMFs originate in water depths between 1000 and 1300 m. Three climatic regions may be identified where SMFs take place: a

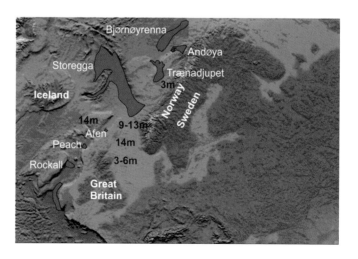

Figure 8.4 Submarine mass failures (SMFs) off northern Europe (after Lee, 2009 and Laberg et al., 2003). Black numbers are run-up heights from the Storegga tsunami of 8200 BP (From Bondevik, et al., 2005).

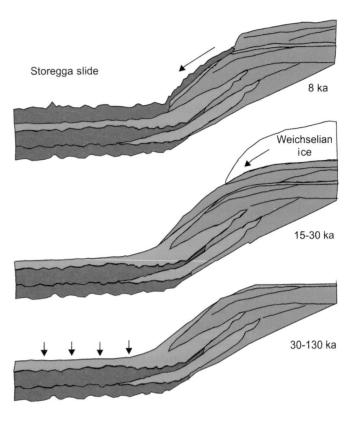

Figure 8.5 Illustration of the cyclic deposition and slide processes in the Storegga area. Arrows in the lower panel indicate marine, hemipelagic deposition, in which the preferred glide planes are found.
Green = glacial clays.
Red = slide deposits.
Blue = marine clays from contour currents. (From Bryn et al., 2005. Explaining the Storegga Slide. Marine and Petroleum Geology, Vol 22, 11–19.)

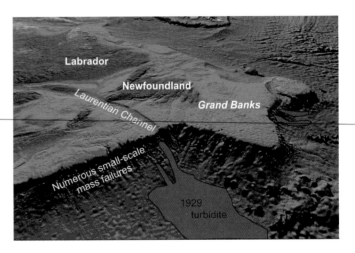

Figure 8.6 Submarine mass failures (SMFs) off Canada. (After Piper & Asku, 1987 and Piper & McCall, 2003.)

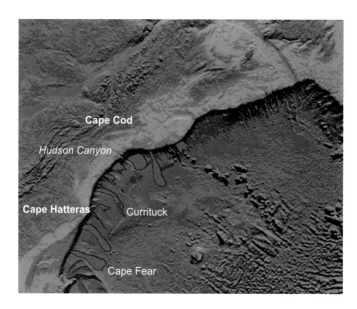

Figure 8.7 The largest submarine mass failure (SMF) off the east coast of the USA. (After Lee, 2009.)

glaciated margin north of 56° N (southern tip of Norway), a 'glacially influenced' margin from 26° N to 56° N and a non-glaciated margin south of 26° N (Weaver et al., 2000) (see Figure 8.3). However, these boundaries reflect present-day climate and have fluctuated as this has changed. During glacial maxima, for example, the limit of glacial margins is located at least 10° farther south than at present.

Glaciated margins

In the eastern Atlantic, four major SMFs border the Norwegian margin: Bjørnøyrenna, Andøya, Trænadjupet and Storegga. Further south, off Britain and Ireland, large SMFs include Peach and Rockall located off Scotland (see Figure 8.4). The Storegga slide is the largest SMF in the Atlantic with a volume of 3500 km³. It generated a tsunami that struck the west coast of Norway with run-ups of up to 20 m and propagated outward striking Scotland and the Faeroes with run-ups of 5–10 m (Bondevik et al., 2005) (see Figure 8.4).

In high-latitude, glaciated regions there is a dynamic evolution of SMFs that is related to sea-level fluctuations associated with ice sheet advance and retreat. Sedimentation rates are highest during glacial periods, with 36–65 m/ka (thousands of years) for the Trænadjupet and Storegga SMFs respectively (Laberg et al., 2003; Hjelstuen et al., 2005). During interglacials, sedimentation rates are much reduced, with less than 1 m/ka being common. These rates reflect the large volumes of sediment produced from ice sheet scouring during glacial episodes, with less sediment delivered during interglacials when ice sheets are smaller and more distant, and sedimentation is mainly from slope parallel contourites. The result of these changing sedimentation regimes is complex sediment preconditioning which includes the development of high pore pressures in the fine-grained glacial sediment, resulting in destabilisation through earthquake loading.

The failure of Storegga is probably representative of similar SMFs along the Norwegian margin (Bryn et al., 2005) (see Figure 8.5). Failure took place at the end of the last glaciation or soon after deglaciation. It was translational and took place along weak layers formed in marine clays subject to strain softening. Before failure, destabilisation was the result of rapid loading from glacial deposits, causing the generation of excess pore pressure and reduction of effective shear strength in the underlying clays. Initiation of failure was at water depths of 1500–2000 m, with retrogressive failure propagating upwards from the base. Climatic processes led to a preconditioning of the sediment mass. During glacial periods, when the sea level was lowered, the ice sheet advanced to the shelf break depositing large volumes of coarse-grained sediment on the slope. During interglacials, higher sea levels resulted in slope-parallel contour currents from which finer-grained sediment was deposited. These fine-grained clays formed weak layers along which failure of the sediment mass took place. Failure was triggered by earthquakes caused by isostatic uplift as the ice sheets melted. Although hydrate destabilisation may contribute locally, this is not regarded as a primary driver of mass failure.

In the western Atlantic, off of Canada, 24 SMFs have been mapped (Piper & McCall, 2003) (see Figure 8.6). The SMFs vary in morphology from slides to retrogressive slumps. They are located in pro-delta settings, on the continental rise and in steep-sided canyons. The oldest is 125 ka, although most are younger than 50 ka. Fourteen failures occurred during the last glacial period. Only two are post-glacial, the best known of which is the Grand Banks event of 1929, when 41 people lost their lives in the resultant tsunami. The 1929 SMF was a debris flow and turbidite released from the Canadian shelf by an earthquake. It is relatively thin (20 m average) and probably retrogressive (Mosher and Piper, 2007). Off Nova Scotia, there are five main episodes of mass wasting over the past 17 ka, at 5–8 ka, 12.7 ka, 13.8 ka, 17.9 ka and 14 ka. Thus SMF took place mainly during glaciation, with a glacial:interglacial ratio of failure of 3 or 4:1. Their broad geographical distribution suggests an earthquake trigger for all these events (Jenner et al., 2007), with earthquakes the result of post-glacial crustal rebound. This readjustment is still ongoing (e.g. Peltier, 2002).

Glacially influenced margins

There are few large SMFs off the European margins where narrow shelves and common submarine canyons mainly funnel turbidite currents to the abyssal plain. Off the US margin, however, 55 SMFs have been identified (Chaytor et al., 2007; Twichell et al., 2009) (see Figure 8.7). Two types of SMFs are recognised: those originating in submarine canyons and those from the open continental slope and rise. They cover 33% of the continental slope and rise of the glacially influenced New England margin, 16% of the sea floor offshore of the fluvially dominated

Middle Atlantic margin, and 13% of the sea floor south of Cape Hatteras. Their distribution is in part controlled by the Quaternary evolution of the margin, resulting from climate controlled eustatic sea-level change. The headwall scarps of open-slope SMFs are mainly located on the lower slope and upper rise; those of the canyon-sourced SMFs lie mostly on the upper slope. SMFs are generally thin (mostly 20–40 m thick) and comprise primarily Quaternary sediment. Volumes of the open-slope SMFs are generally larger (up to $392\,km^3$) than the canyon-sourced ones (up to $10\,km^3$). Largest SMFs along the southern New England margin are located seaward of shelf-edge deltas laid down during the lowered sea levels of the last glaciation. South of Cape Hatteras, SMFs are located adjacent to salt domes that breach the sea floor. The wide spatial distribution of landslides indicates a variety of triggers, although earthquakes are recognised as the most common, probably generated by glacio-isostatic rebound of the glaciated margin, or by salt movement. Other triggering processes, such as fluid over-pressuring, may have contributed to failure by pre-conditioning. The large-volume open-slope landslides have the greatest potential to generate hazardous tsunamis. Few of the SMFs have been dated, but most are considered to have formed during the last glaciation and Early Holocene (Embley & Jacobi, 1986).

The Cape Fear Slide is the largest SMF on the US margin, with a volume of $200\,km^3$ (Lee, 2009). It is dated at between 8 and 25 ka, forming at the transition between the end of the last glaciation and present interglacial. Salt movement, driven by sediment loading, is considered the most likely triggering mechanism, with salt diapirism causing over-steepening of the seabed slope followed by failure. The headwall lies along a major normal fault (Cashman & Popenoe, 1985; Hornbach et al., 2007). The Currituck SMF has a volume of $165\,km^3$ and is formed on a shelf-edge delta (Prior et al., 1986b). It formed between 24 and 50 ka, during glaciation (Locat et al., 2009). Very high pore pressures and/or a strong earthquake are considered the most likely triggering mechanisms. The high pore pressures are attributed to fluid seepage from coastal aquifers, sediment loading from delta construction, local sediment loading, gas hydrates and/or earthquakes. Modelling of the two slides as tsunami sources indicates that they both have the potential to be a hazard today (Hornbach et al., 2007; Geist et al., 2009). However, they formed at lower sea levels, thus making their hazard at the time of deposition much greater. To date, no evidence has been found to indicate that a tsunami was generated by these SMFs.

Non-glaciated margins

Off North Africa there are a number of SMF complexes, although little evidence has been found (or in fact looked for) to indicate associated tsunamis (Figure 8.8). Off the Moroccan margin a complex of turbidites extends over 1700 km to seaward (Weaver et al., 2000; Wynn et al., 2002). The turbidites are up to 200–250 ka. The largest are the Bed 5 event off Morocco and, further south off the western Sahara, the Saharan debris flow. Both are dated at approximately 50–60 ka (Gee et al.,

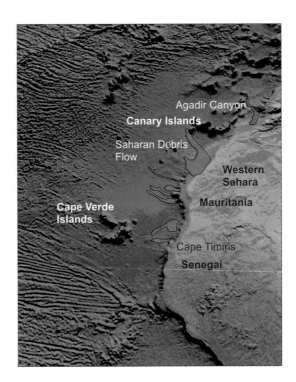

Figure 8.8 Submarine mass failures (SMFs) off Africa. (After Weaver et al., 2000 and Lee, 2009.)

1999; Wynn et al., 2002). Thus both were triggered when global climate was changing between glacial and interglacial conditions, and are associated with changes in eustatic sea level, when this was either rising or falling rapidly. Further south, off Mauritania, turbidite complexes are also clearly climate controlled (Henrich et al., 2009). There is a high frequency of failure during deglaciation, when the sea level was rising, through remobilisation of large-scale aeolian dune fields that had expanded to the shelf edge during glacial sea-level lows. Pervasive, and widespread, fluid escape structures suggest high fluid pore pressures that may have initiated failure (Antobreh & Krastel, 2006). Antobreh and Krastel (2006) also propose that slide formation was preconditioned by uninterrupted deposition of organic-rich sediment, the result of upwelling in an open slope environment. The result was a rapid accumulation of poorly consolidated bedded sediment (turbidites) intercalated with thin weak layers of organic-rich sediment. Turbidite activity is also frequent during glacial periods – an increase that is attributed to rapid sea-level rise during Heinrich events. Henrich et al. (2008) speculate that the presence in one core of a turbidite overlying a debris flow may indicate an associated tsunami. Triggers in this area are undefined, but North Africa is an earthquake-prone area, as recorded in the massively destructive event that struck Agadir in 1960, so earthquake triggers are a probable source. The synchronicity of turbidite events between the Dakar and Timiris canyons also suggests an earthquake trigger.

SMFs have been mapped in the south Atlantic, although to date there is no evidence of any associated tsunamis; again, however, these have not been looked for. Off the Amazon Delta, there are large-volume (up to 2500 km³), catastrophic failures of the continental slope (see Figure 8.3). These failures are sourced from rapidly deposited, under-consolidated sediment laid down on upper-fan levees (Piper et al., 1997; Maslin et al., 1998). The dating is poor, but suggests a late glacial to Early Holocene age. If correct, then the failures may correlate with climate-induced changes in sea level. Failure is attributed to either rapid sea-level fall, which resulted in destabilisation of gas hydrate, or deglaciation in the Andes leading to the large-scale flushing of Amazon River sediment onto the continental slope; this resulted in excessive loading of fan sediment laid down previously during glacially induced sea-level lows. The youngest failures are 14–17 ka (Maslin et al., 1998), with older events at 35 ka and 42–45 ka. Hydrate disassociation is mainly associated with the younger events.

Off Brazil, the Rebelde complex is believed to have been triggered by either an earthquake or high fluid pore pressure (see Figure 8.3). Again the SMF is poorly dated (Ashabranner et al., 2009). Off southern Africa, the Agulhas Slump has a proposed volume of 20,000 km³, although this figure is based on single-beam bathymetry data (Dingle, 1977) (see Figure 8.3).

Other areas

Outside the Atlantic, there are numerous SMFs, although there is little evidence of associated tsunamis, with the notable exception of the Nice event of 1979 in which 12 people died (Dan et al., 2007) (see Figure 8.3). There has been little research on the relationship between SMF and climate. In the Mediterranean, probably the largest SMF is the 'BIG' 95 debris flow off Spain, with an area of 2000 m² and dated at 11.4 ka (Lastras et al., 2004) (see Figure 8.3). Proposed triggering mechanisms include seismicity and over-steepening of the slope due to a volcanic structure underlying the main headwall.

SMFs have been identified off France and in the eastern Mediterranean off Egypt and Israel (see Figure 8.3). Two major SMFs lie off California (see Figure 8.3). The Goleta Slide, in the Santa Barbara Channel, is a compound failure with a total volume of 1.75 km³. It is formed of three main lobes with both surficial slump blocks and mud flows. It is interpreted as Holocene in age (Fisher et al., 2005; Greene et al., 2005). Earthquakes are a likely trigger because historically events of magnitudes ≥5 occur about every approximately 20 years. Modelling of one lobe of the Goleta Slide as a tsunami source produces a local run-up of about 10 m (Greene et al., 2005). Records show that tsunamis struck the area in the nineteenth century, but these appear to be sourced from earthquakes rather than SMFs. The Palos Verdes debris avalanche is located in a submarine canyon offshore to Long Beach (Bohannon & Gardner, 2004). It is the largest late Quaternary SMF, with a volume of 0.34 km³, in the inner California Borderland basins. It is dated at 7.5 ka (Normark et al., 2004). Modelling indicates that it was large enough to generate a significant tsunami that would inundate the adjacent coastline (Locat et al., 2004). However, to date, there is no evidence that such a tsunami struck the coast.

Fjords

SMFs are common in fjords (Syvitski et al., 1986; Hampton et al., 1996) with numerous records of associated tsunamis. In these glacial environments rapid sedimentation results in deposits that are susceptible to failure. Streams that drain glaciers transport and deposit sediment formed of low plasticity rock flour vulnerable to earthquake loading; rapid sedimentation can lead to under-consolidation. Organic matter deposited from rivers may decay to produce methane gas which may lead to elevated pore pressure and further reduce sediment strength. Thus there is a strong climate control on both the location of the fjords and the volume of contributed sediment.

Fjord-head deltas can fail under cyclic loading (Prior et al., 1986a), e.g. at Kitimat Fjord in British Columbia in 1975, a landslide was triggered by a low tide, creating a tsunami up to 8.2 m in height (Prior et al., 1982a; Lee, 1989). Although there was significant damage, no lives were lost. Weak delta sediments can also fail through earthquake loading, as happened during the great Alaskan earthquake of 1964. The resulting tsunamis were enormously destructive, with loss of lives and damage to infrastructure (Plafker et al., 1969; Lee, 1989; Hampton et al., 1993). At Seward a 1-km section of the waterfront collapsed as a result of submarine failure, creating a 10-m-high tsunami (Lemke, 1967). The destruction was compounded by a subsequent earthquake-generated tsunami, also 10 m high, arriving 30 min later. Most of the 13 people who died at Seward were killed by the tsunami. At Valdez, an initial landslide volume of 0.4 km^3 increased to 1 km^3 as it incorporated sediment from the seabed (Coulter & Migliaccio, 1966). The resulting tsunami attained heights up to 52 m, and resulted in the loss of 32 lives.

Convergent margins

Convergent margins, similar to passive margin open continental slopes and rises, are important regions of sediment flux between the land and the sea. Their characterisation into those margins where the sediment flux is significant (sediment rich) and those where it is not (sediment starved) has implications for SMFs, although the relationships are complex (Tappin, 2009). It is not always those margins where sediment flux is large that produce the most hazardous tsunami, e.g. the Papua New Guinea event at Sissano Lagoon in 1998 took place along the New Guinea Trench which is sediment starved (Tappin et al., 2001) (Figures 8.9 and 8.10). Conversely, along the Sunda margin, where there is a much larger accretionary prism than in New Guinea, SMFs are small scale and less of a hazard in sourcing destructive events (Tappin et al., 2007) (Figures 8.11 and 8.12).

SMFs have been mapped along many convergent margins (McAdoo et al., 2004). Their size varies from 'super-scale' in Cascadia (Goldfinger et al., 2000) to

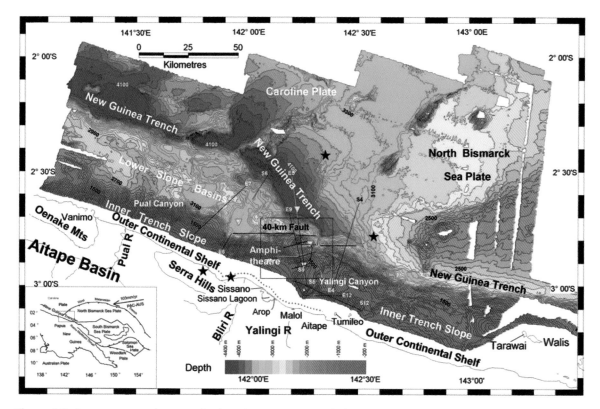

Figure 8.9 Convergent margin tsunami submarine mass failures (SMF) hazard – Papua New Guinea (PNG). The figure shows the regional seabed morphology of the convergent margin area offshore of northern PNG. The box in the centre marks the location of Figure 8.10; the amphitheatre marks the area of the SMF – a slump. (From Tappin et al., 2001. The Sissano Papua New Guinea tsunami of July 1998 – offshore evidence on the source mechanism. Marine Geology, Vol. 175, 1–23.)

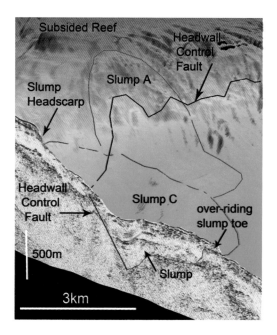

Figure 8.10 Convergent margin tsunami submarine mass failure (SMF) hazard – Papua New Guinea. The figure shows the seabed morphology and sub-seabed seismic structure of the slump C which caused the tsunami of 1998. (From Tappin et al., 2008. The Papua New Guinea tsunami of 17 July 1998: anatomy of a catastrophic event. Nat Hazards Earth Syst. Sci., Vol 8, 243–266.)

Figure 8.11 Convergent margin tsunami submarine mass failure (SMF) hazard – Indian Ocean, Sunda Margin. Map of the margin off Sumatra and the location of the bathymetric data which illustrates the seabed morphology. Boxes are the location of Figure 8.12. (From Tappin et al., 2007. Mass wasting processes offshore Sumatra, pp. 327–336. In Submarine Mass Movements and their Consequences. Lykousis et al, Springer, with kind permission from Springer Science+Business Media B.V.).

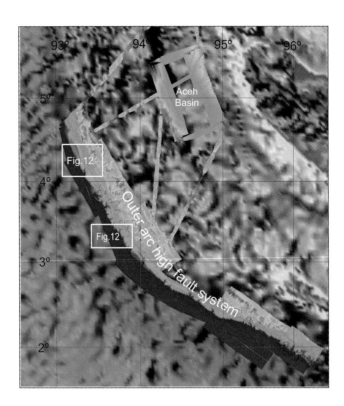

'small' along the Sunda margin in the Indian Ocean (Tappin et al., 2007). Off Japan, there are large SMFs on the upper parts of the Nankai accretionary prism (Kawamura et al., 2009) but the highly eroded lower slopes show little evidence of large, well-preserved events (McAdoo et al., 2004). Along the Makran and Kodiak accretionary margins there is evidence of mass wasting on the upper slopes, with the lower slopes lacking large SMFs. By contrast, along the sediment-starved Sanriku (Japan), Nicaragua and Aleutian margins there are large SMFs. In the Gulf of Alaska, there are a number of large SMFs located on the shelf and the continental slope. Both slumps and debris flows are present and their formation is attributed to glacial processes (Schwab & Lee, 1988; Dobson et al., 1998). During climatic cooling there is increased sediment delivery to the slope through expansion of tide-water glaciers, with resultant formation of point-sourced fans. During interglacials, as sea levels rise, sediment is delivered onto the shelf. Triggering is by either earthquake or storm wave loading.

Since the Sissano, Papua New Guinea tsunami of 1998, research into convergent margin tsunamis from SMFs has increased significantly. Some studies are based on newly acquired multibeam data, e.g., Puerto Rico, 1918 (López-Venegas et al., 2008). Other studies have re-evaluated anomalous tsunamis with run-ups that, to some degree, are too large in relation to their proposed earthquake source, e.g. Messina, 1908 and Alaska, 1946 (Fryer et al., 2004; Billi et al., 2008). Other events, where there are inconsistencies between earthquake magnitude/location and

Figure 8.12 Convergent margin tsunami submarine mass failure (SMF) hazard – Indian Ocean, Sunda Margin – small-scale SMF. Top: deeply eroded toe of the accretionary prism with small-scale debris avalanches and turbidites. Location in Figure 8.11 (Lower box). Bottom: morphology and bathymetry of a blocky debris avalanche in the north of the area. Bathymetric contours in metres, internal dotted lines and numbers within the SMF represent three internal subdivisions. For location see Figure 8.11 (Upper box). (From Tappin et al., 2007. Mass wasting processes offshore Sumatra, pp 327–336. In Submarine Mass movements and Their Consequences. Lykousis et al, Springer, with kind permission from Springer Science+Business Media B.V.)

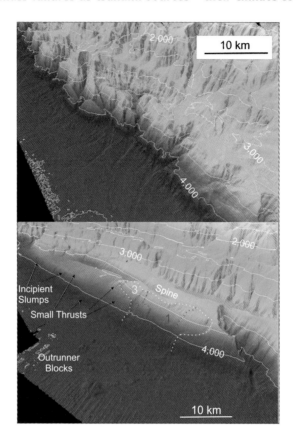

tsunami run-up, include the Makran, Indian Ocean event of 1945 (Rajendran et al., 2008), Sanriku in 1896 (Tanioka & Seno, 2001), Flores Island in 1982 (Imamura et al., 1995) and Java, 2006 (Fritz et al., 2007).

SMFs, tsunamis and climate control

In most, if not all, of the landslide environments described there is evidence of strong climate control on SMF; on initial deposition, post-depositional precon-ditioning of sediment, as well as on triggering. The type of sediment laid down, fine or coarse grained, the rate of deposition and post-depositional alteration are all influenced to a greater or lesser degree by climate. The location of SMFs is also climate controlled through changes in eustatic sea level. Yet there are obvious variations and differences in how climate controls SMF in the different landslide territories. There are both temporal variations over hundreds of thousands of years, associated with global, astronomically forced, climate change, and with geographical variations that are related to present-day climate and the various tectonic environments in which the landslide territories occur. Finally, in the context of tsunamis, there is the evidence on which the relationship between SMF and tsunamis is based, whether this is representative and, if not, why. These interrelations between climate, SMF and tsunami generation are, therefore, complex in temporal, geographical and tectonic frameworks.

Temporal relationships between climate change and SMFs

Evidence from the most intensively studied region of the Atlantic reveals a strong climate control on the mass-wasting systems of the region which is related to the cyclical changes over the past hundreds of thousands of years that relate to glacial and interglacial periods. In the high-latitude, glacially dominated, continental margins off Norway, the thick sedimentary deposits laid down during glacial periods and interbedded, thinner, finer-grained sediments laid down during interglacials are prone to failure, triggered by increased seismicity (caused by isostatic readjustment) during deglaciation. Further south along the glacially influenced margins of the USA, there is also climate control on sedimentation. During the cyclical glacial/interglacial periods changing eustatic sea level controls the location of sediment deposition, as well as the rates of sedimentation. Compared with glaciated margins, sediment delivery is much reduced, thus SMF volumes are smaller by an order of magnitude. Largest SMFs are on the margin of shelf-edge deltas, where rapid sedimentation during glacially lowered sea-levels results in a potential for subsequent catastrophic failure as sea level rises.

Along the low latitude margins off West Africa, although the database is more limited, climate control is reflected in the turbidite mass failures that correlate with climate-induced sea-level change. Off Mauritania, turbidites are sourced from the desert sands that advance to the shelf edge during sea-level lows. Smaller SMFs relate to climate-controlled rates of sedimentation, in association with different types of sediment deposited between glacials and interglacials. Organic-rich fine-grained sediments resulting from upwelling during sea-level highs form weak layers along which failure takes place. Along convergent margins, although the evidence base is again limited, variation in sediment supply in high latitudes, such as along the Alaska margin, is related to glacial/post-glacial sediment variation (e.g. Schwab & Lee, 1988; Dobson et al., 1998).

A major constraint on identifying temporal control is undoubtedly the relatively poor accuracy of landslide dating and the limited number of reliable ages (Table 8.1). Thus climate-induced controls, such as on relatively brief periods of sea-level fall, remain uncertain. Notwithstanding, the dates available indicate that, over the last 20,000 years, there has been a relatively even distribution of large landslides in the period between the last glacial maximum until about 4000 years ago. These failures are in sediments deposited during lowered sea levels, interbedded with weak layers that form during interglacials, and prone to destabilisation during warming and deglaciation. Evidence of two relatively recent tsunamis from SMF (Storegga, 8200 BP, and Grand Banks in 1929) in these glacially dominated

Table 8.1 Distribution of ages of several large submarine landslides

Name of SMF	Age (ka)	Source
East Atlantic		
Afen	3.4	D. Long (personal communication)
Traenadjupet	4	Laberg et al. (2003)
Storegga	8.1	Haflidason et al. (2005)
Faeroe	9.9	Van Weering et al. (1998)
Peach	10.5	Holmes et al. (1998); Maslin et al. (2004)
Rockall	15–16	Flood et al. (1979)
Nyk	16.3	Lindberg et al. (2004)
West Atlantic		
Cape Fear	8–14.5	Embley (1980); Popenoe et al. (1993); Paull et al. (1996); Rodriguez and Paull (2000)
Currituck	25–50	Prior et al. (1986b)
Central Atlantic		
Amazon Shallow East	14–17	Maslin et al. (1998)
Amazon Shallow West	14–17	Maslin et al. (1998)
Amazon Deep Eastern	35	Maslin et al. (1998)
Amazon Deep Western	42–45	Maslin et al. (1998)
Saharan	60	Gee et al. (1999)
Bed 5	50–60	
Mediterranean		
BIG '95	11.4	Canals et al. (2004); Lastras et al. (2004)
California		
Sur slide	1.5–6	Normark and Gutmacher (1988)
Palos Verdes	7.5	Normark et al. (2004)
Goleta west	8	Fisher et al. (2005)
Goleta middle lobe	10	Fisher et al. (2005)

Modified from Lee (2009).

northern regions suggests that the present-day hazards here may be high. However, where triggering of failure is mainly through glacio-isostatic rebound, the decline in seismicity may well result in the hazard today not being as great as perceived (as proposed by Bryn et al., 2005).

Does climate influence preconditioning of submarine sediment sequences to mass failure?

Climate preconditioning of sediment failure is recognised in many regions of the Atlantic. Off Norway, the change in sedimentation between glacials and inter-glacials results in loading of fine-grained sediments, thereby creating high pore pressures, and resulting in sediment susceptible to failure from earthquake loading. In fjords, rapid deposition of organic-rich sediment on fjord head deltas also leads to high pore pressures and gas-rich sediment, resulting in sediment that is sensitive to cyclic loading from storms and earthquake shock. Recent studies from the Gulf of Mexico (Dugan & Stigall, 2009) and Storegga (Kvalstad et al., 2005) show that high pore pressures resulting from rapid sedimentation, can precondition SMFs, but an external instantaneous trigger, such as an glacio-isostatic earthquake, or salt movement is still required. Thus the sediment preconditioning is climate related, although triggering may be mainly from another source.

Climate change as a trigger mechanism for SMFs

There has been much discussion about the instantaneous triggers of SMFs. Earthquakes are undoubtedly the most common triggers. Others include storms, tides and salt movement. More controversial is triggering resulting from instantaneous hydrate destabilisation and pore fluid over-pressuring. There is much circumstantial evidence on the association between hydrates and SMFs, e.g. the relationship between the headwall scarps of SMFs with the intersection of the hydrate stability zone at the seabed. Hydrate dissociation was considered as a trigger of the Storegga slide, but this has been discounted as a major factor (Bryn et al., 2005). Suggestions that rapid methane increase during past interglacials is from SMF have now been disproved from isotopic ice core evidence (Sowers, 2006).

Thus, although hydrate disassociation and sediment fluid over-pressures may precondition sediment for subsequent failure, under present climates an external force, such as from earthquake loading, is still required to actually initiate failure (Bryn et al., 2005; Dugan & Stigall, 2009). Elevated pore pressure is undoubtedly a significant factor in reducing sediment shear strength, but geotechnical analyses have failed to confirm that these can cause instantaneous failure. Most SMFs dated by Lee (2009) from the last 20,000 years were emplaced during stable or rising sea level, a period when hydrate disassociation would be least expected (Figure 8.13). In the future, however, if predictions of rapid global warming are correct, then hydrate stability in shallow ocean waters may be affected. In polar regions, where present models suggest that warming will be most rapid, this may result in an increased potential for hydrate destabilisation (Liggins et al., 2010; Maslin et al., 2010). In high latitudes, SMF triggering by earthquakes during late- to post-glacial periods is well established, with the earthquakes resulting from glacio-isostatic readjustment as ice sheets melt and retreat. In both Canada and northern Europe, rebound seismicity, although still occurring, is much less than in the Early Holocene because the isostatic readjustment is in an asymptotic decline. Triggers, such as cyclic loading by storm waves, are also climatically controlled. Continuing earthquake activity associated with glacio-isostatic readjustment along the Scotia margin suggests that there is on ongoing hazard here.

Figure 8.13 Plot showing the distribution of ages of dated submarine mass failures (SMFs) younger than 20 ka and listed in Table 8.1. These are compared with relative sea level according to corrected ^{14}C dating of samples from Barbados and elsewhere. (From Fairbanks, 1992; from Lee, 2009. Timing of occurence of large submarine landslides on the Atlantic Ocean margin. Marine Geology, Vol 264, 53–64.)

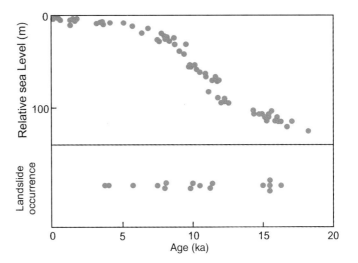

Climate and sea-level change, and the preservation potential of tsunami sediments from SMFs

The identification of so few tsunamis sourced from SMFs is enigmatic. The magnitude of tsunamis from SMFs is determined mainly by the type of failure (cohesive slump or fragmental landslide), SMF volume and water depth (Tappin et al., 2008). Slumps per unit volume are most hazardous, although fragmental failures are more voluminous and, therefore, potentially as hazardous. However, any SMF of sufficient volume and in appropriate water depths has the potential to create a hazardous tsunami. That SMF-sourced tsunamis have been identified only along continental and convergent margins and fjords is thus considered unlikely to be representative, and it is the evidence base that is inadequate to fully identify the hazard.

Limitations on the identification of tsunamis may be due to the limited anecdotal evidence from survivors (for historical events) or, for prehistorically tsunamis, an absence of tsunami sediment laid down on inundation. Of recent SMF generated tsunami, such as those of the Grand Banks, 1929 and Papua New Guinea, 1998, there have been positive identifications of associated sediments. The prehistoric event from Storegga laid down sediment that extends as far afield as the Faeroe Islands. It is less certain that tsunamis were also sourced from Andøya, Trænadjupet and Bjørnøyrenna; the absence of associated sediments from these events is puzzling. The assumption is that any resulting deposits have not been preserved.

Modelling of the Cape Fear and Currituck SMFs indicates that at present sea levels failure would create a hazardous tsunami. As the failures took place during either the late glacial or early interglacial, at the time of failure the tsunami would have made a major impact. The lack of evidence of tsunami from onshore (or in fact offshore-cored sediments) can be explained by the poor preservation potential of tsunami deposits that may lie on the seabed. Reworking during post-glacial periods may have eroded them, however. The absence of deposits on land may be due to the width of the US continental shelf. Given that SMFs took place at sea levels approximately 100 m lower than today, it is unlikely that the tsunami carried completely across the width of the US continental shelf that in places is up to 200 km. Both the large-scale SMF off the Amazon and the smaller SMFs off west Africa have the potential to generate tsunamis, the absence of evidence in these areas may be due to the timing (failure during lowered sea levels) or the lack of investigation in onshore areas to determine whether sediments resulting from these events are present.

The low preservation potential of sediments laid down by tsunamis suggests that the prehistoric record may be unrepresentative. Tsunamis during low stands may not have left a deposit on present-day land, but on the present-day flooded continental shelf, where identification is compromised by difficulties in identification in cores or absence due to sediment reworking during postglacial transgression, e.g. this may explain the lack of evidence of tsunamis sourced from the SMFs off of the eastern USA (e.g. Carrituck and Cape Fear), despite their large volume and modelling that would suggest otherwise.

The absence of evidence for tsunamis on active river deltas is intriguing. The similarity between sedimentation mechanisms on river deltas and other landslide territories, such as continental margins and fjords, in both sediment preconditioning and triggering suggests that SMFs of sufficient volume and at appropriate depths would generate tsunami. The water depths (<1000 m) of known failures in the Gulf are certainly shallow enough to generate tsunamis, if the SMFs are of sufficient volume. On the Gulf Coast the most catastrophic seabed failure was in 1969, when Hurricane Camille struck. Three offshore drilling platforms collapsed as a result of seabed sediment failure which in turn resulted in a change of seabed relief of up to 12 m (e.g. Bea et al., 1983). Thus either the SMFs are too small or the tsunami wave was indistinguishable from the storm surge. In the instance of Katrina in 2005 the surge was between 6 and 9 m, probably of a sufficient magnitude to mask a tsunami.

Conclusions

The hazard and risk from SMF-sourced tsunami have advanced significantly over the past decade. SMFs are generally acknowledged to source hazardous tsunamis; however, the database of well-studied events is still limited. There is a recognisable and strong climate control on SMF that impacts on their potential to source hazardous tsunamis. There are strong climate controls on the type of SMF sediment and its rate of delivery. Climate can precondition sediment instability by introducing 'weak layers' as well as through fluid and gas over-pressuring. Climate influences triggering through earthquake shock and wave loading. There is an obvious latitudinal variation in SMF architecture that suggests tsunami hazard to be greater in northern regions. During glacial periods, rates of sediment delivery are higher and sediment volumes are larger, thus forming larger SMFs that are more prone to failure. Along convergent margins, where earthquakes are more frequent, there is a climate influence in high latitudes.

Any SMF of sufficient volume in an appropriate water depth has the potential to generate a hazardous tsunami. At first sight, as in the North Atlantic, the presence of numerous SMFs suggests a present-day high risk. Consideration of the climate controls on SMF, however, indicates that these SMF took place under different environmental conditions. Their number, therefore, may not reflect present-day hazard. The evidence base for the conclusions here, however, is still small. There are still too few case studies of actual events. More dates of landslide failure are required to test hypotheses on climate control of mass failure processes. The absence of tsunamis off deltas is enigmatic. The general absence of evidence for tsunamis where there are numerous SMFs may be misleading and attributed to limited defining evidence from anecdote, absence of preserved deposits or misinterpretation of tsunami source.

Notwithstanding, consideration of climate controls on SMFs contributes significantly in improving our understanding of SMF mechanisms, allowing an improved assessment of their sources as hazardous tsunami under present climatic conditions. In the future, as climate warms, this improved understanding will underpin prediction of tsunami-sourced SMFs, particularly in regions where climate change will be most rapid, such as in the polar regions.

Acknowledgements

This chapter is published with the permission of the Director of the British Geological Survey, Natural Environment Research Council, UK. Many thanks to Dr Simon Day for a comprehensive and analytical review.

References

Antobreh, A. A. & Krastel, S. (2006) Morphology, seismic characteristics and development of Cap Timiris Canyon, offshore Mauritania: A newly discovered canyon preserved-off a major arid climatic region. *Marine and Petroleum Geology* **23**, 37–59.

Ashabranner, L. B., Tripsanas, E. K. & Shipp, R. C. (2009) Multi-direction flow in a mass-transport deposit, Santos Basin, Offshore Brazil. In: Mosher, D. C., Shipp, R. C., Moscardilli, L., et al. (eds), *Submarine Mass Movements and Their Consequences*. New York: Springer, pp 247–255.

Bea, R. G., Wright, S. G., Sicar, P. & Niedoroda, A. W. (1983) Wave-induced slides in South Pass Block 70, Mississippi delta. *Journal of Geotechnical Engineering* **109**, 619–644.

Billi, A., Funiciello, R., Minelli, L., et al. (2008) On the cause of the (1908) Messina tsunami, southern Italy. *Geophysical Research Letters* **35**.

Biscontin, G., Pestana, J. M. & Nadim, F. (2004) Seismic triggering of submarine slides in soft cohesive soil deposits. *Marine Geology* **203**, 341–354.

Bohannon, R. G. & Gardner, J. V. (2004) Submarine landslides of San Pedro Sea Valley, southwest Long Beach, California. *Marine Geology* **203**, 261–268.

Bondevik, S., Mangerud, J., Dawson, S., Dawson, A. & Lohne, Ø. (2005) Evidence for three North Sea tsunamis at the Shetland Islands between 8000 and 1500 years ago. *Quaternary Science Reviews* **24**, 1757–1775.

Bryn, P., Berg, K., Forsberg, C. F., Solheim, A. & Lien, R. (2005) Explaining the Storegga Slide. *Marine and Petroleum Geology* **22**, 11–19.

Bugge, T. (1983) Submarine slides on the Norwegian continental margin, with special emphasis on the Storegga area. *IKU Report* **110**, 1–152.

Canals, M., Lastras, G., Urgeles, R., et al. (2004) Slope failure dynamics and impacts from seafloor and shallow sub-seafloor geophysical data: case studies from the COSTA project. *Marine Geology* **213**, 9–72.

Cashman, K. V. & Popenoe, P. (1985) Slumping and shallow faulting related to the presence of salt on the continental slope and rise off North Carolina. *Marine and Petroleum Geology* **2**, 260–271.

Chaytor, J. D., Twichell, D. C., ten Brink, U. S., Buczkowski, B. J. & Andrews, B. D., eds (2007) Revisiting submarine mass movements along the U.S. Atlantic continental margin: implications for tsunami hazard. In: *Submarine Mass Movements and their Consequences*. New York: Springer, pp 394–403.

Coulter, H. W. & Migliaccio, R. R. (1966) *Effects of the earthquake of March 27, 1964 at Valdez, Alaska 542-C*. US Geol. Survey Prof., Paper 542-E.

Dan, G., Sultan, N. & Savoye, B. (2007) The 1979 Nice harbour catastrophe revisited: Trigger mechanism inferred from geotechnical measurements and numerical modelling. *Marine Geology* **245**, 40–64.

Dingle, R. V. (1977) Anatomy of a large submarine slump on sheared continentgal margin (southeast Africa). *Jounal of the Geological Society of London* **134**, 293–310.

Dobson, M. R., O'Leary, D. & Veart, M. (1998) Sediment delivery to the Gulf of Alaska: source mechanisms along a glaciated transform margin. *Geological Society, London, Special Publications* **129**, 43–66.

Dugan, B. & Stigall, J. (2009) Origin of overpressure and slope failure in the Ursa Region, Northern Gulf of Mexico. In: Mosher, D. C., Shipp, R. C., Moscardilli, L., et al. (eds), *Submarine Mass Movements and Their Consequences*. New York: Springer, pp 167–178.

Embley, R. W. (1980) The role of mass transport in the distribution and character of deep ocean sediments with special reference to the North Atlantic. *Marine Geology* **38**, 23–50.

Embley, R. W. & Jacobi, R. D. (1986) Mass wasting in the western North Atlantic. In: Vogt, P. R. & Tucholke, B. E. (eds), *The Western North Atlantic Region: Geology of North America*. Boulder, CO: Memoir of the Geological Society of America, pp 479–490.

Fairbanks, R. (1992) *Barbados sea level and Th/U 14C calibrations.* IGBP Pages/ World Data Center for Paleoclimatology Data Contribution Series No. 92-020.

Fisher, M. A., Normark, W. R., Greene, H. G., Lee, H. J. & Sliter, R. W. (2005) Geology and tsunamigenic potential of submarine landslides in Santa Barbara Channel, Southern California. *Marine Geology* **224**, 1–22.

Flood, R. D., Hollister, C. D. & Lonsdale, P. (1979) Disruption of the Feni sediment drift by debris flows from Rockall Bank. *Marine Geology* **32**, 311–334.

Fritz, H. M., Kongko, W., Moore, A., et al. (2007) Extreme runup from the 17 July 2006 Java tsunami. *Geophysical Research Letters* **34**, L12602.

Fryer, G. J., Watts, P. & Pratson, L. F. (2004) Source of the great tsunami of 1 April 1946: a landslide in the upper Aleutian forearc. *Marine Geology* **203**, 201–218.

Gee, M J R, Masson, D. G., Watts, A. B. & Allen, P. A. (1999) The Saharan debris flow: an insight into the mechanics of long runout submarine debris flows. *Sedimentology* **46**, 317–335.

Geist, E. L., Lynett, P. J. & Chaytor, J. D. (2009) Hydrodynamic modeling of tsunamis from the Currituck landslide. *Marine Geology* **264**, 41–52.

Goldfinger, C., Kulm, L. D., McNeill, L. C. & Watts, P. (2000) Super-scale failure of the southern Oregon Cascadia margin. *Pure and Applied Geophysics* **157**, 1189–1226.

Greene, H. G., Murai, L. Y., Watts, P., et al. (2005) Submarine landslides in the Santa Barbara Channel as potential tsunami sources. *Natural Hazards and Earth System Sciences* **6**, 63–88.

Grozic, J. L. H. (2009) Interplay between gas hydrates and submarine slope failure. In: Mosher, D. C., Shipp, R. C., Moscardilli, L., et al. (eds), *Submarine Mass Movements and Their Consequences.* Springer Science + Business Media, pp 11–30.

Haflidason, H., Lien, R., Sejrup, H. P., Forsberg, C. F. & Bryn, P. (2005) The dating and morphometry of the Storegga Slide. *Marine and Petroleum Geology* 123–136.

Hampton, M. A., Lemke, R. W. & Coulter, H. W. (1993) Submarine landslides that had a significant impact on man and his activities: Seward and Valdez, Alaska. In: Schwab, W. C., Lee, H. J. & Twichell, D. C. (eds), *Submarine Landslides: Selected Studies in the US EEZ.* 2002 USGS Bulletin, 123–142.

Hampton, M. A., Lee, H. J. & Locat, J. (1996) Submarine landslides. *Reviews of Geophysics* **34**, 33–59.

Heezen, B. C., Ericsson, D. B. & Ewing, M. (1954) Further evidence of a turbidity current following the 1929 Grand Banks earthquake. *Deep Sea Research* **1**, 193–202.

Henrich, R., Hanebuth, T J J, Krastel, S., Neubert, N. & Wynn, R. B. (2008) Architecture and sediment dynamics of the Mauritania Slide Complex. *Marine and Petroleum Geology* **25**, 17–33.

Henrich, R., Hanebuth, T. J. J., Cherubini, Y., Kraste, S., Pierau, R. & Zühlsdorff, C. (2009) Climate-induced turbidity current activity in NW-African Canyon

Systems. In: Mosher, D. C., Shipp, R. C., et al. (eds) *Submarine Mass Movements and Their Consequences*. New York: Springer.

Hjelstuen, B. O., Sejrup, H. P., Haflidason, H., Nygård, A., Ceramicola, S. & Bryn, P. (2005) Late Cenozoic glacial history and evolution of the Storegga Slide area and adjacent slide flanks regions, Norwegian continental margin. *Marine and Petroleum Geology* **22**, 57–69.

Holmes, R., Long, D. & Doded, L. R. (1998) Large-scale debrites and submarine landslides on the Barra Fan, west of Britain. In: Stoker, M. S., Evans, D. & Cramp, A (eds), *Geological Processes on Continental Margins: Sedimentation, mass-wasting and stability*. Special Publications 129. London: Geological Society, pp 67–79.

Hornbach, M. J., Lavier, L. L. & Ruppel, C. D. (2007) Triggering mechanism and tsunamogenic potential of the Cape Fear Slide complex, U.S. Atlantic margin. *Geochemistry, Geophysics, Geosystems* **8**, Q12008.

Hühnerbach, V., Masson, D. G. et al. (2004) Landslides in the north Atlantic and its adjacent seas: an analysis of their morphology, setting and behaviour. *Marine Geology* **213**, 343–362.

Imamura, F., Gica, E., Takahashi, T. & Shuto, N. (1995) Numerical simulation of the 1992 Flores tsunami: Interpretation of tsunami phenomena in northeastern Flores Island and damage at Babi Island. *Pure and Applied Geophysics* **144**, 555–568.

Jenner, K. A., Piper, D J W, Campbell, D. C. & Mosher, D. C. (2007) Lithofacies and origin of late Quaternary mass transport deposits in submarine canyons, central Scotian Slope, Canada. *Sedimentology* **54**, 19–38.

Jiang, L. & LeBlond, P. H. (1994) Three dimensional modelling of tsunami generation due to submarine mudslide. *Journal of Physical Ocean* **24**, 559–573.

Johnson, R. W. (1987) Large-scale volcanic cone collapse: the 1888 slope failure of Ritter Volcano, and other examples from Papua New Guinea. *Bulletin of Volcanology* **49**, 669–679.

Kawamura, K., Kanamatsu, T., Kinoshita, M., et al. (2009) Redistribution of sediments by submarine landslides on the Eastern Nankai Accretionary Prism. In: Mosher, D. C., Shipp, R. C., Moscardilli, L., et al. (eds), *Submarine Mass Movements and Their Consequences*. New York: Springer, p 28.

Kvalstad, T. J., Andresen, L., Forsberg, C. F., Berg, K., Bryn, P. & Wangen, M. (2005) The Storegga Slide: evaluation of triggering sources and slide mechanics. *Marine and Petroleum Geology* **22**, 245–256.

Laberg, J. S., Vorren, T. O., Dowdeswell, J. A., Kenyon, N. H. & Taylor, J. (2000) The Andøya Slide and the Andøya Canyon, north-eastern Norwegian-Greenland Sea. *Marine Geology* **162**, 259–275.

Laberg, J. S., Vorren, T. O., Mienert, J., Haflidason, H., Bryn, P. & Lien, R. (2003) Preconditions leading to the Holocene Trænadjupet slide offshore Norway. In: Locat, J. & Mienert, J. (eds), *Submarine Mass Movements and their Consequences*. Amsterdam: Kluwer Academic Publishers, pp 247–254.

Lastras, G., Canals, M., Urgeles, R., de Batist, M., Calafat, A. M. & Casamor, J. L. (2004) Characterisation of the recent BIG'95 debris flow deposit on the Ebro margin, Western Mediterranean Sea, after a variety of seismic reflection data. *Marine Geology* **213**, 235–255.

LeBlond, P. H. & Jones, A. (1995) Underwater landslides ineffective at tsunami generation. *Science and Tsunami Hazards* **13**, 25–26.

Lee, H. J. (1989) Undersea landslides: extent and significance in the Pacific Ocean. In: Brabb, E. E. & Harrod, B. L. (eds), *Landslides, Extent and Economic Significance*. Washington, DC: Proceedings of the 28th International Geological Congress: Symposium on landslides, pp 367–380.

Lee, H. J. (2005) Undersea landslides: extent and significance in the Pacific Ocean, an update. *Natural Hazards and Earth System Sciences* **5**, 877–892.

Lee, H. J., et al (2007) Submarine Mass movements on Continental Margins, In: Nittrouer, C. A. et al. (Eds), *Continental Margin Sedementation*. Wiley, pp 213–274.

Lee, H. J. (2009) Timing of occurrence of large submarine landslides on the Atlantic Ocean margin. *Marine Geology* **264**, 53–64.

Lee, H. J., Kayen, R. E., Gardner, J. V. & Locat, J. (2003) Characteristics of several tsunamigenics submarine landslides. In: Locat, J. & Mienert, J (eds), *Submarine Mass Movements and their Consequences*. Amsterdam: Kluwer, pp 357–366.

Lemke, R. W. (1967) *Effects of the earthquake of 27 March 1964, at Seward, Alaska*. US Geological Survey Professional, Paper 542-E.

Liggins, F., Betts, R. A. & McGuire, B. 2010 Projected future climate changes in the context of geological and geomorphological hazards. *Philosophical Transactions of the Royal Society A* **368**, 2347–2368.

Lindberg, B., Laberg, J. S. & Vorren, T. O. (2004) The Nyk Slide – morphology, progression, and age of a partly buried submarine slide offshore northern Norway. *Marine Geology* **213**, 277–289.

Locat, J., Locat, P., Lee, H. J. & Imran, J. (2004) Numerical analysis of the mobility of the Palos Verdes debris avalanche, California, and its implication for the generation of tsunamis. *Marine Geology* **20**, 269–280.

Locat, J., Lee, H., ten Brink, U. S., Twichell, D., Geist, E. & Sansoucy, M. (2009) Geomorphology, stability and mobility of the Currituck slide. *Marine Geology* **264**, 28–40.

López-Venegas, A. M., Brink, U. St. & Geist, E. L. (2008) Submarine landslide as the source for the October 11, 1918 Mona Passage tsunami: Observations and modeling. *Marine Geology* **254**, 35–46.

McAdoo, B., Pratson, G. & Orange, L. F. (2000) Submarine landslide geomorphology, U.S. Continental Slope. *Marine Geology* **169**, 103–136.

McAdoo, B. G., Capone, M. K. & Minder, J. (2004) Seafloor geomorphology of convergent margins: implications for Cascadia seismic hazard. *Tectonics* **23**, TC6008.

Maslin, M., Mikkelsen, N., Vilela, C. & Haq, B. (1998) Sea-level and gas-hydrate-controlled catastrophic sediment failures of the Amazon Fan. *Geology* **26**, 1107–1110.

Maslin, M., Owen, M., Betts, R., et al. (2010) Gas hydrates: past and future geohazard? *Philosophical Transactions of the Royal Society A* **368**, 2369–2394.

Maslin, M., Owen, M., Day, S., Long, D., (2004) Linking continental-slope failures and climate change: Testing the clathrate gun hypothesis. *Geology* **32**(1), 53–56.

Masson, D. G., Wynn, R. B. & Talling, P. J. (2009) Large landslides on passive continental margins: processes, hypotheses and outstanding questions. In: Mosher, D. C., Shipp, R. C., Moscardilli, L., et al. (eds), *Submarine Mass Movements and Their Consequences*. New York: Springer Science, pp 153–165.

Moore, J. G., Normark, W. R. & Holcomb, R. T. (1994) Giant Hawaiian underwater landslides. *Science* **264**, 46–47.

Mosher, D. C. & Piper, D. J. W. (2007) Analysis of multibeam seafloor imagery of the Laurentian Fan and the 1929 Grand Banks landslide area. In: Lykousis, V., Sakellariou, D. & Locat, J (eds), *Submarine Mass Movements and Their Consequences*. New York: Springer, pp 77–88.

Normark, W. R. & Gutmacher, C. E. (1988) Sur submarine slide, Monterey Fan, central California. *Sedimentology* **35**, 629–647.

Normark, W. R., McGann, M. & Sliter, R. (2004) Age of Palos Verdes submarine debris avalanche, southern California. *Marine Geology* **203**, 247–259.

Paull, C. K., Buelow, W. J., Ussler, W. & Borowski, W. S. (1996) Increased continental-margin slumping frequency during sea-level lowstands above gas hydrate-bearing sediments. *Geology* **24**, 143–146.

Peltier, W. R. (2002) Global glacial isostatic adjustment: Palaeogeodetic and space-geodetic tests of the ICE-4G (VM2) model. *Journal of Quaternary Science* **17**, 491–510.

Piper, D. J. W. & Asku, A. E. (1987) The source and origin of the 1929 Grand Banks turbidity current inferred from sediment budgets. *Geo Marine Letters* **7**, 177–182.

Piper, D. J. W. & McCall, C. (2003) A synthesis of the distribution of submarine mass movements on the eastern Canadian Margin. In: Locat, J. & Mienert, J (eds), *Submarine mass movements and their consequences*. Amsterdam: Kluwer Academic Publishers, pp 291–298.

Piper, D. J. W., Pirmez., C., Manley, P. L., et al. (1997) Mass transport deposits of the Amazon Fan. In: Flood, R. D., Piper, D. J. W., Klaus, A. & Peterson, L. C. (eds), *Proceedings of the Ocean Drilling Program, Scientific results*, 155. College Station, Texas (Ocean Drilling Program), pp 109–146.

Plafker, G., Kachadoorian, R., Eckel, E. B. & Mayo, L. R. (1969) *Effects of the earthquake of March 27, 1964 on various communities*. US Geological Survey Professional, Paper 542-G.

Popenoe, P., Schmuck, E. A. & Dillon, W. P. (1993) The Cape Fear landslide: Slope failure associated with salt diapirism and gas hydrate decomposition. In: Schwab, W. C., Lee, H. J. & Twichell, D C (eds), *Submarine Landslides: Selected Studies in the U.S. Exclusive Economic Zone*. 2002 US Geological Survey Bulletin, pp 40–53.

Prior, D. B. & Coleman, J. M. (1982) Active slides and flows on underconsolidated marine sediments on the slopes of the Mississippi delta. In: Saxov, S. & Nieuwenhuis, J. K. (eds), *Marine Slides and Other Mass Movements*. New York: Plenum, pp 21–49.

Prior, D. B., Bornhold, B. D., Coleman, J. M. & Bryant, W. R. (1982a) Morphology of a submarine slide, Kitimat Arm, British Columbia. *Geology* **10**, 588–592.

Prior, D. B., Coleman, J. M. & Bornhold, B. D. (1982b) Results of a known sea-floor instability event. *Geomarine Letters* **2**, 117–122.

Prior, D. B., Bornhold, B. D. & Johns, M. W. (1986a) Active sand transport along a fjord-bottom channel, Bute Inlet, British Columbia. *Geology* **14**, 581–584.

Prior, D. B., Doyle, E. H. & Neurauter, T. (1986b) The Currituck Slide, mid-Atlantic continental slope – Revisited. *Marine Geology* **73**, 25–45.

Rahiman, T I H, Pettinga, J. R. & Watts, P. (2007) The source mechanism and numerical modelling of the 1953 Suva tsunami, Fiji. *Marine Geology* **237**, 55–70.

Rajendran, C. P., Ramanamurthy, M. V., Reddy, N. T. & Rajendran, K. (2008) Hazard implications of the late arrival of the 1945 Makran tsunami. *Current Science* **95**, 1739–1743.

Rodriguez, N. M. & Paull, C. K. (2000) Data report: ^{14}C dating of sediment of the uppermost Cape Fear slide plain: constraints on the timing of this massive submarine landslide. In: Paull, C. K., Matsumoto, R., Wallace, P. J. & Dillon, W. P. (eds), *Proceedings of the Ocean Drilling Program. Scientific Results.* Washington DC: US Government Printing Office, pp 325–327.

Satake, K. (2007) Volcanic origin of the 1741 Oshima-Oshima tsunami in the Japan Sea. *Earth Planets Space* **59**, 381–390.

Satake, K. & Kato, Y. (2001) The 1741 Oshima-Oshima Eruption: Extent and Volume of Submarine Debris Avalanche. *Geophysical Research Letters* **28**, 427–430.

Schwab, W. C. & Lee, H. J. (1988) Causes of two slope-failure types in continental-shelf sediment, northeastern Gulf of Alaska. *Journal of Sedimentary Research* **58**, 1–11.

Siebert, L., Glicken, H. & Ui, T. (1987) Volcanic hazards from Bezymianny- and Bandai-type eruptions. *Bulletin of Volcanology* **49**, 435–459.

Solheim, A., Bryn, P., Sejrup, H. P., Mienert, J. & Berg, K. (2005) Ormen Lange – an integrated study for the safe development of a deep-water gas field within the Storegga Slide Complex, NE Atlantic continental margin; executive summary. *Marine and Petroleum Geology* **22**, 1–9.

Sowers, T. (2006) Late Quaternary Atmospheric CH4 Isotope Record Suggests Marine Clathrates Are Stable. *Science* **311**, 838–840.

Syvitski, J P M, Burrell, D. C. & Skei, J. M. (1986) *Fjords: Processes and products.* New York: Springer.

Tanioka, Y. & Seno, T. (2001) Sediment effect on tsunami generation of the 1896 Sanriku tsunami earthquake. *Geophysical Research Letters* **28**, 3389–3392.

Tappin, D. R. (2009) Mass transport events and their tsunami hazard. In: Mosher, D. C., Shipp, R. C., Moscardilli, L., et al. (eds), *Submarine Mass Movements and Their Consequences.* Springer Science + Business Media, pp 667–684.

Tappin, D. R., Matsumoto, T., Watts, P., et al. (1999) Sediment slump likely caused (1998) Papua New Guinea Tsunami. *EOS, Transactions of the American Geophysical Union* **80**, 329, 334, 340.

Tappin, D. R., Watts, P., McMurtry, G. M., Lafoy, Y. & Matsumoto, T. (2001) The Sissano Papua New Guinea tsunami of July (1998) – offshore evidence on the source mechanism. *Marine Geology* **175**, 1–23.

Tappin, D. R., McNeil, L., Henstock, T. & Mosher, D. (2007) Mass wasting processes – offshore Sumatra. In: Lykousis, V., Sakellarious, D. & Locat, J (eds), *Submarine Mass Movements and Their Consequences*. New York: Springer, pp 327–336.

Tappin, D. R., Watts, P. & Grilli, S. T. (2008) The Papua New Guinea tsunami of 17 July 1998: anatomy of a catastrophic event. *Natural Hazards and Earth Systems Science* **8**, 243–266.

Twichell, D. C., Chaytor, J. D., ten Brink, U. S. & Buczkowski, B. (2009) Morphology of late Quaternary submarine landslides along the U.S. Atlantic continental margin. *Marine Geology* **264**, 4–15.

Urgeles, R., Canals, M., Baraza, J., B. Alonso. & Masson, D. (1997) The most recent megalandslides of the Canary Islands: el Golfo debris avalanche and Canary debris flow, west el Hierro Island. *Journal of Geophysical Research* **102**, 305–320, 323.

Van Weering, T., Nielsen, T., Kenyon, N. H., Akentieva, K. & Kulipers, A. H. (1998) Large submarine slides on the NE Faeroe continental margin. In: Stoker, M. S., Evans, D. & Cramp, A. (eds), *Geological Processes on Continental Margins: Sedimentation, Mass-wasting and Stability*. Special Publications . London: Geological Society, pp 5–17.

Varnes, D. J. (1958) Landslides types and processes. In: Eckel, E. D. (ed.), *Landslides and Engineering Practice*. Washington DC: Highway Research Board, pp 20–47.

Ward, S. N. & Day, S. (2003) Ritter Island Volcano – lateral collapse and the tsunami of 1888. *Geophysical Journal International* **154**, 891–902.

Weaver, P. P. E., Wynn, R. B., Kenyon, N. H. & Evans, J. (2000) Continental margin sedimentation, with special reference to the north-east Atlantic margin. *Sedimentology* **47**(suppl 1), 239–225.

Wynn, R. B., Weaver, P. P. E., Masson, D. G. & Stow, D. A. V. (2002) Turbidite depositional architecture across three inter-connected deep-water basins on the Northwest African Margin. *Sedimentology* **49**, 669–695.

9

High-mountain slope failures and recent and future warm extreme events

Christian Huggel[1], Nadine Salzmann[2] and Simon Allen[3]

[1]*Department of Geography, University of Zurich, Switzerland*
[2]*Department of Geography, University of Zurich, Switzerland and Department of Geosciences, University of Fribourg Switzerland*
[3]*Climate and Environmental Physics, Physics Institute, University of Bern, Switzerland*

Summary

The number of large slope failures in some high mountain regions such as the European Alps has increased over the past two to three decades. There are a number of indications that ongoing climatic changes cause an increase in slope failures, thus possibly further exacerbating future failure events. Although the effects of a gradual temperature rise on glaciers and permafrost have been extensively studied, the impacts of short-term, unusually warm, temperature increases on slope stability in high mountains remain largely unexplored.

We describe several large slope failures in rock and ice in recent years in Alaska, New Zealand and the European Alps, and analyse patterns of meteorological variables in the days and weeks before the failures. Although we did not find one general air temperature pattern, all the studied failures were preceded by unusually warm periods; several happened immediately after temperatures suddenly dropped to freezing level.

We assessed the frequency of warm extremes in future by analysing eight regional climate models (RCMs) from the recently completed EU programme ENSEMBLES for the central Swiss Alps. The models show an increase in the frequency of high-temperature events for the period 2001–2050 compared with a 1951–2000 reference period. The 5-, 10- and 30-day warm events are projected to increase about 1.5–4 times by 2050, and in some models by up to 10 times.

Warm extremes can trigger large landslides in temperature-sensitive, high mountain environments by increasing occurrence of liquid water due to melt of snow and ice, and by rapid thaw of permafrost. In addition to these climate-induced processes, which can reduce slope strength, local geological, glaciological and topographic parameters of a slope also must be considered for comprehensive analyses.

Climate Forcing of Geological Hazards, First Edition. Edited by Bill McGuire and Mark Maslin.
© 2013 The Royal Society and John Wiley & Sons, Ltd. Published 2013 by John Wiley & Sons, Ltd.

Introduction

There is increasing concern that rising air temperatures could affect slope stability in high-mountain areas due to decay of glaciers and degradation of permafrost (Haeberli et al., 1997; Gruber & Haeberli, 2007; Huggel et al., 2008a). Large rock or ice avalanches can be destructive and far reaching, especially when they impact lakes and trigger large outburst floods (Clague & Evans, 2000; Huggel et al., 2004; Salzmann et al., 2004; Kääb et al., 2005; Haeberli & Hohmann, 2008; Romstad et al., 2009).

Although there is a long tradition of research on alpine ice avalanches (Heim, 1932; Alean, 1985; Röthlisberger, 1987), research on rock-slope instability related to permafrost degradation is a more recent development (Wegmann et al., 1998; Gruber & Haeberli, 2007; Harris et al., 2009). Indications of possible effects of climate change on slope stability has come from an increasing number of large-size rock falls and rock avalanches from permafrost areas in the Alps over the past two decades (Barla et al., 2000; Fischer et al., 2006; Gruber & Haeberli, 2007; Fischer & Huggel, 2008; Sosio et al., 2008). Many of these events have not yet been studied in detail, and the causes and trigger processes are not adequately understood. However, understanding of the process and effects of long-term climate change on frozen rock slopes has been improved by systematic observation of near surface and subsurface temperatures, and numerical modelling approaches for simulation of the spatial distribution and temporal evolution of sub-surface temperatures in complex three-dimensional topography in a warming climate scenario (Noetzli et al., 2007; Fischer et al., 2010). Furthermore, recent studies have shown that polythermal glaciers overlying frozen bedrock can induce thermal anomalies in the ground as deep as several tens of metres (Haeberli et al., 1997; Wegmann et al., 1998; Etzelmüller & Hagen, 2005; Huggel et al., 2008a).

Slope failures in high mountains are controlled by a variety of factors, including geology, topography and hydrology, as well as glacier and permafrost, all of which are highly interconnected (Fischer & Huggel, 2008). Due to atmospheric warming glaciers and permafrost are currently changing at the highest rate. An aspect of high-mountain slope instability that has been little studied in this context is the effect of extreme, short-term temperature events of days to several weeks as a potential trigger of rock and ice avalanches. The extreme heat wave in Europe in the summer of 2003 appears to have triggered numerous small rock falls (10^2–10^3 m^3), probably due to rapid thawing and thickening of the active layer (Gruber et al., 2004a). A very large rock-ice avalanche (~5×10^7 m^3) on Mt Steller in Alaska in September 2005 occurred during unusually warm air temperatures, with temperatures far above the freezing point (Huggel, 2009). However, a detailed analysis of meteorological conditions surrounding these events has not yet been completed. In addition, analysis of a larger sample of events is important for understanding the impacts of current and future warm events on slope stability better, especially given that current projections indicate an increase in the number of extreme climatic events with high temperatures (Tebaldi et al., 2006; Meehl et al., 2007).

In this study, we explore meteorological conditions leading up to several large rock and ice avalanches in south–central Alaska, the European Alps and the Southern Alps of New Zealand. We focus on the air temperature history days and weeks before failure. Longer-term ground surface and firn or ice temperatures, geology and topography are also considered but are not the primary focus of the chapter. Based on results from case studies, we define appropriate climatic indicators (time period, temperature range) and use recently completed ENSEMBLES regional climate model runs for the central Alps to assess whether the frequency of events that potentially trigger large slope failures will increase in future. For the analysis of meteorological and climatic conditions of case studies we generally use reference time periods that vary according to available data records. For future projections we consider the period 2001–2050, using the reference period of 1951–2000.

Case studies

Central Southern Alps, New Zealand

Large slope failures are common in New Zealand's Southern Alps, where Carboniferous–Cretaceous greywacke is being rapidly uplifted along an active tectonic plate margin (Allen et al., 2010). Rapid twentieth-century glacial recession is probably implicated in some large landslides in this region (McSaveney, 2002); it has removed the lateral support provided to adjacent steep slopes and exposed previously insulated surfaces to thermal and mechanical erosion. Studies have also begun to consider the possible role of permafrost degradation relating to slope failures occurring about the highest, and steepest terrain of the central Southern Alps (Allen et al., 2009, 2010). Here, we focus on air temperature observations in the months and weeks before the spectacular 1991 summit collapse of New Zealand's highest mountain, Aoraki/Mt Cook, and draw comparisons with two more recent, but smaller, ice-rock avalanches occurring from Vampire Peak (2008) and Mount Dampier (2010).

Aoraki/Mt Cook 1991

Shortly after midnight on 14 December 1991, $12 \times 10^6 \, m^3$ of rock and ice detached from the east face of Aoraki/Mt Cook (3754 m above sea level or asl) (Figure 9.1). The large rock-ice avalanche narrowly missed an occupied alpine hut, completely scoured and entrained a steep icefall, and came to rest after running approximately 70 m up an opposing valley wall. The avalanche travelled 7.5 km, and in the process bulked up to a total volume estimated at $60–80 \times 10^6 \, m^3$ with entrained snow and ice (Figure 9.2) (McSaveney, 2002). It lowered the summit of Aoraki/

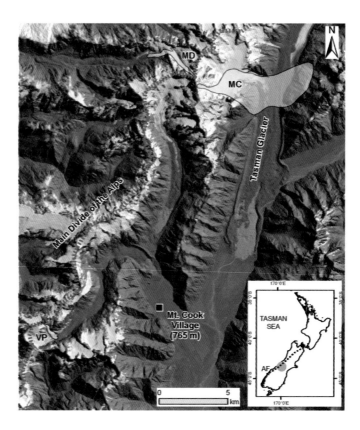

Figure 9.1 Map showing location of the avalanches at Aoraki/Mt Cook (MC), Mt Dampler (MD) and Vampire Peak (VP), together with the meteorological station at Mt Cook village.

Figure 9.2 Rock-ice detachment from Mt Cook (3754 m), 14 December 1991. The failure reduced the elevation of New Zealand's highest mountain by approximately 10 m. (Photo courtesy of I. Owens taken on 16 December 1991.)

Mt Cook and continues to significantly affect the flow of the Tasman Glacier (Quincey & Glasser, 2009).

The Aoraki/Mt Cook rock-ice avalanche initiated with the failure of a large, east- to southeast-facing scarp slope that supported the approximately 15-m-thick summit ice cap. Foliation in closely jointed greywacke and argillite dips steeply in the summit area, and the approximately 60-m-deep failure scar exposed a heavily fractured and dilated rock mass of poor quality (McSaveney, 2002). The detachment zone was steep (50–70°) and included a small hanging glacier that was partially removed during the failure.

The nearest available climate data come from Mount Cook Village about 16 km south of, and about 3000 m below, the Aoraki/Mt Cook summit. High orographic precipitation in this region (>10 m water equivalent precipitation per year) favours the use of a humid free-air lapse rate (0.0055°C/m) to extrapolate air temperature data to the detachment zone of the landslide. Based on data for the years 1961–1990, mean annual air temperature (MAAT) at the base of the detachment zone (2900 m) is estimated to be −3.1°C, decreasing to −7.8°C towards the mountain summit (3754 m). Recent rockwall temperature measurements and solar-radiation-based modelling of mean annual ground surface temperature (MAGST) (Allen et al., 2009) indicates MAGST within the Mt Cook detachment zone ranging from 0°C to −6.3°C at the summit at 3754 m (for a lapse rate of 0.0055°C/m), with variations dependent on both elevation and aspect. In combination, MAGST and MAAT values suggest that most of the detachment zone supported cold permafrost; only the lowermost part of the detachment zone was likely in a warm or marginal state.

After a particularly cold winter, the summit of Aoraki/Mt Cook experienced repeated periods of daily maximum temperatures far above 0°C in November and early December of 1991, interspersed with periods of cooling where temperatures remained below 0°C (Figure 9.3). During the week immediately before the

Figure 9.3 Daily maximum air temperature range between the top (3754 m) and bottom (2900 m) of the Mt Cook rock-ice detachment zone, extrapolated from Mt Cook Village (765 m). Dashed lines indicate the longer-term (1961–1990) 95th percentile temperatures extrapolated to these same elevations. The timing of the failure is indicated by the red line.

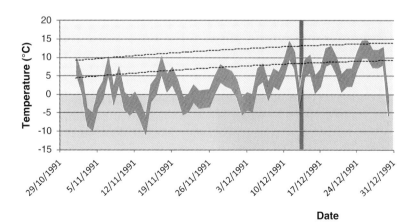

rock-ice avalanche, temperatures warmed again, culminating in estimated daily maximum temperatures of 14.4°C and 9.7°C at, respectively, the base and top of the detachment zone on 11 December. It is notable that this was the first time during the summer in which the entire detachment area experienced extremely warm seasonal temperatures (>95th percentile), and, although only mid-December, these temperatures were within the warmest 92nd percentile of maximum daily temperatures for all the summer months (December–February) as recorded for the period 1961–1990. Prolonged melting would have been enabled within the detachment zone, followed by rapid cooling to sub-freezing temperatures 24 hours before the failure.

Given the presence of a small glacier and an extensive cover of seasonal snow within the detachment zone, the unusually warm temperatures would have melted firn and produced abundant melt water. McSaveney (2002) suggested that refreezing of this melt water on a clear cold night may have been sufficient to trigger the failure. Freezing could have blocked movement of melt water in the bedrock, causing rapid pore pressure variations.

As the lower part of the detachment zone was probably characterised by warm or discontinuous permafrost, melt water may have penetrated discontinuities in bedrock that were previously ice filled, but thawing as a result of twentieth-century warming in the region (Salinger et al., 1995).

Vampire Peak 2008 and Mount Dampier 2010

Similar to the much larger rock-ice avalanche occurring from Aoraki/Mt Cook, more recent failures from Vampire Peak and Mt Dampier (see Figure 9.1) also exhibit detachment zones likely to be characterised by permafrost in a warm or marginal state, particularly towards the lowermost parts of these zones. The failure of Vampire Peak from a steep (>70°), shaded southeast-facing scarp slope of the Main Divide Fault Zone (Figure 9.4a) was inferred from seismic records to have initiated as two distinct pulses of rockfall late on 7 January 2008 (Cox & Allen, 2009). There was no glacial ice within the detachment area, yet running water was clearly visible seeping from the failure zone, positioned between 2380 and 2520 m (Figure 9.4b). A rock failure volume of 150,000 m^3 was calculated, doubling in volume as the avalanche travelled 1.7 km down through an icefall, depositing onto the Mueller Glacier as a series of distinct lobes (Cox & Allen, 2009). Based on a lapse rate of 0.0055°C/m, MAGST is estimated to have ranged from −0.3°C to −1.1°C, dependent on aspect and solar exposition across the failure zone (modified from Allen et al., 2009). The Mount Dampier failure occurred from a much higher-elevation (3200–3400 m), sunny, northwest-facing slope. A large proportion of hanging glacial ice detached in the initial failure (Figure 9.4c), and subsequent erosion of steep glacial ice occurred throughout the 4.2-km avalanche path (Figure 9.4d). The timing of the failure is less certain, with satellite images confirming it occurred between 19 February and 7 April 2010. Helicopter pilots first noticed and photographed the avalanche deposit on 8 April and, given the regularity with which tourist flights operate in this area, it is considered highly likely that

Figure 9.4 (a) Cumulative Vampire peak rock avalanche depositing across the Mueller Glacier, with 2008 detachment zone indicated with the yellow arrow, and adjacent scar from the 2003 failure indicated with the red arrow. (b) Enlarged view of the 2008 detachment zone, showing water seeping from rock discontinuities. (c) View of the 2010 failure from Mt Dampier (note the area of hanging glacier removed at the base of the failure). (d) Failure path showing substantial erosion down through a steep icefall. (Photos courtesy of D. Bogie (a,b) and T. Delaney (c,d).)

Figure 9.5 Maximum daily temperature trends from the Mt Cook village climate station (765 m) extrapolated to the elevation at the bottom of the failure zones on Mt Cook (2900 m), Vampire Peak (2380 m) and Mt Dampier (3200 m) recorded in the 6–8 weeks before these large rock-ice avalanches. Dashed lines represent extremely warm (95th percentile) maximum daily temperature limits established from a 1961–1990 reference period extrapolated to these same elevations. The timing of the avalanches is indicated by the arrows. Qualitative accounts strongly suggest the Mt Dampier event occurred in the week before 8 April, but a smaller dashed arrow indicates the wider possible time frame that has so far been constrained by available quantitative data. Note that for the Mt Cook and Mt Dampier events temperatures immediately before failure probably dropped to freezing levels after days to weeks of unusually warm conditions.

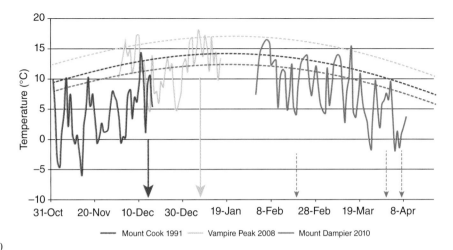

the failure occurred in this first week of April. The failure removed a large (~200 m high × 100 m wide) slab of rock, dipping parallel to the 50–60° slope. Based on an average bedding thickness of 20 m, a rock failure volume of 500,000 m³ is estimated. MAGST for exposed bedrock in this instance is estimated to have ranged from −0.2°C at the base (3200 m) to −1.8°C at the top (3400 m) of the detachment zone. This generalised thermal regime is complicated by the presence of a polythermal glacier at the base of the detachment zone, with associated latent heat from surface melting probably producing temperatures near 0°C at the ice–rock interface.

Comparing the air temperature patterns before the Mt Dampier, Vampire Peak and Mt Cook failures reveals that extremely warm temperatures (>95th percentile) preceded all three instances (Figure 9.5). Although the exact timing of the Mt Dampier failure cannot be confirmed, extremely warm temperatures were experienced throughout February and March 2010, with mean daily maximum air temperature during these months being 4.1°C and 2.6°C, respectively, above the longer-term (1961–1990) average. In contrast, the Vampire Peak, and Mt Cook failures have occurred within 12–48 hours after much shorter-term periods of extreme temperature. The fact that the Mt Dampier failure did not occur similarly as a more immediate response to extreme temperatures in early February highlights the difficulties in quantitatively linking warming with instability. On different timescales, warming may operate as a gradual precursor or a more direct trigger of slope failure (see Discussion – Figure 9.15), yet the physical processes involved are not yet sufficiently understood, e.g. a rock mass with numerous large open discontinuities should facilitate much more rapid warming at depth relative to a less fractured rock mass. If the Mt Dampier failure did in fact occur within the first week of April, it should be noted that over 300 mm of precipitation was recorded between 22 March and 4 April, followed by rapid temperature cooling to sub-freezing temperatures. Hence, rain or snowmelt may have been the trigger, following a gradual reduction in rock strength as ice-filled joints thawed during the extremely warm preceding months.

Alaskan case histories

Some researchers have highlighted recent widespread landsliding in several mountain ranges in Alaska, some related to earthquakes (Jibson et al., 2006), others to volcanic activity (Caplan-Auerbach and Huggel, 2007; Huggel et al., 2007) and still others with no obvious external trigger (Arsenault and Meigs, 2005; Huggel et al., 2008b). Our focus here is a series of recent, large rock and ice avalanches in the Bagley Ice Field region of south-east Alaska which have intriguing relations to air and ground temperature conditions.

Mt Steller 2005

The largest of a number of rock and ice avalanches in the Bagley Ice Field occurred at Mt Steller on 14 September 2005. Mt Steller (3236 m asl) is part of Waxell Ridge, which separates the Bagley Ice Field from Bering Glacier (Figure 9.6). The avalanche initiated from the south face of Mt Steller, close to the summit. Observations made during an overflight a day after the failure suggest that the initial failure was in bedrock, between 2500 and 3100 m asl. The south flank of Mt Steller is formed of Tertiary sedimentary rocks which dip sub-parallel to the slope. Significant parts of the ice mass on the summit ridge also broke off or were entrained by the rock slope failure. The initial volume of failed rock has been estimated to be $10\text{--}20 \times 10^6\,\text{m}^3$; an additional $3\text{--}4.5 \times 10^6\,\text{m}^3$ of glacier ice also failed (Huggel et al., 2008b). The avalanche travelled 9 km horizontally and 2430 m vertically. It entrained up to a few tens of millions of cubic meters of glacier ice, snow and debris along its path, resulting in a total volume of $40\text{--}60 \times 10^6\,\text{m}^3$.

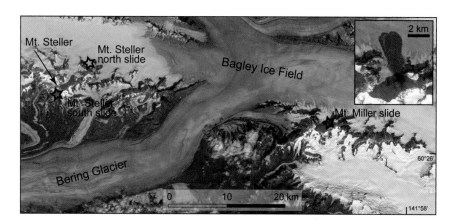

Figure 9.6 Landsat satellite image (9 September 2007) showing the Bagley Ice Field regions and the locations of the landslides in 2005 and 2008.

Evidence of water flowing on exposed bedrock in the detachment zone immediately after the failure raised the question of to what extent the slope failure was triggered by melting of glacier ice or thawing of permafrost. Analysis of daily temperatures derived from radiosonde measurements at Yakutat, Alaska indicates that temperatures in the summit region of Mt Steller were above freezing for about 10 days before the avalanche (Huggel, 2009). An assessment of mean annual ground surface temperatures indicates that permafrost on the south face of Mt Steller should be cold. However, modelling studies show that the summit ice cap is probably polythermal, in which case it induced a thermal anomaly to the ground, warming the underlying bedrock by several degrees Celsius (Huggel et al., 2008b).

Although the trigger of the avalanche is not known with certainty, the thermal state of the hanging glacier and bedrock, in combination with extremely warm air temperatures, could have generated melt water that infiltrated fractures in the summit rock mass, destabilising it. Melt water flowing at the base of the summit glacier probably reduced the shear strength of the rock mass and promoted its failure.

Mt Steller 2008

Two landslides occurred from Mt Steller in July 2008 (Figure 9.6). One detached from the north face of the ridge that extends east from the summit. The north flank of Mt Steller is heavily glacierised, with few bedrock outcrops (Figures 9.7 and 9.8). The landslide initiated in steep ice about 100 m below the summit ridge at about 2350 m asl. The failure extended into bedrock beneath the ice. High-resolution satellite images taken on 21 June 2008 show a developing slope instability in the bedrock beneath the steep glaciers on the north face of the mountain, as well as a precursory smaller rock avalanche (Figure 9.8). The main landslide occurred before 9 July, when an overview flight of Mt Steller was made (Figure 9.9).

Figure 9.7 Two rock-ice avalanches from Mt Steller region running out on the Bagley Ice Field. View is towards southeast. (Photo courtesy of C. Larsen taken on 24 July 2008.)

Figure 9.8 Left: three-dimensional view based on a high-resolution satellite image of the source area of the 2008 Mt Steller rock-ice avalanche taken on 21 June 2008 (© GoogleEarth), showing a precursory rock slide. Right: the main failure (24 July 2008). (Photo courtesy of C. Larsen.)

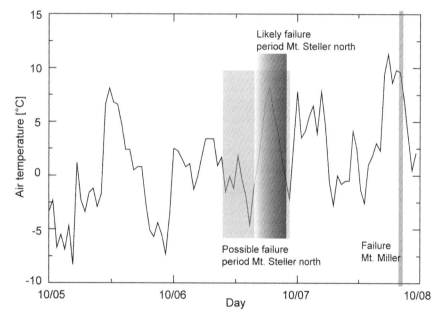

Figure 9.9 Air temperature record for the period between 10 May and 10August 2008, based on radiosonde data from Yakutat, extrapolated to 2000 m asl, the approximate elevation of the failure zones of the Mt Steller north and Mt Miller landslides. The Mt Miller slide was on 6 August 2008, whereas the exact day of failure of the Mt Steller north landslides is unknown.

It is difficult to assess from post-event imagery the depth of the detachment surface. The upper part of the scarp has a near-vertical failure plane in sedimentary rock. The exposed rock of the scarp is highly jointed and several large joint systems run across the scarp. Given the evidence for instabilities in bedrock before the main failure, it is likely that the landslide had its origin in a rock-slope instability, rather than in glacier ice. Nevertheless, thermal instability in rock at the base of the ice may have played an important role (Haeberli et al., 1997; Huggel, 2009).

We estimate that the volume of ice that failed and became entrained in the avalanche to be 400,000–700,000 m^3, based on a scarp length of 400 m and width of 200 m and an estimated ice thickness of 5–10 m (see Figure 9.8). The amount of bedrock that failed is difficult to assess, but a rough estimate can be derived from the avalanche deposit volume, which is about 1×10^6 m^3. Deposits showed predominantly shattered rock with single large blocks of several metres' diameter.

A second landslide occurred at about 1950 m asl, about 80 m below the crest of a rock ridge on the north-east-facing flank of Mt Steller and less than 2 km from the first landslide. It overrode the distal part of deposits of the first avalanche. The rock slope failed in two stages: a larger failure occurred before 9 July and involved the upper part of the failure zone; a smaller one involved detachment of a rock mass from the lower part of the failure zone and occurred between 20 and 24 July. The maximum run-out of the landslide is 2.2 km, and its deposit area is about 0.45 km^2. The estimated volume is $1–1.5 \times 10^6$ m^3, based on field observations of deposit thickness of 1–3 m and 3–5 m, respectively, in the lower and middle parts of the deposit. Only small amounts of glacier ice were involved in this landslide.

MAGST at the 2008 failure sites on the north flank of Mt Steller can be estimated using the thermal approach of Huggel et al. (2008b). MAAT derived from troposphere temperature data provided by the Yakutat radiosonde, 220 km south-east of the landslides, was used, together with a vertical gradient of 0.0065°C/m, to calculate MAGST. MAAT derived from the radiosonde data was increased by 1°C for northern aspects and by 3°C for southern aspects (Haeberli et al., 2002; Gruber et al., 2004b; Huggel et al., 2008b). The analysis yields MAGST of −2.5°C at 2000 m asl on the north-facing slope at the location of the first landslide (Figure 9.10). Exposed bedrock surface temperatures at the location of the first slide may have been in the range −1.5 to −3.5°C. However, the failure zone was largely covered by glacier ice, which was probably polythermal given the prevailing air temperatures. Thus, at least some parts of the failure area were close to the freezing point. Modelling studies have shown the thermal effect of polythermal glaciers on underlying bedrock can be complex (Haeberli et al., 1997; Wegmann et al., 1998; Huggel et al., 2008b), but water was probably present at the base of the glacier and possibly within fractures in the underlying rock. Close-up photographs of the failure zone a few days after the avalanche show extensive water flowing on the scarp that was probably produced in part by melting of ice and snow, but may to a lesser extent have its origin in rock fractures. Liquid water has similarly been observed on exposed detachment surfaces of landslides at high elevations in the Alps (Gruber & Haeberli, 2007; Huggel, 2009; Fischer et al., 2010).

The inferred thermal conditions at the site of the second landslide are different. The upper end of the failure zone is some 300 m lower than that at the first site and the site has a north-east aspect. Based on the analysis outlined above, MAGST of the failure zone of this landslide may be between −1 and −2°C, with warm permafrost close to thawing. Here bedrock is largely exposed, rather than covered by glacier ice, although the thermal effect of the snow cover must also be taken into account. As in the case of the first landslide, water was visible on the exposed scar and may have flown from rock fractures.

To assess the thermal effects of weather conditions days and weeks before the Mt Steller avalanches, we extrapolated 700 mbar level temperatures (~3000 m asl) of the Yakutat radiosonde to a 2000 m asl level, which is the height of the failure zones of the two landslides. The record shows an approximately 10-day period of very warm temperature (up to 8°C) at the end of May (Figure 9.9), temperature fluctuations around the freezing point in June, pronounced warming at the beginning of July with a maximum temperature of >8°C on July 4, and a sudden drop in temperature culminating in temperatures below 0°C 5 days later. Only 2 days later, on 11 July, temperatures were again up to approximately 8°C (Figure 9.9). If our analysis of the date of the landslide is correct (i.e. one to a few days before 9 July), failure occurred either during the very warm period with temperatures far above freezing, or during the sudden temperature drop at the end of the first week of July. The pattern is similar to that at the 1991 Aoraki/Mt Cook, the 2008 Vampire Peak and the 2010 Mt Dampier events (compare Figure 9.5).

Mt Miller 2008

A large rock-ice avalanche occurred on the north slope of a ridge near Mt Miller within 1 month of the 2008 Mt Steller failures (Figure 9.10). The summit ridge elevation at the site of the landslide is about 2400 m, and the failure zone extends from about 2200 m to 1600 m. The avalanche ran out 4.5 km on to the Bagley Ice Field, coming to rest at an elevation as low as 1290 m (Figure 9.11) The Fahrböschung (ratio of horizontal run-out to vertical drop – H/L) is 0.2, which is low but within the range of similarly sized ice and rock avalanches (Legros, 2002; Huggel

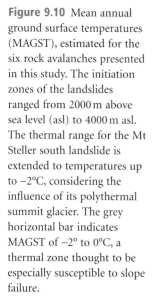

Figure 9.10 Mean annual ground surface temperatures (MAGST), estimated for the six rock avalanches presented in this study. The initiation zones of the landslides ranged from 2000 m above sea level (asl) to 4000 m asl. The thermal range for the Mt Steller south landslide is extended to temperatures up to −2°C, considering the influence of its polythermal summit glacier. The grey horizontal bar indicates MAGST of −2° to 0°C, a thermal zone thought to be especially susceptible to slope failure.

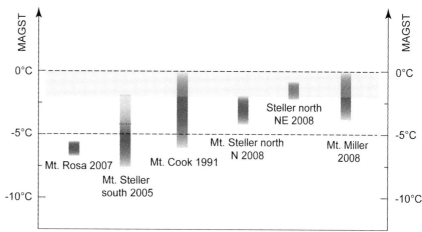

Figure 9.11 (a) The large rock-ice avalanche from the Mt Miller region on the Bagley Ice Field. (b) Close-up view of the failure zone showing the thickness of glacier ice involved in the rock-ice avalanche. (photos courtesy of C. Larsen).

et al., 2007). The landslide was recorded at about 20:25 UTC (11.25am local time) on 6 August 2008, at a seismic station 10 km to the east of the slide. The site was visited a few days after the landslide, and photos of the failure and deposit were taken from the ground and air. The deposit had a mean thickness of 3–5 m, which yields a total volume of $16–28 \times 10^6 \, m^3$.

Inspection of the scarp confirmed that large amounts of glacier ice and bedrock were involved in the landslide, with initial failure probably being in bedrock. The steepest section of the scarp flank revealed 50–80 m of glacier ice overlying exposed bedrock. The scarp extends several tens of metres into the bedrock which consists of relatively intact basalt.

MAGST was estimated in the same way as for the Mt Steller landslides using the Yakutat radiosonde data, 180 km to the south-east of the landslide site. The failure zone is north–north-west facing and therefore the MAGST of −2.5°C at 2000 m asl determined for the north-facing slope at Mt Steller should be applicable to Mt Miller. The range of MAGST of the failure zone (about 2200–1800 m asl) is −3.7 to −1.3°C, suggesting that relatively warm permafrost conditions existed in the lower part of the zone. However, the exact elevation range of the failure zone is difficult to determine because significant snow and ice were entrained along the path. Even more difficult is identification of the precise area where the failure started. Notwithstanding these difficulties, it is likely that areas at the transition of frozen to non-frozen bedrock were involved in the failure.

No liquid water was observed on the exposed bedrock of the Mt Miller landslide scarp at the time of the field visit, during which air temperatures were around the freezing point at the elevation of the scarp. Photographs taken 4 days after the landslide suggested that the upper end of the failure zone was frozen to the glacier bed.

We based our analysis of meteorological conditions days and weeks before the Mt Miller failure on an extrapolation of the 700 mbar level temperatures from the Yakutat radiosonde to an elevation of 2000 m, which we consider to represent thermal conditions in the middle to upper part of the failure zone. Temperature increased from −2.5°C on July 27 to over 11°C on 2 August 2008. This very warm period was maintained until the day of the failure on 6 August, but then dropped to freezing. The air temperature data thus indicate that the entire slope where the landslide occurred was in a melting state for several days before failure.

The long-term effect of the glacier on the underlying bedrock must also be considered. Based on the long-term radiosonde record, MAAT at the elevation of the failed glacier is about −5°C, and accordingly firn temperatures should be temperate. These are conditions where seasonal melting is possible and latent heat effects from refreezing of melt water can significantly warm the ice (Suter et al., 2001). Thermal modelling studies have shown that the thermal anomaly produced by freezing of water at the base of a glacier can penetrate tens of metres into underlying bedrock (Wegmann et al., 1998; Huggel et al., 2008b).

Monte Rosa, Alps

The east face of Monte Rosa extends from about 2200 m to over 4600 m asl and was the site of two spectacular avalanches in 2005 and 2007 (Fischer et al., 2006). The slope increases upward, reaching over 55° in exposed gneissic bedrock sections and over 40° in sections with glaciers (Figure 9.12). Studies based on sequential historical photographs have shown that the ice cover on the east face of the mountain changed little during the 20th century until about 1980, when it began to rapidly decrease (Haeberli et al., 2002; Fischer et al., 2012). Slope instability involving both ice and rock increased around 1990 and has continued to the present (Fischer et al., 2006). Instability culminated in two large avalanches, one on 25 August 2005, and the other on 21 April 2007.

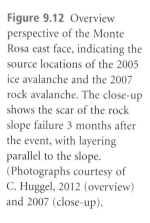

Figure 9.12 Overview perspective of the Monte Rosa east face, indicating the source locations of the 2005 ice avalanche and the 2007 rock avalanche. The close-up shows the scar of the rock slope failure 3 months after the event, with layering parallel to the slope. (Photographs courtesy of C. Huggel, 2012 (overview) and 2007 (close-up).

The 2005 event was a large ice avalanche ($1.1 \times 10^6 \, m^3$) that started from a steep glacier terminating at 3500 m asl and reached the foot of the face, where a large supraglacial lake had formed in 2002 but had drained in 2003 (Figure 9.12). Had the lake still existed, the avalanche would have generated a displacement wave with catastrophic consequences for the downstream community of Macugnaga. The avalanche occurred at night, which probably prevented injuries to tourists who often spend daylight hours on the pasture that was affected.

The 2007 event was a rock avalanche that detached from exposed bedrock at approximately 4000 m asl near the top of the east face of Monte Rosa (Figure 9.12). It involved about $0.3 \times 10^6 \, m^3$ of rock that fell to the base of the slope, again impacting the area of the former supraglacial lake.

Two meteorological stations at elevations over 3000 m are near Monte Rosa – a station at Testa Grigia, Italy (3488 m asl), 15 km to the west, and the Swiss station at Gornergrat (3130 m asl), 9 km to the north-west. The Testa Grigia station has a temperature record for 1951–2000 but measurements unfortunately were not continued after 2000, whereas the station at Gornergrat came into operation only in 1994 and has been working properly since then. Based on these data, temperature extrapolations yield a MAAT of −5° to −6°C at the lower end of the ice avalanche failure zone, suggesting a cold glacier front, but probably polythermal to temperate conditions at some distance behind the front. Previous studies show that the lower limit of permafrost is at about 3000 m asl on north- to north-east-facing slopes, and up to 500 m higher on east-facing sections (Zgraggen, 2005; Fischer et al., 2006; Huggel, 2009).

Estimates of MAGST for the 2007 rock slope failure, based on data from the aforementioned meteorological stations and rock temperature loggers deployed on the east face (Zgraggen, 2005), suggest temperatures of about −6°C. These conditions compare with those on Mt Steller south, but, unlike Mt Steller, the Monte Rosa site was not thermally perturbed by overlying glacier ice. A more likely destabilization factor is the enormous loss of ice at the base of the failure zone over the past 20 years with a volume of more than $20 \times 10^6 \, m^3$, which probably caused significant changes to the stress and temperature fields (Fischer et al., 2012). In addition, the dip of the foliation in the gneissic bedrock is parallel to the surface slope, adversely affecting slope stability.

Weather in the days and weeks before the landslides was very different for the 2005 and 2007 slides. Several warm periods of 5–10 days' duration occurred in June and July 2005. Temperatures rose to 5°C above the 1994–2009 average (of the Gornergrat station record). The warm periods were interrupted by temperature drops, with several freeze–thaw cycles during the 20 days before the failure (Figure 9.13). After the last freezing event 4 days before the landslide, temperatures again increased to 5°C on the day of failure. Much melt water was produced during the warm periods and possibly penetrated to the base of the steep glacier, lowering the strength at its contact with the underlying bedrock. The repeated cycles of melt and refreezing may have also destabilised the bedrock.

Temperatures in April 2007 were extraordinarily warm in central Europe, producing a spring heat wave. April temperatures at Jungfraujoch at 3580 m asl were

Figure 9.13 Summary of air temperatures up to 40 days before the failures. All temperatures are extrapolated to the elevation of the respective failure zone, using regional lapse rates (0.55–0.65°C/100 m). Air temperatures for all landslides except the April 2007 Monte Rosa slide fluctuated around the freezing point, with extended warm periods. Several of the slides show a rapid drop in temperature after a warm period and immediately before failure. The 2007 Monte Rosa slide occurred under considerably lower temperatures than the other slides. However, the last approximately 10 days before this failure are over 1 standard deviation above the 1994–2009 mean.

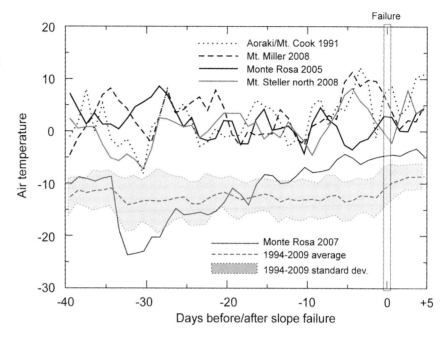

up to 5°C above one standard deviation of mean of the previous 49 years. The temperature of −3.5°C, 1 day before the landslide, is in the 98th to 99th percentile of the long-term April record (for Jungfraujoch available since 1958) (Figure 9.14). The Jungfraujoch climate record correlates highly with the record at the Gornergrat station, which is the nearest station above 3000 m asl to the failure site and reveals interesting thermal patterns during the weeks before the failure. Exactly 1 month before the landslide, temperatures dropped to an estimated −24°C at the failure site, which is about 10°C lower than the long-term average (see Figure 9.13). After this unusually cold period, temperature rose steadily to about −5°C near the date of the failure. However, radiation on the east face of Monte Rosa in April is high and cloud cover was generally low in April 2007. We thus infer that snow and ice melted at the surface in spite of the sub-freezing air temperature. In addition, it is possible that thermal energy during particularly warm summer months in 2003 and 2006, when the 0°C isotherm was above 4000 m asl, penetrated into bedrock some metres deep at the level of the rock slope failure.

Assessing changes in warm event frequencies based on RCM simulations

Several studies indicate that warm extremes increased during the twentieth century (Alexander et al., 2006) and are likely to further increase during the

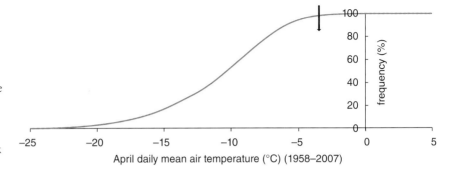

Figure 9.14 Frequency distribution of daily mean temperatures for April 1958–2007, Jungfraujoch station (3580 m above sea level). The mean temperature for April 2007 was 5°C warmer than that of the previous 49 years. The temperature of −3.5°C (black arrow) for 20 April 2007 (1 day before failure) plots in the 98th to 99th percentiles of the long-term record.

twenty-first century (Beniston et al., 2007). The Intergovernmental Panel on Climate Change (IPCC) defines an 'extreme weather event' as a rare event lying outside the 90th, 95th or 99th (or 10th, 5th, 1st, respectively) percentile of a statistical reference distribution (Trenberth et al., 2007). Extreme events also can be defined in terms of their intensity or severity of damage (Beniston et al., 2007). In terms of temperature, analyses of trends often use indices such as the number of daily maximum or minimum temperatures, or monthly to seasonal maximum temperatures (Aguilar et al., 2005; Schär et al., 2004).

Here, we consider whether the frequency of unusually warm periods of a few days to a couple of weeks might increase in the future. We explore this issue using the results from the most recent regional climate model (RCM) simulations. RCMs are currently among the most comprehensive tools to project climate on regional scales. The recently completed EU programme ENSEMBLES (van der Linden & Mitchell, 2009) ran a large number of RCMs over Europe which now serve the climate-impacted community with an ensemble of state-of-the-art regional climate simulations. To evaluate and express uncertainty, the ENSEMBLES simulations include a large number of RCMs, driven by different general circulation models (GCMs) with identical boundary settings. The RCMs were run with 25- or 50-km horizontal resolution for the period 1951–2050 (some until 2100). The SRES Scenario A1B (Nakicenovic & Swart, 2000) was applied for all model runs. Scenario A1B represents future conditions of rapid economic growth and introduction of more efficient technologies, with a balance between fossil and non-fossil energy production.

For the present study, we analysed results for air temperature 2 m above ground from eight ENSEMBLES RCM simulations. We chose a mix of different RCMs, driven with different GCMs, to provide a representative selection:

- HIRHAM_ARPEGE (from Danish Meteorological Institute – DMI)
- HIRHAM_ECHAM5 (from DMI)
- ETHZ-CLM_HADCM3Q (from Swiss Federal Institute of Technology – ETHZ)
- HIRHAM_HADCM3Q (from Norwegian Meteorological Institute – METNO)
- HIRHAM_BCM (from METNO)
- HIRAC_BCM (from Swedish Meteorological Institute – SMHI)
- HIRHAM_HADCM3Q (from SMHI).

The first part of each acronym represents the RCM, whereas the second part refers to the driving GCM.

Mean daily temperature results were analysed for anomalously warm temperature events in both the past and the future. The analysis is based on the one-grid box that represents the longitudes and latitudes of the Jungfrau region. The Jungfrau region was chosen based on the availability of long-term high-elevation observational time series (since 1958) from the Swiss Federal Institute of Meteorology and Climatology (MeteoSwiss), which enables performance analysis and de-biasing of the RCM simulations. Typically, measured air temperatures at high-altitude climate stations are highly correlated. The temperature records at the Jungfraujoch and Gornergrat stations have a correlation coefficient of 0.98 for the common period of record. Therefore, we assume that the Jungfraujoch data also are representative for conditions at or near Gornergrat (including Monte Rosa).

A horizontal resolution of 25 km, which is now the standard for many RCM simulations, is too coarse to represent the topography of a high mountain region realistically. The selected grid box is referenced to an elevation of 2244 m asl and therefore needed to be adjusted to the level of Jungfraujoch at 3580 m asl, applying a basic lapse-rate correction of 0.006°C/m (as an average over the year; Rolland, 2003) to adjust the air temperature of the RCM grid box with an elevation of 2244–3580 m, the elevation of the Jungfraujoch climate station.

In addition to the elevation adjustment, we applied a bias correction to each RCM time series (Salzmann et al., 2007a, 2007b). The bias relates to the difference between the mean annual temperature observed at Jungfraujoch for the available time period 1960–2000, and the respective temperature of the RCM time series (Table 9.1). The bias correction involves only one temperature value per RCM; different values for different seasons or months were not used.

Table 9.1 Summary of changes (expressed as factor) in warm air temperature anomalies between 1951–2000 and 2001–2050, based on eight regional climate models (RCMs)

Regional climate model	Δ30-day events	Δ10-day events	Δ5-day events	Bias (°C)
DMI-HIRHAM5_ARPEGE	5.5	1	0.9	−5.5
DMI-HIRHAM_ECHAM5	[a]	[a]	[a]	−3
ETHZ-CLM_HADCM3Q	10	2.2	1.5	−3
METNO-HIRHAM_HADCM3Q	[a]	2.5	1.2	−4.5
METNO-HIRHAM_BCM	[a]	[a]	[a]	−1
SMH-HIRAC_BCM	8	3.7	2.1	−2.5
SMH-HIRHAM_HADCM3Q	[a]	1.7	3.7	−6.5
MPI-REMO_ECHAM5	[a]	8.3	2.2	−5

A factor of 1 means that no change is projected between the two time periods, a factor of 10 foresees a 10 times increase for the future 2001–2050 period. The column 'Bias' refers to the difference between the observed mean annual temperature of Jungfraujoch for the available time period 1960–2000, and the respective temperature of the RCM time series.
[a]No event indicated either in the past or in the future.

The following analyses are based on the RCM time series with elevation adjustment and bias correction as described above. We studied eight RCM time series to identify periods with air temperature continuously exceeding a threshold of +5°C for periods of 5, 10 and 30 days, thus representing significant melting conditions. These thresholds are based on the case studies described above, where warm air temperature anomalies of many days' duration were observed before each failure. We analysed the change in frequency of these events for the period 2001–2050, as compared with the reference period 1951–2000. The results are shown in Table 9.1.

Results from the eight models show a clear increase in the frequency of warm air temperature events in the next several decades compared with the second half of the twentieth century (Table 9.1). For a matrix comprising eight models and three event types, only one model (DMI-HIRHAM5_ARPEGE) produces a slight decrease (~10%) of the 5-day events. The differences among the model outputs are large, but most models show increases in frequency of extreme events of about 1.5–4 times. Large increases in the frequency of warm extremes, by a factor of 8–10, are projected by three models. In two of these three cases, the increase is for 30-day events. On the other hand, five models show no 30-day extreme warm events, for either the past or the future.

Discussion and conclusions

Our analysis of temperature records days and weeks before several major high-mountain rock and ice avalanches provides new insights to a little investigated aspect of slope instability. A clear predictive thermal trigger is scarcely discernable for all events. However, examination of the similar temperature pattern before failure for our case studies suggests that some thermal mechanisms are repeated at locations as different and distant from one another as Alaska, New Zealand and the European Alps: (1) unusually warm air temperatures over several days during the weeks or days before failure and (2) sudden drops of temperatures, typically below freezing level, after warm periods and hours to days before failure.

All the studied events had warm temperatures, far above freezing level, before failure, except the 2007 Monte Rosa landslide. In most of the cases, temperatures above freezing level during summer months are not exceptional, although the observed temperatures were far above normal. At Mt Cook, for example, the peak temperature 3 days before failure was 8.5°C above the long-term average. Temperatures in the days before the April 2007 Monte Rosa rockslide were up to 4–5°C above one standard deviation of the long-term record. Such unusually warm periods initiate melt of surface snow and ice. The liquid water from melting processes can infiltrate rock slopes via fractures and joints, increasing hydrostatic pressures and thus reducing shear strength (Watson et al., 2004; Huggel et al., 2008b; Fischer et al., 2010). Photographs taken after the 2008 Mt Steller north slides show large amounts of water on the landslide scar, and some of the water appears to have seeped from bedrock discontinuities. Melt water can also penetrate to the base of steep glaciers and initiate gliding and/or failure processes.

Temperature pattern 2 has been inferred at the sites of the Mt Cook and the 2005 and 2008 Mt Steller events (Huggel, 2009), as well as at the site of a rock slide in permafrost in the eastern Swiss Alps, not discussed further here (Fischer et al., 2010). It did not occur, however, at the 2005 Monte Rosa ice avalanche and the 2007 Monte Rosa landslide. A sudden lowering of temperature may favour slope failure by refreezing the surface after infiltration of melt water into bedrock during the preceding warm period. Such 'lock-off' situations are difficult to quantify due to a lack of on-site measurements with piezometers and other instruments (Watson et al., 2004; Willenberg et al., 2008), but have been invoked in similar conditions (Fischer et al., 2010). In this context, we should also consider the importance of rainfall or melt of fresh snow as potential sources of infiltrating water and slope destabilisation. Accurate measurements of precipitation at sites of high-mountain slope failures rarely exist because of low technical practicability. The Gornergrat meteorological station is less than 10 km from Monte Rosa, but we did not use its precipitation data for this study because precipitation is highly variable in this region (Machguth et al., 2006). There is more certainty at Mount Cook, where total precipitation of 48 mm and 11 mm, respectively, was measured windward and leeward of Mount Cook in the week before failure; this amount is not significant in a region with common, much larger daily precipitation.

Another climate variable that was not examined in this study, but may have an important effect on surface warming and melting, is radiation. Short-wave radiation constitutes a major portion of the energy available for melt during summer and may be close to $200 \, W/m^2$ in alpine conditions (Oerlemans, 2001; Ohmura, 2001). In several of the cases presented here, short-wave radiation may have played a role in melting snow and ice, e.g. air temperatures reached only $-5°C$ before the failure at the site of the 2007 Monte Rosa landslide, but clear sky conditions resulted in high short-wave radiation that may have melted surface snow, with possible infiltration of water into the highly fractured bedrock. Radiation may also have played a role in the 2008 Steller and Miller landslides.

Our analysis concentrated on temperature aspects of large high-mountain slope failures, but this is only one component of a highly complex physical system which, in response to gradual and sudden changes in external and internal controls, produces a slope failure. Several geological factors, including structure and rock type, glaciation, permafrost, topography and seismicity, are important determinants of slope stability in this environment. It is fundamental in this context to consider the timescales involved in causative and trigger factors. Figure 9.15 shows two hypothetical histories of slopes before failure. Slope 1 has a higher initial shear strength corresponding, for example, in more stable geological or topographic conditions. Geology and topography are typical predisposing factors. The histories of both slopes are characterised by processes that gradually reduce

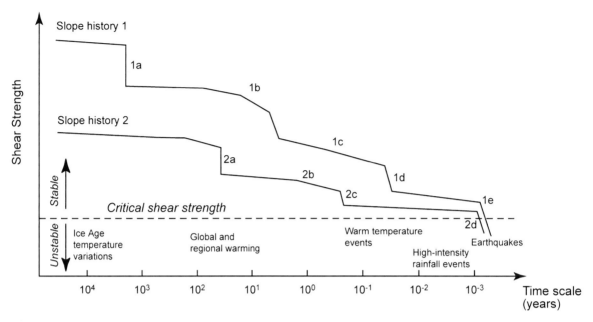

Figure 9.15 A hypothetical sketch of two slope histories, shown over a relevant period of time before slope failure. The dashed line indicates the critical shear strength threshold below which the slope is unstable and failure would occur. Both slope histories are characterised by processes that produce a gradual decrease in shear strength over long periods of time (e.g. warming at the close of the Pleistocene), and abrupt reductions in shear strength, due, for example, to seismic activity. Slope 2 has a lower initial shear strength due, for example, to rock type or structure. Processes that can cause abrupt reductions in shear strength, including warm extremes or high-intensity rainfall, act as slope failure triggers only if the shear strength is sufficiently low.

their shear strength over periods of decades to millennia. Short-term events, operating over days to weeks, such as warm extremes, high-intensity rainfall or earthquakes, can rapidly reduce the strength of the slope. However, these events may not necessarily trigger failure, depending on their impact and the shear strength of the slope when they occur. Slope failure only occurs if the potential triggering event reduces the shear strength below a critical threshold. These events will trigger a landslide only if slope stability is already low and near the threshold of failure.

The concept of effective timescales is particularly important when considering bedrock permafrost and slope stability. Thermal perturbations resulting from twentieth-century warming, for example, have now penetrated to depths of a few decametres in high rock slopes (Haeberli et al., 1997; Noetzli et al., 2007), whereas short warm extremes may have an effect at a few metres' depth only one or several years later. Open fractures can, however, facilitate infiltration of water into rock slopes and thus contribute to a much more immediate effect on slope stability.

Based on an analysis of the ENSEMBLE RCM simulations, we have found that short periods with very high temperatures may increase in frequency by a factor of 1.5–4 times the next several decades compared with the 1951–2000 reference period. Short-duration warm periods may produce a critical input of liquid water into slopes. The projected increase in extended warm periods of up to 1 month is also of concern because such events can lead to substantial thermal perturbation of subsurface hydrology.

The large range in the RCM results suggests that it is difficult to provide robust regional climate projections, even when using the same SRES scenario and the most advanced RCMs. Nevertheless, the projected increase in the frequency of short-term temperature extremes is consistent with earlier findings on heat waves on the global (Meehl et al., 2007) and European scales at the end of the twenty-first century, based on the earlier generation of PRUDENCE RCMs (Beniston et al., 2007). Although not investigated here, several studies have documented an increase in warm extremes, including summer heat waves, during the twentieth century, on global and regional scales (Alexander et al., 2006; Hegerl et al., 2007). However, the limited availability of high-quality and homogenised climate station data is an important constraint on a more detailed analysis of extreme events (Hegerl et al., 2007).

We have performed the analysis of future warm extremes for the Alps, motivated by the recent completion of the ENSEMBLES project which made RCM data available for a range of models and thus allows for a better consideration of uncertainties involved. Results found for Europe cannot directly be transformed to our other study regions in Alaska and New Zealand. However, global assessments show similar tendencies (Meehl et al., 2007), and we therefore assume that our conclusions are broadly valid for these regions as well. Furthermore, similar RCM programmes such as ENSEMBLES are currently ongoing in other parts of the world: North America: NARCCAP (Mearns et al., 2009); South America: CREAS (Marengo & Ambrizzi, 2006). In addition, with the World Climate Research Programme's CORDEX initiative, RCM simulations will be available for all continents and allow regional-scale scenario analysis worldwide.

Several aspects of the role of warm extremes in high-mountain slope stability remain unresolved, but this study will hopefully stimulate further discussion and research. As argued in Figure 9.15, not every warm extreme will trigger a large slope failure. However, we think that large slope failures will increase in temperature-sensitive, high-mountain areas as the number of warm extreme events increases.

The landslides described in the case studies did not cause any major damage to people or infrastructure, due partly to fortunate circumstances, partly to the remote location of the landslides. In densely populated and developed mountain regions such as the European Alps, however, serious consequences have to be considered from large slope failures. Cascading processes (e.g. landslides impacting natural or artificial lakes, generating outburst floods) are of particular concern. With the Monte Rosa case study it has been indicated that similar landslides as in 2005–2007 would probably have resulted in a major disaster had they occurred during the existence of a large glacier lake in 2002–2003.

Acknowledgements

We gratefully acknowledge comments by, and collaboration with Jackie Caplan-Auerbach, Luzia Fischer, Isabelle Gärtner-Roer, Wilfried Haeberli, Ruedi Homberger, Chris Larsen, Bruce Molnia, Gianni Mortara, Demian Schneider, Manuela Uhlmann and Rick Wessels, including provision of photographs, satellite images. Information on New Zealand slope failures has been compiled in collaboration with Simon Cox. John Clague and Bill McGuire made very useful comments on the manuscript.

References

Aguilar, E., Peterson, T. C., Ramírez Obando, P., et al. (2005) Changes in precipitation and temperature extremes in Central America and northern South America 1961–2003. *Journal of Geophysical Research* **110**, D23107.

Alean, J. (1985) Ice avalanches, some empirical information about their formation and reach. *Journal of Glaciology* **31**, 324–333.

Alexander, L. V., Zhang, X., Peterson, T. C., et al. (2006) Global observed changes in daily climate extremes of temperature and precipitation. *Journal of Geophysical Research* **111**, D05109.

Allen, S. K., Gruber, S. & Owens, I. (2009) Exploring steep bedrock permafrost and its relationship with recent slope failures in the Southern Alps of New Zealand. *Permafr P P* **20**, 345–356.

Allen, S., Cox, S. & Owens, I. (2010) Rock-avalanches and other landslides in the central Southern Alps of New Zealand: A regional assessment of possible climate change impacts. *Landslides*. DOI 10.1007/s10346–010–0222-z

Arsenault, A. M. & Meigs, A. J. (2005) Contribution of deep-seated bedrock landslides to erosion of a glaciated basin in southern Alaska. *Earth Surface* **30**, 1111–1126.

Barla, G., Dutto, F. & Mortara, G. (2000) Brenva glacier rock avalanche of 18 January 1997 on the Mount Blanc range, northwest Italy. *Landslide News* **13**, 2–5.

Beniston, M., Stephenson, D. B., Christensen, O. B., et al. (2007) Future extreme events in European climate, an exploration of regional climate model projections. *Climate Change* **81**, 71–95.

Caplan-Auerbach, J. & Huggel, C. (2007) Precursory seismicity associated with frequent, large ice avalanches on Iliamna volcano, Alaska, USA. *Journal of Glaciology* **53**, 128–140.

Clague, J. J. & Evans, S. G. (2000) A review of catastrophic drainage of moraine-dammed lakes in British Columbia. *Quaternary Science Reviews* **19**, 1763–1783.

Cox, S. C. & Allen, S. K. (2009) Vampire rock avalanches of January 2008 and 2003, Southern Alps, New Zealand. *Landslides* **6**, 161–166.

Etzelmüller, B. & Hagen, J. O. (2005) Glacier permafrost interaction in arctic and alpine environments – examples from southern Norway and Svalbard. In: Harris, C. & Murton, J. (eds), *Cryospheric systems – Glaciers and Permafrost*. British Geological Society, Special Publication 242, pp 11–27.

Fischer, L. & Huggel, C. (2008) Methodical design for stability assessments of permafrost-affected high-mountain rock walls. In: Kane, D. L. & Hinkel, K. M. (eds), *Proceedings of the 9th International Conference on Permafrost*, pp 439–444.

Fischer, L., Kääb, A., Huggel, C. & Noetzli, J. (2006) Geology, glacier retreat and permafrost degradation as controlling factors of slope instabilities in a high-mountain rock wall, the Monte Rosa east face. *Natural Hazards and Earth Systems Sciences* **6**, 761–772.

Fischer, L., Amann, F., Moore, J. & Huggel, C. (2010) The 1988 Tschierva rock avalanche (Piz Morteratsch, Switzerland), An integrated approach to periglacial rock slope stability assessment. *Engineering Geology* **116**, 32–43.

Fischer, L., Eisenbeiss, H., Kääb, A., Huggel, C. & Haeberli, W. (2012) Detecting topographic changes in steep high-mountain flanks using combined repeat airborne LiDAR and aerial optical imagery – a case study on climate-induced hazards at Monte Rosa east face, Italian Alps. *Permafr P P.*

Gruber, S. & Haeberli, W. (2007) Permafrost in steep bedrock slopes and its temperature-related destabilization following climate change. *Journal of Geophysical Research* **112**, F02S18.

Gruber, S., Hoelzle, M. & Haeberli, W. (2004a) Permafrost thaw and destabilization of Alpine rock walls in the hot summer of 2003. *Geophysical Research Letters* **31**, L13504.

Gruber, S., Hoelzle, M. & Haeberli, W. (2004b) Rock-wall temperatures in the Alps, modelling their topographic distribution and regional differences. *Permafr P P* **15**, 299–307.

Haeberli, W. & Hohmann, R. (2008) Climate, glaciers and permafrost in the Swiss Alps 2050, scenarios, consequences and recommendations. In: Kane, D. L. & Hinkel, K. M. (eds), *Proceedings of the 9th International Conference on Permafrost*, pp 607–612.

Haeberli, W., Wegmann, M. & Vonder Mühll, D. (1997) Slope stability problems related to glacier shrinkage and permafrost degradation in the Alps. *Eclogae geologicae Helvetia* **90**, 407–414.

Haeberli, W., Kääb, A., Paul, F., et al. (2002) A surge-type movement at Ghiacciaio del Belvedere and a developing slope instability in the east face of Monte Rosa, Macunaga, Italian Alps. *Norsk Geografisk Tidsskrift* **56**, 104–111.

Harris, C., Arenson, L. U., Christiansen, H. H., et al. (2009) Permafrost and climate in Europe, Monitoring and modelling thermal, geomorphological and geotechnical responses. *Earth-Science Reviews* **92**, 117–171.

Hegerl, G., Zwiers, F., Braconnot, P., et al. (2007) Understanding and attributing climate change. In: Solomon, S., D., Qin, M., et al. (eds), *Climate Change 2007, The Physical Science Basis*. Contribution of Working Group I to the Fourth Assessment Report of the Intergovernmental Panel on Climate Change. Cambridge: Cambridge University Press, pp 663–745.

Heim, A. (1932) Bergsturz und Menschenleben. *Beiblatt zur Vierteljahrschrift der Naturorschenden Gesellschaft in Zürich* **77**, 218 pp.

Huggel, C. (2009) Recent extreme slope failures in glacial environments, effects of thermal perturbation. *Quaternary Science Reviews* **28**, 1119–1130.

Huggel, C., Haeberli, W., Kaab, A., Bieri, D. & Richardson, S. (2004) An assessment procedure for glacial hazards in the Swiss Alps. *Canadian Geotechnical Journal* **41**, 1068–1083.

Huggel, C., Caplan-Auerbach, J., Waythomas, C. F. & Wessels, R. L. (2007) Monitoring and modeling ice-rock avalanches from ice-capped volcanoes. A case study of frequent large avalanches on Iliamna Volcano, Alaska. *Journal of Volcanology and Geothermal Research* **168**, 114–136.

Huggel, C., Caplan-Auerbach, J. & Wessels, R. (2008a) Recent Extreme Avalanches, Triggered by Climate Change. *Eos Transactions AGU* **89**, 469–470.

Huggel, C., Caplan-Auerbach, J., Gruber, S., Molnia, B. & Wessels, R. (2008b) The 2005 Mt Steller, Alaska, rock-ice avalanche, A large slope failure in cold permafrost. In: Kane, D. L. & Hinkel, K. M. (eds), *Proceedings of the 9th International Conference on Permafrost*, pp 747–752.

Jibson, R. W., Harp, E. L., Schulz, W. & Keefer, D. K. (2006) Large rock avalanches triggered by the M 7.9 Denali Fault, Alaska, earthquake of 3 November 2002. *Engineering Geology* **83**, 144–160.

Kääb, A., Reynolds, J. M. & Haeberli, W. (2005) Glaciers and permafrost hazards in high mountains. In: Huber, U. M., Burgmann, H. K. H. & Reasoner, M. A. (eds), *Global Change and Mountain Regions – An overview of current knowledge*. Dordrecht: Springer, pp 225–234.

Legros, F. (2002) The mobility of long-runout landslides. *Engineering Geology* **63**, 301–331.

Machguth, H., Eisen, O., Paul, F. & Hoelzle, M. (2006) Strong spatial variability of snow accumulation observed with helicopter-borne GPR on two adjacent Alpine glaciers. *Geophysical Research Letters* **33**, L13503.

McSaveney, M. J. (2002) Recent rockfalls and rock avalanches in Mount Cook National Park, New Zealand. In: Evans, S. G. & DeGraff, J. V. (eds), *Catastrophic Landslides, Effects, Occurrence and Mechanisms*. Boulder, CO: Geological Society of America, Reviews in Engineering Geology, pp 35–70.

Marengo, J. A. & Ambrizzi, T. (2006) Use of regional climate models in impacts assessments and adaptation studies from continental to regional and local scales – The CREAS (Regional Climate Change Scenarios for South America) initiative in South America. Proceedings 8 ICSHMO, Foz do Iguaçu, Brazil, April 24–28 2006, INPE, pp 291–296.

Mearns, L. O., Gutowski, W., Jones, R., et al. (2009) A regional climate change assessment program for North America. *Eos Transactions AGU* **90**, 36, 311.

Meehl, G., Stocker, T., Collins, W., et al. (2007) Global climate projections. In: Solomon, S. D., Qin, M., Manning, Z., et al. (eds), *Climate Change 2007, The Physical Science Basis*. Contribution of Working Group I to the Fourth Assessment Report of the Intergovernmental Panel on Climate Change. Cambridge: Cambridge University Press.

Nakicenovic, N. & Swart, R. (2000) *Special Report on Emissions Scenarios, a special report of Working Group III of the Intergovernmental Panel on Climate Change.* Cambridge: Cambridge University Press.

Noetzli, J., Gruber, S., Kohl, T., Salzmann, N. & Haeberli, W. (2007) Three-dimensional distribution and evolution of permafrost temperatures in idealized high-mountain topography. *Journal of Geophysical Research* **112**, F02S13.

Oerlemans, J. (2001) *Glaciers and Climate Change.* Lisse: A.A. Balkema Publishers.

Ohmura, A. (2001) Physical basis for the temperature-based melt-index method. *Journal of Applied Meteorology* **40**, 753–761.

Quincey, D. J. & Glasser, N. F. (2009) Morphological and ice-dynamical changes on the Tasman Glacier, New Zealand 1990–2007. *Global and Planetary Change* **68**, 185–197.

Rolland, C. (2003) Spatial and seasonal variations of air temperature lapse rates in Alpine regions. *Journal of Climate* **16**, 1032–1046.

Röthlisberger, H. (1987) Sliding phenomena in a steep section of Balmhorngletscher, Switzerland. *Journal of Geophysical Research* **92**(B9), 8999–9014.

Romstad, B., Harbitz, C. & Domaas, U. (2009) A GIS method for assessment of rock slide tsunami hazard in all Norwegian lakes and reservoirs. *Natural Hazards and Earth System Sciences* **9**, 353–364.

Salinger, J. M., Basher, R. E., Fitzharris, B., et al. (1995) Climate trends in the South-West Pacific. *International Journal of Climatology* **15**, 285–302.

Salzmann, N., Kääb, A., Huggel, C., Allgöwer, B. & Haeberli, W. (2004) Assessment of the hazard potential of ice avalanches using remote sensing and GIS-modelling. *Norsk Geografisk Tidsskrift-Norwegian Journal of Geography* **58**, 74–84.

Salzmann, N., Frei, C., Vidale, P. & Hoelzle, M. (2007a) The application of Regional Climate Model output for the simulation of high-mountain permafrost scenarios. *Global and Planetary Change* **56**, 188–202.

Salzmann, N., Nötzli, J., Hauck, C., Gruber, S., Hoelzle, M. & Haeberli, W. (2007b) Ground surface temperature scenarios in complex high-mountain topography based on regional climate model results. *Journal of Geophysical Research* **112**, F02S12.

Schär, C., Vidale, P. L., Lüthi, D., et al. (2004) The role of increasing temperature variability in European summer heatwaves. *Nature* **427**, 332–336.

Sosio, R., Crosta, G. & Hungr, O. (2008) Complete dynamic modeling calibration for the Thurwieser rock avalanche (Italian Central Alps). *Engineering Geology* **100**, 11–26.

Suter, S., Laternser, M., Haeberli, W., Frauenfelder, R. & Hoelzle, M. (2001) Cold firn and ice of high-altitude glaciers in the Alps, measurements and distribution modelling. *Journal of Glaciology* **47**, 85–96.

Tebaldi, C., Hayhoe, K., Arblaster, J. M. & Meehl, G. A. (2006) Going to the extremes. *Climatic Change* **79**, 185–211.

Trenberth, K., Jones, P., Ambenje, P., et al. (2007) Observations, Surface and Atmospheric Climate Change. In: Solomon, S. D., Qin, M., Manning, Z., et al.

(eds), *Climate Change 2007, The Physical Science Basis.* Contribution of Working Group I to the Fourth Assessment Report of the Intergovernmental Panel on Climate Change. Cambridge: Cambridge University Press, pp 235–336.

van der Linden, P. & Mitchell, J. (2009) ENSEMBLES, Climate Change and its Impacts, Summary of research and results from the ENSEMBLES project, Met Office Hadley Centre, FitzRoy Road, Exeter EX1 3PB, UK.

Watson, A. D., Moore, D. P. & Stewart, T. W. (2004) Temperature influence on rock slope movements at Checkerboard Creek. In: *International Symposium on Landslides, Balkema, Rio de Janeiro.* pp 1293–1298.

Wegmann, M., Gudmundsson, G. H. & Haeberli, W. (1998) Permafrost changes in rock walls and the retreat of Alpine glaciers, a thermal modelling approach. *Permafr PP* **9**, 23–33.

Willenberg, H., Evans, K. F., Eberhardt, E., Spillmann, T. & Loew, S. (2008) Internal structure and deformation of an unstable crystalline rock mass above Randa (Switzerland), Part II – Three-dimensional deformation patterns. *Engineering Geology* **101**, 15–32.

Zgraggen, A. (2005) Measuring and modeling rock surface temperatures in the Monte Rosa East face. Master thesis, ETH Zurich/University of Zurich.

10

Impacts of recent and future climate change on natural hazards in the European Alps

Jasper Knight[1], Margreth Keiler[2] and Stephan Harrison[3]

[1]School of Geography, Archaeology and Environmental Studies, University of the Witwatersrand, Johannesburg, South Africa
[2]Geographical Institute, University of Bern, Switzerland
[3]College of Life and Environmental Sciences, University of Exeter, Penryn, UK

Summary

Climate and environmental changes associated with anthropogenic global warming are being increasingly identified in the European Alps and other mountain areas worldwide. Evidence for this is seen by changes in long-term high-alpine temperature, precipitation, glacier and snow cover, and permafrost thickness and temperature. These changes can be linked to a combination of both long-term climatic amelioration following the European Little Ice Age (about AD 1550–1850), and an acceleration of this warming trend caused by anthropogenic global warming. In turn, these changes impact on land surface stability and lead to increased frequency and magnitude of mountain natural hazards, including rock-falls, debris flows, landslides, avalanches and floods. These hazards also impact on ecosystems, infrastructure, and socioeconomic and cultural activities in mountain regions. This chapter presents two case studies from the European Alps (2003 heatwave, 2005 floods) that demonstrate some of the interlinkages between physical processes and human activity in climatically sensitive Alpine regions that are responding to ongoing climate change. Based on this evidence we outline future implications of climate change on mountain environments and its impact on hazards and hazard management in paraglacial mountain systems.

Introduction

Mountain systems are particularly sensitive to climate change. This comes about largely as a result of feedbacks associated with seasonal variations in high-elevation snow cover, whereby high albedo during winter leads to increased cooling, and

Climate Forcing of Geological Hazards, First Edition. Edited by Bill McGuire and Mark Maslin.
© 2013 The Royal Society and John Wiley & Sons, Ltd. Published 2013 by John Wiley & Sons, Ltd.

low albedo during summer leads to increased warming (Haeberli et al., 2007; Vavrus, 2007). Such climate amplification does not so strongly affect the seasonal climate variations and heat budgets of lowland areas, or Alpine valleys within the mountains where most human activity in these regions is concentrated. As a result mountain climates are highly spatially and temporally complex, and exhibit greater variability than the climates of surrounding lowlands (Calanca et al., 2006). Such climate amplification has meant that temperatures in the European Alps have increased twice as much as the global average since the late nineteenth century (Brunetti et al., 2009). In addition, precipitation and other variables have also changed non-linearly, with significant regional and seasonal differences, and differences by elevation and aspect (Auer et al., 2007; Haeberli et al., 2007; Schmidli et al., 2007). Regional changes in the distribution of snow, rainfall and temperature have implications for snow-cover thickness and duration (which also affect subsurface temperatures), and catchment run-off (Beniston et al., 2003; Vanham et al., 2008). Temperature and precipitation changes can be demonstrably linked to changes in glacier mass balance (including equilibrium line altitude) and terminus position, in particular at high elevations where climate amplification is most strong when snowblow and precipitation changes are most significant (Zemp et al., 2006; Lambrecht & Kuhn, 2007; Huss et al., 2008b; Steiner et al., 2008; Nemec et al., 2009). Permafrost monitoring sites throughout the Alps also show changes in Alpine permafrost distribution, subsurface temperature profile and active layer thickness that are specifically linked to climate amplification (Harris et al., 2003; Luetschg et al., 2008). Hazardous events related to variations in precipitation and temperature, such as snow avalanches, upland river floods and mass movements, have also increased in frequency and magnitude in many areas (e.g. Beniston et al., 2003, 2007; Kääb, 2008; Baggi & Schweizer, 2009; Hilker et al., 2009; Höller, 2009; Bollschweiler & Stoffel, 2010). This pattern of climate and geomorphic change is consistent across most mountain blocks worldwide.

Although the direct effects of climate forcing on these cryospheric systems have now been monitored for several decades in the European Alps (e.g. Laternser & Schneebeli, 2003; Paul et al., 2004; Beniston, 2005; Lambrecht & Kuhn, 2007; Huss et al., 2008a; Abermann et al., 2009), the indirect effects on geomorphological processes and on sedimentary systems are less well known. Based on records of past events, Alpine geomorphological hazards such as landslides, debris flows, mudflows and other expressions of slope instability can result from aspects of human activity and climate variability either alone or in combination (e.g. Gehrig-Fasel et al., 2007; Stoffel et al., 2008), e.g. debris flows can be initiated directly by heavy rainfall events, but also where the land surface is potentially vulnerable to such events by land use change, warm permafrost temperatures, soil saturation and sediment availability (Marchi et al., 2009; Bollschweiler & Stoffel, 2010). Although human activity and land management are important, it is now being acknowledged that ongoing climate change is exerting a more significant role in the generation of geomorphological hazards by influencing the operation of all landscape elements (e.g. Agrawala, 2007; López-Moreno et al., 2008; Stoffel et al., 2008; Liggins et al., 2010). Cryospheric responses to climate change in particular can give rise to 'downstream' geomorphological impacts such as hazardous events that represent periods of decreased land surface stability (Gude & Barsch, 2005; Stoffel & Beniston, 2006), e.g. a negative glacier mass balance and ice retreat contribute an increasing amount of glacigenic sediment and meltwater into outflowing streams, thereby increasing flood hazard risk. Furthermore, geomorphological processes in high-relief areas are strongly influenced by slope angle and aspect, sediment availability and slope moisture supply, and these processes (and their capacity to lead to hazards) evolve in a downslope direction, leading to high spatial and temporal variability in process domain and thus hazard risk (Marchi et al., 2009; Huggel et al., 2010).

As a result, the dynamics of and controls on geomorphological hazards under the effects of ongoing climate change are presently a major issue for landscape management and planning in the European Alps and similar mountain settings worldwide (Agrawala, 2007; López-Moreno et al., 2008; Huggel et al., 2010), particularly for areas of high population density in Alpine valleys. There is therefore an imperative for accurate monitoring and modelling of Alpine geomorphological processes and for hazard risk mapping in order to minimise potential impacts on human activity within sensitive Alpine landscapes both at the present time and into the future.

Future climate change can be predicted by global climate models (GCMs), but the downscaled results from different model scenarios highlight regional uncertainty in these predictions (e.g. Wanner et al., 2006; Allan & Soden, 2007; Schmidli et al., 2007; Reichler & Kim, 2008). For the European Alps in particular, GCMs cannot account for the climatological effects of high relief due to the coarse spatial resolution of the climate models and their insensitivity to sub-grid scale variability in relief and albedo (Calanca et al., 2006). In addition, climate feedbacks are significant in mountain settings and have implications for local patterns of snow preservation and melt and maintenance of permafrost (Vavrus, 2007; Slaymaker & Embleton-Hamann, 2009). These are significant areas of concern that highlight the complexity of mountain landscape systems and their land surface (geomorphological) responses to future climate change (Beniston, 2006).

Aims and structure of this chapter

This chapter considers some of the impacts of recent and future climate change on geomorphological processes and natural hazards in the European Alps, which is a climatically sensitive location with respect to changing patterns of temperature and precipitation. These probable impacts of climate change are inferred from observations of those geomorphological processes and natural hazards that took place during the 2003 heatwave and 2005 summer floods, both of which had diverse impacts across the European Alps. These recent events are chosen because their processes and impacts were well monitored and GCMs predict that under global warming there will be an increased frequency of temperature and precipitation extremes over central Europe (Frei et al., 2005). This means that geomorphological responses to these recent events are likely to be indicative of wider landscape responses to be expected under future climate change. More widely, these climatic responses in the European Alps are likely to be typical of many mountain regions worldwide.

In detail the chapter has three main aims: (1) to outline the present climatic regime of the European Alps and recent changes in glacier, Alpine permafrost and river systems; (2) to describe the climate events and geomorphological impacts of the pan-Alpine 2003 heatwave and 2005 floods; and (3) to discuss these geomorphological impacts within the wider context of ongoing climate change taking place in sensitive paraglacial (glacier-influenced) Alpine landscapes. This forward projection of likely impacts has important implications for planning, policy and management of sensitive Alpine landscapes and environments worldwide.

Climate and environment of the European Alps

The European Alps is a climatically transitional region, located at the interface between major atmospheric circulation source areas of the Atlantic Ocean, Mediterranean Sea and continental Europe west of the Urals. As such, there are major north–south and east–west climatic gradients (Auer et al., 2005). The European Alps are also split into eastern and western sectors by the Rhine and Splügen pass; the eastern sector, which is the geographical focus of this chapter, has Atlantic-influenced climate in its northern part, with a continental influence in the east and Mediterranean influence in the south (Figure 10.1). Monitoring of temporal changes in temperature and precipitation patterns across the Alps has already highlighted the effects of ongoing climate changes on temperature and precipitation anomalies, winter snow depth and duration, and river discharge (e.g. Beniston et al., 2003; Laternser & Schneebeli, 2003; Auer et al., 2005). Therefore future patterns of these meteorological phenomena will reflect synoptic-scale reorganisation of atmospheric circulation cells and air mass source areas that are a likely outcome of anthropogenic global warming (Lionello et al., 2008).

In detail, precipitation varies markedly across the Alps by both yearly and seasonal totals, and in response to variations in moisture source, wind direction and microclimate (Frei & Schär, 1998; Casty et al., 2005; North et al., 2007; Van der Schrier et al., 2007). Precipitation is highest on the northern side of the Alps, and decreases from west to east with distance from Atlantic source regions. Temperature variability across the Alps is due mainly by elevation, with valley floors substantially warmer than adjacent mountain tops. Temperature inversions are common during winter (Agrawala, 2007). Variation in Alpine climate is also linked closely to the North Atlantic Oscillation, which in part determines the trajectory of precipitation-bearing storm tracks across Europe, as well as temperature anomalies (Beniston & Jungo, 2002; Casty et al., 2005; Auer et al., 2007; Bartolini et al., 2009).

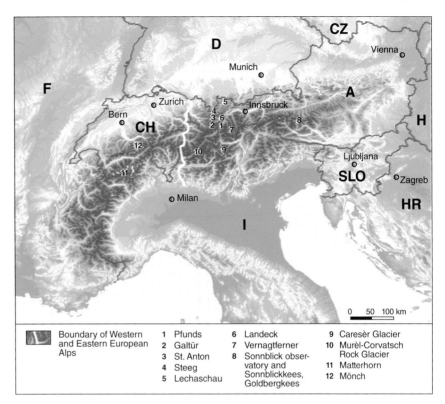

Boundary of Western and Eastern European Alps	1 Pfunds	6 Landeck	9 Caresèr Glacier
	2 Galtür	7 Vernagtferner	10 Murèl-Corvatsch Rock Glacier
	3 St. Anton	8 Sonnblick observatory and Sonnblickkees, Goldbergkees	11 Matterhorn
	4 Steeg		12 Mönch
	5 Lechaschau		

Figure 10.1 Location and topography of the European Alps and places named in the text.

The distribution and thickness of alpine permafrost closely follow these temperature and precipitation patterns (Harris et al., 2003; Luetschg et al., 2008). On north-facing slopes permafrost occurs at altitudes above 2600 m, on south-facing slopes above 3000 m, and in zones with long-lasting avalanche snow at altitudes several hundred metres lower (Lieb, 1998; Luetschg et al., 2008). Permafrost is also sensitive to snow cover, topography, aspect and ground surface material (Damm, 2007). Recent studies show ongoing permafrost warming in the European Alps: during the twentieth century, permafrost warmed by 0.5–0.8°C in the upper tens of metres (Gruber et al., 2004b), particularly at higher elevations, with accompanying thickening of the seasonal active layer (Harris et al., 2003, 2009). In Alpine environments frozen water bodies are also present in rock glaciers, and within talus-foot deposits and bedrock fractures (Gruber et al., 2004a; Lambiel & Pieracci, 2008). Permafrost degradation and ice melt is therefore a trigger for slope instability and mass movement hazards (Kääb, 2008; Avian et al., 2009). Many slope instability events take place towards the end of the permafrost melt season.

Recent climate change effects are also recorded in Alpine glaciers (Haeberli et al., 2007; Huss et al., 2008a) which, by the 1970s, had lost 35% of their 1850 value of total area, and almost 50% of their 1850 value by 2000 (Paul et al., 2004; Zemp et al., 2006). In the Ötzal Alps (Austria) glaciated area decreased by 8% in the period 1997–2006 (Abermann et al., 2009), which is similar to the area of ice loss in Switzerland over the same period (Farinotti et al., 2009). The Pasterze glacier in Austria retreated continuously by an average of 18.9 m/year in the period 1984–2000 (Hall et al., 2001) based on satellite and field data and field photos (Figure 10.2). Across the Alps, however, there is wide variability in glacier response, attributed to changes in macroscale synoptic climate (Beniston & Jungo, 2002) as

Figure 10.2 Photos of the Pasterze Glacier, Austria, at different dates. From top to bottom – 1989 (photo courtesy of Herwig Wakonigg), 2003 (photo courtesy of Gerhard Hohenwarter), 2010 (photo courtesy of Gerhard Lieb).

well as microclimate effects that lead to variations in temperature and precipitation (Lambrecht & Kuhn, 2007). These recent changes are superimposed upon a longer-term decrease in glacier area since the Little Ice Age (Ivy-Ochs et al., 2009). Changes in snowmelt and glacier ablation have implications for the amount and timing of soil water saturation as well as mountain river discharge, e.g. increased volume and earlier onset of spring snowmelt lead to increased flood hazard downstream, as well as increased seasonality in water resource provision (Laternser & Schneebeli, 2003).

Future climate patterns in the European Alps

Modelled predictions of future climate changes in the European Alps, including spatial and temporal patterns of temperature and precipitation anomalies relative to present conditions, have been undertaken on both continental and regional scales. Following Déqué et al. (2007) it is argued that the temperature in the European Alps will increase by 0.3–0.45°C per decade to 2100 from 1961 to 1990 mean values, with a higher expected increase in summer and autumn and increased frequency of summer heatwaves (Beniston, 2004). This averaged value will be substantially modified, however, by altitudinal and other local effects. Changes in precipitation in Europe will result in a higher north–south gradient, with increased precipitation in the north (especially in winter) and a strong decrease in the south (especially in summer). In the European Alps, winter precipitation is expected to increase in the north-west and decrease in the south and east, with increased precipitation intensity in all regions (Frei et al., 2005). Despite uncertainties related to regional climate simulations of precipitation in complex terrain, Beniston (2006) suggested that mean and extreme precipitation values may undergo a seasonal shift, with more spring and autumn heavy precipitation events than at present, and fewer in summer.

Snow is a key feature of environmental change in Alpine areas (Laternser & Schneebeli, 2003; Vavrus, 2007). Under current climatic conditions, a shift in snow amount and snow-cover duration can already be observed (Laternser & Schneebeli, 2003). Due to warmer temperatures in the next decades, the snow volume may respond with reduction at mid-elevation sites (1000–2000 m) by 90–50% and at high-elevation sites (>2000 m) by 35% (Beniston et al., 2003). Furthermore, the duration of snow cover will be sharply reduced mainly because of earlier spring snowmelt (Beniston et al., 2003; Laternser & Schneebeli, 2003). Changes in snow-cover thickness and duration also affect subsurface temperatures and permafrost distribution (Gruber et al., 2004b; Harris et al., 2009); therefore decreasing snowfall will promote permafrost degradation irrespective of changes in air temperature. A thick snow cover acts to insulate the ground surface, whereas a thinner snow cover is associated with enhanced heat loss.

Modelled changes in glacier volume and extent show a clear relationship to increasing temperature. Warming of 3°C in summer air temperature would lead to an upward shift in the glacier equilibrium line elevation (Vincent, 2002) and reduce current Alpine glacier cover by some 80% (Zemp et al., 2006). In the event of a 5°C temperature increase, the Alps would become almost completely ice free. The impacts of precipitation changes are deduced with less certainty because of its co-variation with temperature and role of seasonality (Zemp et al., 2006; Steiner et al., 2008). In addition, glacier decay also leads to climate feedback through changes in surface albedo, areal downwasting and stagnation, and formation of debris-covered glacier margins (Paul et al., 2004; Kellerer-Pirklbauer et al., 2008).

In order to investigate the likely impacts and interlinkages between climate forcing and geomorphological responses in the European Alps, we present two case studies of recent climatological events: the 2003 heatwave and 2005 floods.

The 2003 heatwave and its impacts

The summer of 2003 was characterised by the hottest temperatures in the last 500 years in Central Europe (Luterbacher et al., 2004) and the warmest summer in a 1250-year-long record for the European Alps (Büntgen et al., 2006; Rebetez et al., 2006). The mean summer temperatures (June-August) in a large area of the European Alps exceeded the 1961–1990 mean by 3–5°C, and showed high spatial anomalies (Schär et al., 2004). At the Hoher Sonnblick observatory (3106 m asl) located in the central Austrian Alps a mean air temperature of 4.7°C was recorded during summer 2003, which is 4.4 times the standard deviation of the long-term mean (1886–2000) (Koboltschnig et al., 2009). Summer precipitation in Austria and Switzerland was only 50% of average (Patzelt, 2004; Schmidli & Frei, 2005), which had also been preceded by a dry spring (February–June). This situation was similar across Europe (Black et al., 2004; Schär & Jendritzky, 2004). The extreme temperature and precipitation conditions corresponded to persistent anticyclonic conditions during May to August 2003 (Black et al., 2004; Rebetez et al., 2006) which reinforced dry soil conditions (Fischer, E. M. et al., 2007; Fennessy & Kinter, 2009). Rapid glacier mass loss and ice-margin retreat associated with the 2003 heatwave also changed river discharge and initiated mass movement events in the surrounding mountains during this period and afterwards (Gruber et al., 2004b).

The summer 2003 heatwave triggered a record Alpine glacier loss that was three times above the 1980–2000 average (Haeberli et al., 2007), continuing a long-term and accelerating pattern of mass loss (Zemp et al., 2006) (Figure 10.3). Mass loss from Swiss glaciers was 3.5% of total mass in 2003 alone (Farinotti et al., 2009). Main factors for this remarkable 2003 loss were a thinner snowpack due to the low springtime precipitation, and melting of firn in the accumulation area of small- and medium-sized glaciers, resulting in a long-term albedo feedback. After the melt of the spring snow cover, the bare glacier surface had a lower albedo, which was additionally depressed due to deposition of dark, wind-blown dust during the dry summer (Paul et al., 2005; Haeberli et al., 2007; Koboltschnig et al., 2009). In 2003, Austrian glaciers (88 measured) retreated by an average of

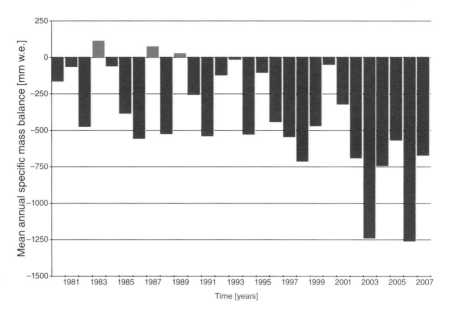

Figure 10.3 Mean annual specific mass balance of reference glaciers in Austria (www.wgms.ch/mbb/mbb10/sum07.html). (Reproduced with permission of the World Glacier Monitoring Service.)

23 m and with a maximum distance of 73 m (Patzelt, 2004), exposing large, debris-covered areas. According to the long-term glacier mass balance time series (e.g. Vernagtferner in Austria from 1965 to present), the mass balance of 2002–2003 was the most negative on record (Commission for Glaciology, 2009).

The response of permafrost temperature and the thickness of the active layer varied considerably in summer, 2003 (Harris et al., 2009). In 2003, the active layer of ice-rich frozen debris at Murtèl-Corvatsch in the Swiss Alps was deeper than previously recorded but the effects of the higher temperature were restricted by greater latent heat demands (Harris et al., 2009). In contrast, the thaw depth in permafrost on bedrock slopes was twice the average of previous years and indicates a strong coupling between atmospheric and ground temperatures (Gruber et al., 2004b). Permafrost degradation set into action by this warming was reflected in increased rock fall activity throughout the Alps during summer 2003 (Gruber et al., 2004b; Fischer, L. et al., 2006). The response of permafrost to the atmospheric warming generally takes place at different scales of time and depth. In particular, localities near the lower elevational limit of the discontinuous permafrost are very sensitive to changes, and respond with a short time lag (of days to months), e.g. a large rockfall event at the elevation of the lower permafrost boundary on the Matterhorn (Switzerland), July 2003, revealed weakened interstitial ice at the back of the rockfall scar (Nötzli et al., 2004). Large-scale responses to warming permafrost such as deep-seated slope instabilities may show a delay of decades to centuries (Gruber et al., 2004b; Harris et al., 2009).

Further effects of the 2003 heatwave are complex hydrological responses of rivers, both within Alpine basins and in their lower catchments (Zappa & Kan, 2007). Apart from the physical characteristics of the basins, run-off is also influenced by anomalies in air temperature, potential evapotranspiration, snow accumulation and ice melt (Zappa & Kan, 2007). Meltwater run-off is a function of the degree of glacierisation of the basin; a historical minimum of summer discharge results from glacierisation of less than 1% of basin area, whereas in basins of up to 10% glacierisation summer discharge was about 70–80% of average (Zappa & Kan, 2007). Koboltschnig et al. (2009) presented similar results from the glacierised Goldbergkees basin in the Austrian Alps, where glacier melt during August 2003 contributed 81% of total run-off. Only in basins where glaciers covered more than 15% of the basin area did increased glacier melt compensate for reduced run-off in ice-free areas of the basin. In 2003, runoff yield in such heavily glaciated basins was close to summer and annual averages, but sub-basins exhibited strong positive and negative runoff anomalies.

The impacts of the 2003 heatwave on hazardous events and society are multi-faceted. Rockfalls increased in frequency in high-elevation areas due to degradation of mountain permafrost (Gruber et al., 2004b). Rockfalls and also debris flows endangered climbing and hiking routes in high-elevation areas, and some of these paths had to be closed or diverted (Nötzli et al., 2004). In Switzerland, some climbers and tourists were evacuated from popular sites including the Matterhorn and Mönch due to rockfall hazard. Geotechnical problems with infrastructure and buildings arose as a result of enhanced permafrost melt, e.g. the Hoher Sonnblick observatory is becoming destabilised, requiring considerable infrastructural investment (Krainer, 2007). The high temperature and drought conditions experienced in summer 2003 contributed directly to increased mortality across Europe by between 35,000 (Schär & Jendritzky, 2004; Fischer et al., 2007) and 70,000 (Gómez & Souissi, 2008). Mortality in Switzerland was 7% above average between June and August 2003, with the increase tracking that of temperature rise (Grize et al., 2005). Temperature and precipitation changes lengthened the growing season in mountain ecosystems (Jolly et al., 2005) and gave rise to the earliest Swiss grape harvest since records began in 1480 (Meier et al., 2007). Elsewhere, crop failure increased and animal fodder production sharply decreased, and forest fires were more common. The strongly reduced discharge in many rivers also affected downstream navigation and industry, including decreased hydroelectric plant production and effects on ecosystems (Beniston, 2004).

The 2005 flood event and its impacts

In August 2005, central Europe and especially the Alpine region were affected by severe floods accompanied by river bank erosion and sediment transport, as well as debris flows, rockfalls and landslides in the smaller catchments (Rickenmann et al., 2008). These events caused the most catastrophic flood damage in the last 100 years with respect to loss of life and damage to infrastructure, communication routes and agriculture (Beniston, 2006; Frei, 2006). In Switzerland, the August 2005 event alone caused a quarter of all damage by floods, debris flows, landslides and rockfalls recorded since 1972 (Hilker et al., 2008). In the Valais Alps (Switzer-

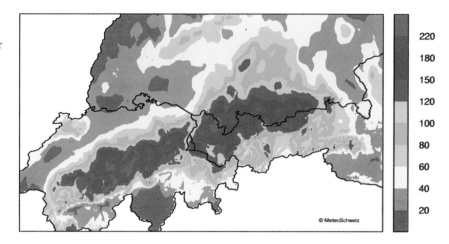

Figure 10.4 Map of 72-hour total precipitation (in millimetres) from 21 August 2005 to 24 August 2005 (starting with 06 Universal Time Coordinated [UTC] to 06 UTC) in the European Alps region (Frei, 2006). Outline of Switzerland and Austria shown for scale and location. (Reproduced with permission of MeteoSwiss, Switzerland.)

land), the timing of debris flows that disturb surrounding trees has been reconstructed using dendrochronology (Stoffel et al., 2008). Based on this evidence, the size of the debris flow event produced at this location by the 2005 floods was comparable to events in 1993 and 1987, but to no other events in the record from AD 1793 onwards.

The major factor for the occurrence of the flooding event was heavy precipitation on 20–23 August 2005. The most intense rainfall was located on the northern slopes of the Alpine ridge (Figure 10.4), which was caused by Vb-type atmospheric patterns, similar to that which had caused previous flood events in Austria (Amt der Vorarlberger Landesregierung, 2005). A low-pressure system situated over the Gulf of Genoa (Adriatic Sea) first brought heavy precipitation in the south and south-east of Austria, then moved slowly to the east and circled back to the Alps. The warm and moist air was lifted by both a cold air mass located over Bavaria (southern Germany) and the topographic effect of the Alpine ridge (Amt der Vorarlberger Landesregierung, 2005).

Several factors apart from the high precipitation rate (30 hours of rainfall with intensity of >10 mm/h for several hours) contributed to the high run-off in 2005: (1) the intense rainfall occurred across a large area and covered whole catchments and valleys; (2) the soil was already pre-saturated due to a very wet situation in July and August, leading to rapid and high surface run-off; and (3) because of high summer air temperatures the freezing level was above 2900–3200 m asl, so only a small amount of precipitation could be buffered as snow. In addition glaciers were already saturated with meltwater due the high summer temperatures (BMLFUW, 2006a).

In the eastern European Alps (Switzerland, western Austria [states of Vorarlberg and Tyrol] and southern Germany) this climatological situation resulted in large and long-lasting river discharges (see Table 10.1 for the Tyrol). The high discharge was accompanied by intense sediment transport related to high erosion rates and subsequent sediment deposition mainly on floodplains. In smaller catchments a

Table 10.1 Discharge of 2005 flood peak and frequency of the event in selected catchments in the state of Tyrol, Austria (modified from BMLFUW, 2006a)

Catchment name	Gauge location	Catchment area (km²)	Flood peak (m³/s)	Frequency in years	Comments
Lech river	Steeg	247.9	361	~5000	
Lech river	Lechaschau	1012.2	943	≫500	
Rosanna river	St Anton	130.6	(≫54)	(≫100)	Gauge destroyed 22 August 2005 (21:30h)
Trisanna river	Galtür	97.6	>141	~5000	
Sanna river	Landeck	727.0	514	~5000	

high number of debris flow events occurred: in Austria alone, 115 debris flows and 111 other landslides were recorded (Internationale Forschungsgesellschaft Interpraevent, 2009), mainly with medium-to-high sediment volumes. A few debris flow events transported more than 100,000 m³ of sediment (Rickenmann et al., 2008). A substantial part of the sediment load (debris and woody debris) accumulated as torrent fans in the main valleys. Some of the debris flows delivered their sediment load into mountain rivers and contributed to the high sediment transport in the river downstream, in addition to the sediment produced locally (Rickenmann et al., 2008). The strong geomorphological activity during the flood event resulted in changes to channel courses, enhanced bank overtopping and sediment deposition outside the main channels (Stoffel et al., 2008). This led to substantial flood damage on inhabited areas and infrastructure alongside the river channels (BMLFUW, 2006a, 2006b; Rickenmann et al., 2008).

The municipality of Pfunds, in the Tyrol (Austria), was one of the communities heavily affected by the 2005 floods. Pfunds is located on the fan of the Stubenbach river, which is a left tributary of the Inn river with a catchment area of 30 km² between 525 m and 3035 m asl (Figure 10.5). A nearby meteorological station (at Ladis-Neuegg) recorded 127% of normal precipitation during August 2005 with over a third of the total (60 mm) falling on 22 August alone. During the flood event about 65,000 m³ of debris was deposited and reached up to 6 m deep (1 m on average). The debris was derived from sediments upstream and by partial erosion of the vegetation cover when reworking the Pleistocene valley fill (BMLFUW, 2006b). Therefore, it is assumed that the state of the catchment system changed due to this extreme event and started to rework the sediments that were in quasi-storage within the upper catchment.

The Pfunds community consists of 2500 residents and 675 buildings. The 2005 event caused severe damage to those parts of the village located directly on the fan where 89 buildings were damaged or destroyed (Figures 10.6 and 10.7). Losses for this community were estimated at €11 million, including damage to buildings, roads, bridges and streets that were blocked for several days (BMLFUW, 2006b).

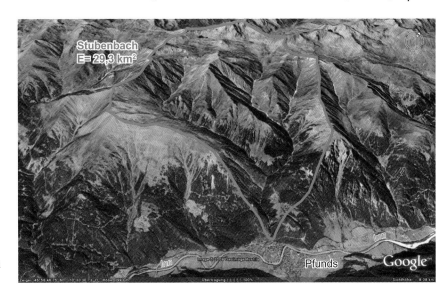

Figure 10.5 Oblique GoogleEarth image (looking north-west) of the village of Pfunds and the high-relief catchment of the Stubenbach river. © Google Earth.

Figure 10.6 View of debris accumulation on the Stubenbach fan during the 2005 flood event in the municipality of Pfunds. The red circle shows the location of the building in Figure 10.7. (Photo courtesy of www.alpinesicherheit.com.)

Figure 10.7 Debris accumulation and damage on buildings and infrastructure following the 2005 flood event in Pfunds. The lorry is located on the bridge and the digger in the former riverbed. (Photo courtesy of Johannes Hübl.)

More widely, the 2005 event caused losses of up to €555 million across Austria, of which €442 million occurred in the western states of Austria (BMLFUW, 2006b). Apart from the damage to buildings and industrial areas due to inundation, lateral erosion of the rivers and loose sediments caused the highest losses, including 100 days of discontinued service on the railway connecting Tyrol and Vorarlberg (the main line between Vienna and Zürich) (Amt der Vorarlberger Landesregierung, 2005; BMLFUW, 2006b).

Discussion

The geomorphological events that took place in the European Alps as a direct consequence of the 2003 heatwave and 2005 floods demonstrate the impact of meteorological variability in driving landscape-change processes over short times-cales and in climatically sensitive regions, and with wider impacts on ecosystems and human activity. There are four main points to be noted from these responses in the Alps:

1. There is considerable spatial (and, less, temporal) variation in river response to both flood and drought events, including where there is greatest sediment erosion, deposition and flood risk, and both across the Alps as a whole and within individual catchments, including the transition from headwaters to immediately adjacent Alpine valleys. Responses to the 2003 and 2005 events show this clearly, where floods had the greatest geomorphic impact in upstream and droughts greatest human impact in downstream locations. Antecedent conditions may be less important in flashy, upland catchments, but river response is strongly modified by catchment relief and sediment availability, which have potential to inhibit downstream water and sediment transport. Enhanced sediment evacuation from upland storage during the 2005 floods also reduced sediment yield subsequently.

2. In both events, glacier response was very rapid (such as high ice mass loss during the 2003 event), but this can be best described as a transient response superimposed on a longer time scale climate-driven signal (Ivy-Ochs et al., 2009). Direct glacier response is measured by variations in meltwater production and mass balance (Haeberli et al., 2007), but other measures of glacier health, including position of the ice margin and structural integrity of the glacier as a whole (including presence of crevasses), should also be used. In addition, glacier responses to summer (temperature) and winter (precipitation, including snow) anomalies have their greatest and most immediate impacts on different parts of the glacier system, with high-elevation source areas more strongly precipitation limited, and lower ablations areas more temperature limited. Rapid temperature-driven retreat of the ice margin can also be followed, counterintuitively, by glacier thickening and positive mass balance (e.g. Hughes, 2008). Simple climatic forcing by temperature or precipitation alone, therefore, does not fully explain glacier dynamics.

3. Responses of Alpine permafrost are difficult to quantify over short timescales because of, first, time-lag effects due to the slow penetration into the sediment pile of the effects of climate forcing at the surface, and, second, the meteorological events that take place over short (sub-seasonal) timescales do not necessarily affect the following freezing–thawing season. This is clearly shown by responses to, in particular, the 2003 event (Harris et al., 2009). In addition, the role of permafrost is uncertain: permafrost monitoring stations are generally widely spaced, and local effects of sediment type, microtopography and microclimate are likely significant controls that cannot be resolved spatially by the current monitoring stations. Although the effects of land surface heating can be effectively modelled, the role of increased surface wetness on permafrost stability is less well known but may also include increased ice lens thickness.

4. Implications for, and interactions between, these geomorphological processes and the biosphere are not well understood, and although individual events such as the 2003 and 2005 events do not have the temporal extent to trigger large-scale biosphere impacts, they form part of a long-term response to ongoing climate change. Biosphere effects are of particular importance for aspects of rock surface weathering and stability of vegetated slopes, and have long-term implications for ecosystem viability including Alpine refugia. Such biosphere responses may buffer effects of climate forcing of geomorphological hazards on warming mountain slopes.

Hazardous events in the European Alps, as in other mountain blocks worldwide, are linked directly to glacial, periglacial, snow and river processes (Korup & Clague, 2009). A number of studies have examined magnitude–frequency relationships of different mountain hazard types, hazard triggering and flow mechanisms, role of slope angle and aspect, and relationship to climate forcing, e.g. Huggel et al. (2004) and Werder et al. (2010) consider the role of changes in glacier front position and meltwater supply as a control on formation of proglacial lakes and jökulhlaup hazard. Kääb (2008) and Avian et al. (2009) use remote sensing data to identify locations where mass movement hazards are most likely to be generated in permafrost terrain, near the lowermost permafrost limit. These studies show that mountain hazards, and therefore the risks associated with them,

Figure 10.8 Photo of debris flow and landslide hazards in the Southern Alps, New Zealand. (Photo courtesy of Jasper Knight.)

are unevenly distributed and are strongly contingent on localized and antecedent conditions that are difficult to monitor and model. Increased hazard frequency associated with both the 2003 and 2005 events impacted negatively on human activity, in both direct and indirect ways. These different responses are highly variable spatially and temporally, and in high-relief areas are also dependent on elevation and aspect. The geomorphological and wider landscape impacts of the 2003 and 2005 events can be considered analogous to the likely hazardous impacts of future climate changes in the European Alps, under which meteorological variability will be more common (Beniston, 2006; Rebetez et al., 2008).

Similar geomorphological processes, hazards and hazard risks are also observed in other mountainous settings worldwide, e.g. in the Southern Alps (New Zealand) geomorphological mapping of past debris flow distribution has been used as a tool to identify where future glacigenic hazards (including debris flows, lake outburst floods and ice avalanches) may be located (Allen et al., 2009). These show considerable spatial variability, where debris flow paths are topographically directed, building flow cones and alluvial fans in scarp-foot locations (Figure 10.8). Many rockfall events in New Zealand are triggered by earthquakes (Bull & Brandon, 1998) but rockfalls and rock avalanches can also lead to unanticipated hazards include landslide dam formation and failure, flooding, erosion and alluviation that may take place some hours to months after the initial event (Hancox et al., 2005).

Implication for natural hazard and risk management

Geomorphological events become natural hazards when they conflict with the human environment and lead to damage. The countries that include the European Alps in their territory have a long tradition of hazard mitigation, but climate and socioeconomic change are major future challenges (Bründl et al., 2009). Population and socioeconomic trends in the European Alps in the recent century (Bätzing, 2003; Slaymaker & Embleton-Hamann, 2009) reflect a shift from agriculture to service and leisure-oriented industries (Bätzing, 1993; Keiler, 2004). This trend is also observed in the development of tourist infrastructure in hazard-prone sites (Fuchs & Bründl, 2005; Keiler et al., 2005), which can result in high and increasing losses if extreme events occur, as illustrated by the 2005 floods and as has happened in the Alps during recent decades (e.g. avalanche events in 1999 [SLF, 2000]; flood events in 2002 [Hilker et al., 2009]). In contrast, the short-term geomorphic impact and economic damage of the 2003 event in the European Alps was relatively low (Nötzli et al., 2004).

As highlighted above, geomorphic responses to climate change are complex and highly variable spatially and temporally, but will have hazard-specific implications for risk management. The observed increase in periglacial and glacial hazards, as during the 2003 event, is concentrated in high-altitude areas where it impacts on small settlements and mountaineering/skiing infrastructure (paths, huts). This may be of limited economic significance nationally but is important locally (Agrawala, 2007). Public awareness of these mountain hazards (as opposed to floods and snow avalanches) is generally low, and hazard management has often focused on build-up of mountain monitoring systems (Kääb, 2008). Despite this low direct effect on society, these high-mountain hazards increase the availability of easily mobilised debris in upland catchments and so increase future hazard risk (Korup & Clague, 2009). The 2005 event showed the connectivity between process chains or cascades resulting from such mountain hazard events that leads in turn to high sediment transport through landslides and other processes, and their impacts on valley settlements.

The traditional hazard assessment does not meet the challenges of the ongoing and future environmental and geomorphic change. The traditional approach includes determining the hazard potential and its probability of occurrence by studying historical events, modelling, and assessing individual processes and defined design events. The considered design events, which are the basis for delimitation of hazard zones and mitigation strategies, vary across the Alps (Keiler, 2004; Fuchs et al., 2005), e.g. Austria uses design events with a 100-year recurrence interval for floods and 150 years for debris flows. With reference to Table 10.1, the 2005 event was far beyond these design events and therefore historical experience, meaning that the record of past events is no longer sufficient for hazard assessment. This reflects changes in the spatial pattern of climatic parameters as well as the connectivity between different geomorphic processes that may alter magnitude–frequency relationships of hazards and damaging events. The role of climate change in the Alps, which the 2003 and 2005 events probably prefigure, has significant implications for hazard triggering and mitigation.

The model of risk management, conceptualised in terms of a 'risk cycle' (e.g. Carter, 1991; Alexander, 2000; Kienholz et al., 2004, with respect to natural hazards in the Alps), extends a traditional hazard management framework that focuses on emergency management responses and procedures to hazardous events (Figure 10.9). Consideration of these risks, including the use of mitigation measures and restrictions in land use, are part of an integrated risk management approach with

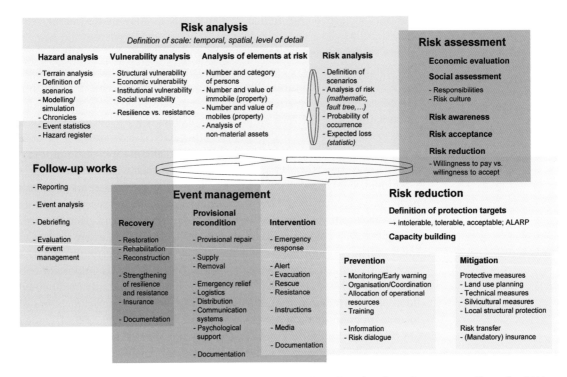

Figure 10.9 A 'risk cycle' model of integrated risk management. (Based on data from Carter, 1991; Alexander, 2000; Kienholz et al., 2004.)

an ex-ante and scenario perspective. This model integrates natural science and social science approaches with the aim of creating disaster-resilient communities through inductive learning (Fuchs, 2009). Risk analysis and subsequent risk assessment are aimed at the evaluation and reduction of hazard risk, whereby prevention and mitigation are targeted at modification of the hazard or the modification of vulnerability of the human environment. This risk reduction includes prevention (monitoring and early warning, but also training and information) and mitigation (by protective measures and by risk transfer) (Bründl et al., 2009). Intervention in the risk cycle by event analysis, debrief and future risk reduction strategies, referred to as follow-up works in Figure 10.9, facilitate an enhanced future risk analysis. The risk cycle approach focuses on land-use regulation, risk transfer and information to the public to build up awareness (Holub & Fuchs, 2009). Land use planning activities such as hazard maps, based on recurrence intervals, may change due to variable climatic conditions. This, therefore, sets in place a responsive framework for the management of future changes in sensitive mountain environments as a result of climate change.

Conclusions and wider implications

The European Alps, in common with many mountains worldwide, are being disproportionately affected by ongoing climate change which is itself superimposed upon a longer-term paraglacial signal that corresponds to processes of landscape readjustment/relaxation (Church & Ryder, 1972) following the last glacial event (the Würmian glaciation in the European Alps). Rockfalls, landslides and debris flows are significant processes whereas deglaciated slopes remain steep, unvegetated and water saturated (Stoffel et al., 2008). The frequency and magnitude of these events decrease exponentially over the paraglacial period, reaching background interglacial values up to 10,000 years after initial ice retreat (Ballantyne, 2002). A long-term consequence of paraglacial relaxation in mountains is that these locations can be considered as 'disturbed landscapes' which are in transition from a glacial to a non-glacial state (Slaymaker, 2009). Two important characteristics of this transition period are glacier retreat and melting of alpine permafrost, both of which are known from field studies across the European Alps and show consistent warming patterns over several decades (Haeberli et al., 2007; Abermann et al., 2009; Harris et al., 2009). Ice retreat and increased active layer thickness in higher-elevation settings are genetically associated with land surface instability and enhanced sediment delivery to mountain slopes and valley bottoms (Otto et al., 2009). As such, it can be anticipated that ongoing and accelerating ice loss in the European Alps, over the next decades to centuries as a result of global warming, will have significant impacts on hazard type, location and frequency, and slope sediment supply (Knight & Harrison, 2009; Huggel et al., 2010). Furthermore, increased sediment supply to mountain valleys will in turn result in future floodplain aggradation and increased flood hazards downstream, outside the Alps region (Knight & Harrison, 2009; Otto et al., 2009).

The European Alps is sensitive to such changes because: (1) it is located at the boundary between different moisture source regions; (2) it is a paraglacial landscape in long-term response to post-glacial environmental change; (3) its glaciers and permafrost are in decay due to climate change; and (4) hazardous events can readily impact on areas of high population density in Alpine valleys. Future climate changes in the Alps will probably have unforeseen outcomes on physical processes and natural hazards related to ongoing changes to the cryosphere caused by increasing temperatures (Knight & Harrison, 2009; Huggel et al., 2010). It is unlikely that coeval changes in precipitation alone can offset temperature-driven changes to Alpine glaciers. Increased natural hazard frequency and/or magnitude is a likely signature of climate forcing on unstable land surfaces. Understanding future climate impacts in these alpine areas is, however, hampered by problems of GCM downscaling in areas of complex local relief, microclimate and sediment supply (Calanca et al., 2006). Future research on geomorphic processes, and monitoring of land surface systems, are needed to establish the sensitivity of these systems to climate forcing. Furthermore, this knowledge will lead to an improvement of hazard and risk management in the European Alps and in similar mountain settings worldwide.

References

Abermann, J., Lambrecht, A., Fischer, A. & Kuhn, M. (2009) Quantifying changes and trends in glacier area and volume in the Austrian Ötztal Alps (1969–1997–2006). *The Cryosphere* **3**, 205–215.

Agrawala, S., ed. (2007) *Climate Change in the European Alps. Adapting winter tourism, and natural hazards management.* Paris: OECD, 127pp.

Alexander, D. (2000) *Confronting catastrophe.* Oxford: Oxford University Press, 282pp.

Allan, R. P. & Soden, B. J. (2007) Large discrepancy between observed and simulated precipitation trends in the ascending and descending branches of the tropical circulation. *Geophysical Research Letters* **34**, L18705. doi:10.1029/2007GL031460.

Allen, S. K., Schneider, D. & Owens, I. F. (2009) First approaches towards modelling glacial hazards in the Mount Cook region of New Zealand's Southern Alps. *Natural Hazards and Earth System Sciences* **9**, 481–499.

Amt der Vorarlberger Landesregierung, eds (2005) Starkregen- und Hochwasserereignis August 2005 in Vorarlberg, 52pp.

Auer, I., Böhm, R., Jurkovic, A., et al. (2005) A new instrumental precipitation dataset in the greater alpine region for the period 1800–2002. *International Journal of Climatology* **25**, 139–166.

Auer, I., Böhm, R., Jurkovic, A., et al. (2007) HISTALP – Historical instrumental climatological surface time series of the greater Alpine region 1760–2003. *International Journal of Climatology* **27**, 17–46.

Avian, M., Kellerer-Pirklbauer, A. & Bauer, A. (2009) LiDAR for monitoring mass movements in permafrost environments at the cirque Hinteres Langtal, Austria, between 2000 and 2008. *Natural Hazards and Earth System Sciences* **9**, 1087–1094.

Baggi, S. & Schweizer, J. (2009) Characteristics of wet-snow avalanche activity: 20 years of observations from a high alpine valley (Dischma, Switzerland). *Natural Hazards* **50**, 97–108.

Ballantyne, C. K. (2002) A general model of paraglacial landscape response. *The Holocene* **12**, 371–376.

Bartolini, E., Claps, P. & D'Odorico, P. (2009) Interannual variability of winter precipitation in the European Alps: relations with the North Atlantic Oscillation. *Hydrology and Earth System Sciences* **13**, 17–25.

Bätzing, W. (1993) *Der sozio-ökonomische Strukturwandel des Alpenraums im 20. Jahrhunder.* Bern: Geographica Bernensia, p 26.

Bätzing, W. (2003) *Die Alpen: Geschichte und Zukunft einer europäischen Kulturlandschaft.* München: Ch. Beck.

Beniston, M. (2004) The 2003 heat wave in Europe: A shape of things to come? An analysis based on Swiss climatological data and model simulations. *Geophysical Research Letters* **31**, L02202. doi:10.1029/2003GL018857.

Beniston, M. (2005) Mountain climates and climatic change: an overview of processes focusing on the European Alps. *Pure and applied Geophysics* **162**, 1567–1606.

Beniston, M. (2006) August 2005 intense rainfall event in Switzerland: Not necessarily an analog for strong convective events in a greenhouse climate. *Geophysical Research Letters* **33**, L05701. doi:10.1029/2005GL025573.

Beniston, M. & Jungo, P. (2002) Shifts in the distributions of pressure, temperature and moisture and changes in the typical weather patterns in the Alpine region in response to the behavior of the North Atlantic Oscillation. *Theoretical and Applied Climatology* **71**, 29–42.

Beniston, M., Keller, F. & Goyette, S. (2003) Snow pack in the Swiss Alps under changing climatic conditions: an empirical approach for climate impacts studies. *Theoretical and Applied Climatology* **74**, 19–31.

Beniston, M., Stephenson, D. B., Christensen, O. B., et al. (2007) Future extreme events in European climate: an exploration of regional climate model projections. *Climate Change* **81**, 71–95.

Black, E., Blackburn, M., Harrison, G., Hoskins, B. & Methven, J. (2004) Factors contributing to the summer 2003 European heatwave. *Weather* **59**, 217–223.

Bollschweiler, M. & Stoffel, M. (2010) Changes and trends in debris-flow frequency since AD 1850: Results from the Swiss Alps. *The Holocene* **20**, 907–916.

Bründl, M., Romang, H. E., Bischof, N. & Rheinberger, C. M. (2009) The risk concept and its application in natural hazard risk management in Switzerland. *Natural Hazards and Earth System Sciences* **9**, 801–813.

Brunetti, M., Lentini, G., Maugeri, M., et al. (2009) Climate variability and change in the Greater Alpine Region over the last two centuries based on multi-variable analysis. *International Journal of Climatology* **29**, 2197–2225.

Bundesministerium für Land- und Forstwirtschaft, Umwelt und Wasserwirtschaft (BMLFUW), eds (2006a) *Hochwasser 2005 – Ereignisdokumentation, Teilbericht des Hydrographischen Dienstes*. Vienna: BMLFUW, 26pp.

Bundesministerium für Land- und Forstwirtschaft, Umwelt und Wasserwirtschaft (BMLFUW), eds (2006b) *Hochwasser 2005 – Ereignisdokumentation, Teilbericht der Wildbach- und Lawinenverbauung*. Vienna: BMLFUW, 126pp.

Bull, W. B. & Brandon, M. T. (1998) Lichen dating of earthquake-generated regional rockfall events, Southern Alps, New Zealand. *GSA Bulletin* **110**, 60–84.

Büntgen, U., Frank, D. C., Nievergelt, D. & Esper, J. (2006) Summer temperature variations in the European Alps: AD 755–2004. *Journal of Climate* **19**, 5606–5623.

Calanca, P., Roesch, A., Jasper, K. & Wild, M. (2006) Global warming and the summertime evapotranspiration regime of the Alpine region. *Climate Change* **79**, 65–78.

Carter, W. (1991) The disaster management cycle. In: Carter, W. (ed.), *Disaster Management: A disaster manager's handbook*. Manila: Asian Development Bank, pp 51–59.

Casty, C., Wanner, H., Luterbacher, J., Esper, J. & Böhm, R. (2005) Temperature and precipitation variability in the European Alps since 1500. *International Journal of Climatology* **25**, 1855–1880.

CFG (Commission for Glaciology) (2009) *Current data collection for the mass balance determination of the Vernagtferner 1964 to 2008*. Bavarian Academy of Sciences and Humanities. Available at: www.lrz-muenchen.de/~a2901ad/webserver/webdata/massbal/index.html (accessed 15 November 2009).

Church, M. & Ryder, J. M. (1972) Paraglacial sedimentation: a consideration of fluvial processes conditioned by glaciation. *Bulletin of the Geological Society of America* **83**, 3059–3071.

Damm, B. (2007) Temporal Variations of Mountain Permafrost Creep: Examples from the Eastern European Alps. In: Kellerer-Pirklbauer, A., Keiler, M., Embleton-Hamann, C. & Stötter, J. (eds), *Geomorphology for the Future*. Innsbruck: Innsbruck University Press, pp 81–88.

Déqué, M., Rowell, D. P., Lüthi, D., et al. (2007) An intercomparison of regional climate simulations for Europe: assessing uncertainties in model projections. *Climatic Change* **81**, 53–70.

Farinotti, D., Huss, M., Baunder, A. & Funk, M. (2009) An estimate of the glacier ice volume in the Swiss Alps. *Global and Planetary Change* **68**, 225–231.

Fennessy, M. J. & Kinter III, J. L. (2009) *Climatic Feedbacks during the 2003 European Heatwave*. COLA Technical Report 282, Center for Ocean-Land-Atmosphere Studies, Calverton, MD 20705, 32pp.

Fischer, E. M., Seneviratne, S. I., Lüthi, D. & Schär, C. (2007) Contribution of land-atmosphere coupling to recent European summer heat waves. *Geophysical Research Letters* **34**, L06707. doi:10.1029/2006GL029068.

Fischer, L., Kääb, A., Huggel, C. & Noetzli, J. (2006) Geology, glacier retreat and permafrost degradation as controlling factors of slope instabilities in a high-mountain rock wall: the Monte Rosa east face. *Natural Hazards and Earth System Sciences* **6**, 761–772.

Frei, C. (2006) *Eine Länder übergreifende Niederschlagsanalyse zum August Hochwasser 2005*. Ergänzung zu Arbeitsbericht 211. Arbeitsberichte der Meteo-Schweiz, 213, 10pp.

Frei, C. & Schär, C. (1998) A precipitation climatology of the Alps from high-resolution rain-gauge observations. *International Journal of Climatology* **18**, 873–900.

Frei, C., Schöll, R., Fukutome, S., Schmidli, J. & Vidale, P. L. (2005) Future change of precipitation extremes in Europe: an intercomparison of scenarios from regional climate models. *Journal of Geophysical Research* **111**, D06105. doi: 10.1029/2005JD005965.

Fuchs, S. (2009) Susceptibility versus resilience to mountain hazards in Austria – paradigms of vulnerability revisited. *Natural Hazards and Earth System Sciences* **9**, 337–352.

Fuchs, S. & Bründl, M. (2005) Damage potential and losses resulting from snow avalanches in settlements in the Canton of Grisons, Switzerland. *Natural Hazards* **34**, 53–69.

Fuchs, S., Keiler, M., Zischg, A. & Bründl, M. (2005) The long-term development of avalanche risk in settlements considering the temporal variability of damage potential. *Natural Hazards and Earth System Sciences* **5**, 893–901.

Gehrig-Fasel, J., Guisan, A. & Zimmermann, N. E. (2007) Tree line shifts in the Swiss Alps: Climate change or land abandonment? *Journal of Vegetation Science* **18**, 571–582.

Gómez, F. & Souissi, S. (2008) The impact of the 2003 summer heat wave and the 2005 late cold wave on the phytoplankton in the north-eastern English Channel. *Comptes Rendus Biologies*, **331**, 678–685.

Grize, L., Huss, A., Thommen, O., Schindler, C. & Braun-Fahrländer, C. (2005) Heat wave 2003 and mortality in Switzerland. *Swiss Medical Weekly* **135**, 200–205.

Gruber, S., Hoelzle, M. & Haeberli, W. (2004a) Rock-wall temperatures in the Alps: modelling their topographic distribution and regional differences. *Permafrost and Periglacial Processes* **15**, 299–307.

Gruber, S., Hoelzle, M. & Haeberli, W. (2004b) Permafrost thaw and destabilization of Alpine rock walls in the hot summer of 2003. *Geophysical Research Letters* **31**, L13504. doi:10.1029/2004GL020051.

Gude, M. & Barsch, D. (2005) Assessment of geomorphic hazards in connection with permafrost occurrence in the Zugspitze area (Bavarian Alps, Germany). *Geomorphology* **66**, 85–93.

Haeberli, W., Hoelzle, M., Paul, F. & Zemp, M. (2007) Integrated monitoring of mountain glaciers as key indicators of global climate change: the European Alps. *Annals of Glaciology* **46**, 150–160.

Hall, D. K., Bayr, K. J., Bindschadler, R. A. & Schoner, W. (2001) Changes in the Pasterze Glacier, Austria, as measured from the ground and space. *58th Eastern Snow Conference, Ottawa, Ontario, Canada, 2001*. Available at: www.easternsnow.org/proceedings/2001/Hall_1.pdf (accessed 4 October 2010).

Hancox, G. T., McSaveney, M. J., Manville, V. R. & Davies, T. R. (2005) The October 1999 Mt Adams rock avalanche and subsequent landslide dam-break flood and effects in Poerua River, Westland, New Zealand. *New Zealand Journal of Geology and Geophysics* **48**, 683–705.

Harris, C., Vonder Muhll, D., Isaksen, K., et al. (2003) Warming permafrost in European mountains. *Global and Planetary Change* **39**, 215–225.

Harris, C., Arenson, L. U., Christiansen, H. H., et al. (2009) Permafrost and climate in Europe: Monitoring and modelling thermal, geomorphological and geotechnical responses. *Earth-Science Reviews* **92**, 117–171.

Hilker, N., Hegg, C. & Zappa, M. (2008) Flood and landslide caused damage in Switzerland 1972–2007 – with special consideration of the flood in August 2005. *INTERPRAEVENT 2008, Conference Proceedings* **1**, 99–110.

Hilker, N., Badoux, A. & Hegg, C. (2009) The Swiss flood and landslide damage database 1972–2007. *Natural Hazards and Earth System Sciences* **9**, 913–925.

Höller, P. (2009) Avalanche cycles in Austria: an analysis of the major events in the last 50 years. *Natural Hazards* **48**, 399–424.

Holub, M. & Fuchs, S. (2009) Mitigating mountain hazards in Austria – legislation, risk transfer, and awareness building. *Natural Hazards and Earth System Sciences* **9**, 523–537.

Huggel, C., Haeberli, W., Kääb, A., Bieri, D. & Richardson, S. (2004) An assessment procedure for glacial hazards in the Swiss Alps. *Canadian Geotechnical Journal* **41**, 1068–1083.

Huggel, C., Salzmann, N., Allen, S., et al. (2010) Recent and future warm extreme events and high-mountain slope stability. *Philosophical Transactions of the Royal Society, London, Series A* **368**, 2435–2459.

Hughes, P. D. (2008) Response of a Montenegro glacier to extreme summer heatwaves in 2003 and 2007. *Geografiska Annaler* **90A**, 259–267.

Huss, M., Bauder, A., Funk, M. & Hock, R. (2008a) Determination of seasonal mass balance of four Alpine glaciers since 1865. *Journal of Geophysical Research* **113**, F01015. doi:10.1029/2007JF000803.

Huss, M., Farinotti, D., Baunder, A. & Funk, M. (2008b) Modelling runoff from highly glacierized alpine drainage basins in a changing climate. *Hydrological Processes* **22**, 3888–3902.

Internationale Forschungsgesellschaft Interpraevent, eds (2009) *Alpine Naturkatastrophen: Lawinen Muren Felsstürze Hochwässer*. Graz: Stocker, 120pp.

Ivy-Ochs, S., Kerschner, H., Maisch, M., Christl, M., Kubik, P. W. & Schlüchter, C. (2009) Latest Pleistocene and Holocene glacier variations in the European Alps. *Quaternary Science Reviews* **28**, 2137–2149.

Jolly, W. M., Dobbertin, M., Zimmermann, N. E. & Reichstein, M. (2005) Divergent vegetation growth responses to the 2003 heat wave in the Swiss Alps. *Geophysical Research Letters* **32**, L18409. doi:10.1029/2005GL023252.

Kääb, A. (2008) Remote sensing of permafrost-related problems and hazards. *Permafrost and Periglacial Processes* **19**, 107–136.

Keiler, M. (2004) Development of the damage potential resulting from avalanche risk in the period 1950–2000, case study Galtür. *Natural Hazards and Earth System Sciences* **4**, 249–256.

Keiler, M., Zischg, A., Fuchs, S., Hama, M. & Stötter, J. (2005) Avalanche related damage potential – changes of persons and mobile values since the mid-twentieth century, case study Galtür. *Natural Hazards and Earth System Sciences* **5**, 49–58.

Kellerer-Pirklbauer, A., Lieb, G. K., Avian, M. & Gspurning, J. (2008) The response of partially debris-covered valley glaciers to climate change: the example of the Pasterze Glacier (Austria) in the period 1964 to 2006. *Geografiska Annaler* **90A**, 269–285.

Kienholz, H., Krummenacher, B., Kipfer, A. & Perret, S. (2004) Aspects of integral risk management in practice – Considerations with respect to mountain hazards in Switzerland. *Öster-reichische Wasser- und Abfallwirtschaft* **56**, 43–50.

Knight, J. & Harrison, S. (2009) Sediments and future climate. *Nature Geoscience* **2**, 230.

Koboltschnig, G. R., Schöner, W., Holzmann, H. & Zappa, M. (2009) Glaciermelt of a small basin contributing to runoff under the extreme climate conditions in the summer of 2003. *Hydrological Processes* **23**, 1010–1018.

Korup, O. & Clague, J. J. (2009) Natural hazards, extreme events, and mountain topography. *Quaternary Science Reviews* **28**, 977–990.

Krainer, K. (2007) Permafrost und Naturgefahren in Österreich. *Ländlicher Raum 1. Online-Fachzeitschrift des Bundesministeriums für Land- und Forstwirtschaft. Umwelt und Wasserwirtschaft.* Austria. Available at: www.laendlicher-raum.at/filemanager/download/19380 (accessed 15 November 2009).

Lambiel, C. & Pieracci, K. (2008) Permafrost distribution in talus slopes located within the Alpine periglacial belt, Swiss Alps. *Permafrost and Periglacial Processes* **19**, 293–304.

Lambrecht, A. & Kuhn, M. (2007) Glacier changes in the Austrian Alps during the last three decades, derived from the new Austrian glacier inventory. *Annals of Glaciology* **46**, 177–184.

Laternser, M. & Schneebeli, M. (2003) Long-term snow climate trends of the Swiss Alps (1931–99). *International Journal of Climatology* **23**, 733–750.

Lieb, G. K. (1998) High-Mountain permafrost in the Austrian Alps (Europe). In: Lewkowicz, A. G. & Allard, M. (eds), *7th International Conference on Permafrost.* Proceedings, Collection Nordicana 57, Centre d'Etudes Nordiques, Université Laval, Yellowknife, Canada 7th International Permfrost Conference, Yellowknife (Canada), pp 663–668.

Liggins, F., Betts, R. A. & McGuire, B. (2010) Projected future climate changes in the context of geological and geomorphological hazards. *Philosophical Transactions of the Royal Society, London, Series A* **368**, 2347–2367.

Lionello, P., Boldrin, U. & Giorgi, F. (2008) Future changes in cyclone climatology over Europe as inferred from a regional climate simulation. *Climate Dynamics* **30**, 657–671.

Luetschg, M., Lehning, M. & Haeberli, W. (2008) A sensitivity study of factors influencing warm/thin permafrost in the Swiss Alps. *Journal of Glaciology* **54**, 696–704.

Luterbacher, J., Dietrich, D., Xoplaki, E., Grosjean, M. & Wanner, H. (2004) European seasonal and annual temperature variability, trends, and extremes since 1500. *Science* **303**, 1499–1503.

López-Moreno, J. I., Beniston, M. & García-Ruiz, J. M. (2008) Environmental change and water management in the Pyrenees: Facts and future perspectives for Mediterranean mountains. *Global and Planetary Change* **61**, 300–312.

Marchi, L., Cavalli, M., Sangati, M. & Borga, M. (2009) Hydrometeorological controls and erosive response of an extreme alpine debris flow. *Hydrological Processes* **23**, 2714–2727.

Meier, N., Rutishauser, T., Pfister, C., Wanner, H. & Luterbacher, J. (2007) Grape harvest dates as a proxy for Swiss April to August temperature reconstructions back to AD 1480. *Geophysical Research Letters* **34**, L20705. doi:10.1029/2007 GL031381.

Nemec, J., Huybrechts, P., Rybak, O. & Oerlemans, J. (2009) Reconstruction of the annual balance of Vadret da Morteratsch, Switzerland, since 1865. *Annals of Glaciology* **50**, 126–134.

North, N., Kljun, N., Kasser, F., et al. (2007) *Klimaänderung in der Schweiz. Indikatoren zu Ursachen, Auswirkungen, Massnahmen.* Umwelt-Zustand Nr. 0728. Bundesamt für Umwelt, Bern, 77 S.

Nötzli, J., Gruber, S. & Hoelzle, M. (2004) Permafrost und Felsstürze im Hitzesommer 2003. *GEOforum actuel* **20**, 11–14.

Otto, J-C., Schrott, L., Jaboyedoff, M. & Dikau, R. (2009) Quantifying sediment storage in a high alpine valley (Turtmanntal, Switzerland). *Earth Surface Processes and Landforms* **34**, 1726–1742.

Patzelt, G. (2004) Gletscherbericht 2002/2003. *Mitteilungen des Österreichischen Alpenvereins* **59**, 8–15.

Paul, F., Kääb, A., Maisch, M., Kellenberger, T. & Haeberli, W. (2004) Rapid disintegration of Alpine glaciers observed with satellite data. *Geophysical Research Letters* **31**, L21402. doi:10.1029/2004GL020816.

Paul, F., Machguth, H. & Kääb, A. (2005) On the impact of glacier albedo under conditions of extreme glacier melt: The summer of 2003 in the Alps. *EARSeL eProceedings* **4**, 139–149.

Rebetez, M., Mayer, H., Dupont, O., et al. (2006) Heat and drought 2003 in Europe: a climate synthesis. *Annals of Forestry Science* **63**, 569–577.

Rebetez, M., Dupont, O. & Giroud, M. (2008) An analysis of the July 2006 heatwave extent in Europe compared to the record year of 2003. *Theoretical and Applied Climatology* **95**, 1–7.

Reichler, T. & Kim, J. (2008) How well do coupled models simulate today's climate? *Bulletin of the American Meteorological Society* **89**, 303–311.

Rickenmann, D., Hunzinger, L. & Koschni, A. (2008) Flood events and sediment transport during the rainstorm of August 2005 in Switzerland. *INTERPRAEVENT 2008, Conference Proceedings*, **1**, 465–476.

Schär, C. & Jendritzky, G. (2004) Climate change: Hot news from summer 2003. *Nature* **432**, 559–560.

Schär, C., Vidale, P. L., Lüthi, D., et al. (2004) The role of increasing temperature variability in European summer heatwaves. *Nature* **427**, 332–335.

Schmidli, J. & Frei, C. (2005) Trends of heavy precipitation and wet and dry spells in Switzerland during the 20th century. *International Journal of Climatology* **25**, 753–771.

Schmidli, J., Goodess, C. M., Frei, C., et al. (2007) Statistical and dynamical downscaling of precipitation: an evaluation and comparison of scenarios for the European Alps. *Journal of Geophysical Research* **112**, D04105. doi:10.1029/2005JD007026.

Slaymaker, O. (2009) Proglacial, periglacial or paraglacial? In: Knight, J. & Harrison, S. (eds), *Periglacial and Paraglacial Processes and Environments* Special Publications 320. London: Geological Society, pp 71–84.

Slaymaker, O. & Embleton-Hamann, C. (2009) Mountains. In: Slaymaker, O., Spencer, T. & Embleton-Hamann, C. (eds), *Geomorphology and Global Environmental Change*. Cambridge: Cambridge University Press, pp 37–70.

SLF (ed) (2000) *Der Lawinenwinter 1999.* SLF (Eidgenössisches Institut für Schnee- und Lawinenforschung), Davos.

Steiner, D., Pauling, A., Nussbaumer, S. U., et al. (2008) Sensitivity of European glaciers to precipitation and temperature – two case studies. *Climatic Change* **90**, 413–441.

Stoffel, M. & Beniston, M. (2006) On the incidence of debris flows from the early Little Ice Age to a future greenhouse climate: A case study from the Swiss Alps. *Geophysical Research Letters* **33**, L16404. doi:10.1029/2006GL026805.

Stoffel, M., Bollschweiler, M., Leutwiler, A. & Aeby, P. (2008) Tree-Ring Reconstruction of Debris-Flow Events Leading to Overbank Sedimentation on the Illgraben Cone (Valais Alps, Switzerland). *Open Geology Journal* **2**, 18–29.

Van der Schrier, G., Efthymiadis, D., Briffa, K. R. & Jones, P. D. (2007) European Alpine moisture variability for 1800–2003. *International Journal of Climatology* **27**, 415–427.

Vanham, D., Fleischhacker, E. & Rauch, W. (2008) Seasonality in alpine water resources management – a regional assessment. *Hydrology and Earth System Sciences* **12**, 91–100.

Vavrus, S. (2007) The role of terrestrial snow cover in the climate system. *Climate Dynamics* **29**, 73–88.

Vincent, C. (2002) Influence of climate change over the 20th Century on four French glacier mass balances. *Journal of Geophysical Research* **107**, 4375. doi: 10.1029/2001JD000832.

Wanner, H., Grosjean, M., Röthlisberger, R. & Xoplaki, E. (2006) Climate variability, predictability and climate risks: a European perspective. *Climatic Change* **79**, 1–7.

Werder, M. A., Bauder, A., Funk, M. & Keusen, H-R. (2010) Hazard assessment investigations in connection with the formation of a lake on the tongue of Unterer Grindelwaldgletscher, Bernese Alps, Switzerland. *Natural Hazards and Earth System Sciences* **10**, 227–237.

Zappa, M. & Kan, C. (2007) Extreme heat and runoff extremes in the Swiss Alps. *Natural Hazards and Earth System Sciences* **7**, 375–389.

Zemp, M., Haeberli, W., Hoelzle, M. & Paul, F. (2006) Alpine glaciers to disappear within decades? *Geophysical Research Letters* **33**, L13504. doi:10.1029/2006GL026319.

11 Assessing the past and future stability of global gas hydrate reservoirs

Mark Maslin[1], Matthew Owen[1], Richard A. Betts[2], Simon Day[3], Tom Dunkley Jones[4] and Andrew Ridgwell[5]

[1]Department of Geography, University College London, London, UK
[2]Met Office Hadley Centre, Exeter, UK
[3]Aon Benfield UCL Hazard Centre, Department of Earth Sciences, University College London, UK
[4]School of Geography, Earth and Environmental Sciences, University of Birmingham, Birmingham, UK
[5]Department of Geography, Bristol University, Bristol, UK

Summary

Gas hydrates are ice-like deposits containing a mixture of water and gas, the most common naturally occurring gas being methane. Gas hydrates are stable under high pressures and relatively low temperatures and are found underneath the oceans and in permafrost regions. Estimates range from 500 GtC to 10,000 GtC (best current estimate 1600–2000 GtC or giga-tonnes of carbon) stored in ocean sediments and approximately 400 GtC in Arctic permafrost. Gas hydrates may pose a serious geohazard in the near future due to the adverse effects of global warming on the stability of gas hydrate deposits in both ocean sediments and permafrost. It is still unknown whether future ocean warming could led to significant methane release, because thermal penetration of marine sediments to the clathrate–gas interface could be slow enough to allow a new equilibrium to occur without any gas escaping. Even if methane gas does escape from the sediments it is still unclear how much of this could be oxidised in the overlying ocean, preventing it from ever reaching the atmosphere. Models of the global inventory of hydrates and trapped methane bubbles suggest a global 3°C warming could release between 35 and 940 GtC, which could add up to an additional 0.5°C to global warming. The destabilisation of gas hydrate reserves in permafrost areas is more certain as climate models predict that high latitude regions will be disproportionately effected by global warming with temperature increases of over 12°C predicted for much of North America and northern Asia. Our current estimates of gas hydrate storage in the Arctic region are, however, extremely poor and non-existent for Antarctica. The shrinking of both the Greenland and Antarctic ice sheets in response to regional warming may also lead to destabilisation of gas hydrates. As ice sheets shrink the weight removed allows the coastal region and adjacent continental slope to rise through isostacy. This removal of hydrostatic pressure could destabilise gas hydrates, leading to massive slope failure and may increase the risk of tsunamis.

Introduction

The importance of gas (methane) hydrates in the global climate systems has only been realised in the last two decades. Gas hydrates form when there is an adequate supplies of water and methane at relatively low temperatures and high pressures. They are affected by the sediment type and textures as well as seawater salinity. Increasing temperatures and/or lowering pressures can cause gas hydrates to break down, releasing the trapped methane. Gas hydrates were first recognised nearly 200 years ago, by the British scientist Sir Humphrey Davy, who discovered that, if a mixture of gas and water is exposed to high pressure and low temperature, it turns into a solid compound called gas hydrate or clathrate (Henriet & Mienert, 1998). He found this in 1810 by producing an ice-like substance (chlorine hydrate) by effervescing chlorine gas through water under elevated pressure. However, during the following century, gas hydrate was perceived as nothing more than a chemical oddity, and as such it did not receive much attention. In the 1930s, however, the oil and gas industry found that the formation of gas hydrate, or as they called it dirty ice, clogged natural gas pipelines during cold weather. In the 1950s the first two structures of gas hydrates were described. But it was not until the 1970s that Russian scientists, based on theoretical models, postulated that there must be vast quantities of natural gas hydrate deposits on our planet (e.g. Vasil'ev et al., 1970; Trofimuk et al., 1973, 1975, 1979; Tucholke et al., 1977). This theory was confirmed when seafloor samples were recovered in the Black Sea by Russian vessels (Yefremova & Zhizhchenko, 1974), Blake Ridge (Harrison & Curiale, 1982; Shiply & Didyk, 1982) and off Central America by the Deep Sea Drilling Project (DSDP; Stoll et al., 1971). It was also observed that methane was the predominant gas trapped in naturally occurring gas hydrates (Figures 11.1 and 11.2). Since then many countries have set up national research programmes for the investigation of gas hydrate, due to their geohazard and economic implications (Haq, 1998). The most important projects are being run by Japan, Canada, the USA, Germany and India (Henriet & Mienert, 1998).

Gas hydrate structure

Gas hydrates are non-stoichiometric compounds, which means that water molecules (so-called structural molecules) form cage-like structures in which gas molecules are enclosed as guest molecules (Figure 11.1). For this reason, they are also called cage compounds or clathrates (Latin *clatratus* for cage). Generally speaking, a gas hydrate can contain different types of gas molecules in separate cages, depending on the mixture of gas molecules in the direct environment. In addition to CH_4 in naturally occurring gas hydrates. these gases will mostly be H_2S, CO_2 and, less frequently, other hydrocarbons.

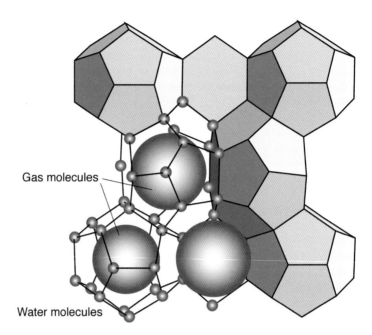

Gas molecules

Water molecules

Figure 11.1 Typical structure of gas hydrate with water molecules linked together to form a cage trapping gas molecules such as methane within.

Hydrates normally form in one of three different repeating crystal structures (Figure 11.2). Structures SI and SII both crystallise to a cubic (isometric) system, whereas the third structure (also denominated H) crystallises to a hexagonal system, similar to ice (von Stackelberg & Müller, 1954; Sloan, 1998). The structure of gas hydrate can be seen as a packing of polyhedral cages. All three structures occur naturally. Structure I is most frequent. Its unit cell consists of eight cages of two different types. Structure I cages can enclose gas molecules that are smaller in diameter than propane molecules, such as CH_4, CO_2 or H_2S. Therefore, the natural occurrence of this crystal structure mainly depends on the presence of biogenic gas, as commonly found in sediments of the ocean floor. A unit cell of structure II consists of 24 cages, i.e. 16 small cages and 8 large ones. The latter are larger than those of structure I. Hence, structure II contains natural mixtures of gases with molecules bigger than ethane and smaller than pentane. It is usually confined to areas where a thermogenic formation of gas takes place in the sediment. H is a more complicated structure (Sloan, 1998). Apart from smaller cages, it contains a cage type, which requires very large gas molecules such as methyl cyclohexane.

Where are gas hydrates found?

We have compiled a map of the currently published possible locations of gas hydrate deposits based on the original dataset of Kvenvolden and Lorenson (2001),

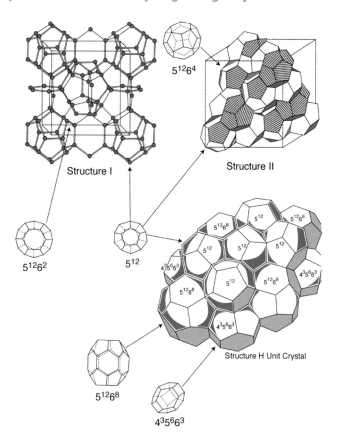

Figure 11.2 Illustration of the three different gas hydrate cage structures; within each cage one gas molecule can be held. (Adapted from Sloan, 1998.)

Ginsbury and Soloviev (1998) and Lorenson and Kvenvolden (2007). The gas hydrate locations are based on both physical samples and seismic evidence taken over the last 20 years (Figure 11.3). Samples of gas hydrate have been taken at approximately 20 different sites, although at another 80 sites the existence of gas hydrate has been suggested by seismic evidence, in the form of bottom-simulating reflectors (BSRs). BSRs are seismic reflectors with a negative reflection coefficient (Henriet & Mienert 1998; Paull & Dillon, 2001; Haacke et al., 2008). They occur at the interface between sediment containing methane hydrate and sediment containing free methane gas; it is the gas that produces the negative reflector. This is because below a certain depth in ocean sediment the geothermal gradient makes the sediment too warm to support the solid gas hydrates, so any methane produced below this depth is trapped as a layer of gas beneath the solid gas hydrate layer. The BSR structures exist roughly parallel to the seafloor morphology along isotherms, hence their name (Shipley et al., 1979). They do not necessarily follow the trend of stratigraphic horizons, but may intersect them. Haacke et al. (2008) provide an excellent model of the formation of the BSRs. The BSRs can occur at depths of several hundred metres below the seafloor, indicating the lower

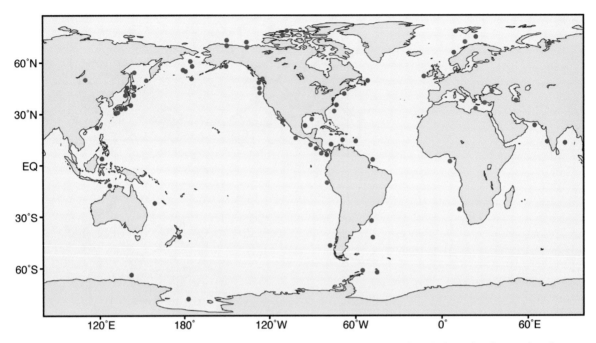

Figure 11.3 Global map of where gas hydrates reserves have been found compiled using data in Kvenvolden and Lorenson (2001), Ginsburg and Soloviev (1998) and Lorenson and Kvenvolden (2007). Red points are locations where the presence of gas hydrates has been inferred, e.g. by the presence of a bottom-simulating reflector (BSR), whereas the blue points are locations where actual samples of gas hydrates have been recovered.

boundary of gas hydrate stability. Consequently, gas hydrate can be assumed to exist above the BSR, otherwise the free gas below the BSR would have migrated upwards. However, gas chimneys have been found in which large quantities of gas can exist surrounded by gas hydrate with no BSR signature (Wood et al., 2002). Figure 11.3 indicates that gas hydrates are found along most continental shelf and slope regions and in many permafrost areas.

Seafloor samples have confirmed the existence of gas hydrates in sediments above BSRs (Abegg et al., 2007, 2008). Most gas hydrate samples have been obtained either by drill-ships (e.g. DSDP, Ocean Drilling Program [ODP] and Integrated Ocean Drilling Program [IODP]), or by research vessels from shallow sediment (Kvenvolden, 1995, 1998). Once exposed to atmospheric temperature and pressure conditions, gas hydrate is unstable and dissociates. For this reason it is likely that the presence of gas hydrate was not noticed in samples taken in the past. The only way to study gas hydrate samples in the laboratory is either to use pressurised sampling devices or to ensure extremely quick sampling and processing on board, preserving the gas hydrate in liquid nitrogen. Gas hydrate can occur in various forms, ranging from finely dispersed lumps to massive, pure layers, which are several centimetres thick (Kvenvolden, 1998). Other structures may consist of layers that are wedge shaped, interbeddings with thicknesses only in the 1- to 10-mm or irregularly branching gas hydrate that completely dissolves the original sedimentary structure and leads to a formation of clastic sediments (Kvenvolden, 1995, 1998). Gas bubbles migrating upward from the seafloor have been observed by submersibles (Sauter et al., 2006) and have also been recorded in the water column by means of echo sounder systems (Westbrook et al., 2009); these have been taken as evidence of near-ocean floor gas hydrate deposits.

How much gas hydrate is there?

The original estimates for the total amount of gas hydrate stored on Earth were between 10,000 and 11,000 GtC (2×10^{16} m^3) (MacDonald, 1990; Kvenvolden, 1998). If correct this would mean that there was over 10 times the amount of carbon stored in gas hydrates as in the atmosphere and would exceed by far the amount of carbon stored in other fossil fuel reservoirs. However, despite the size of the oceanic methane hydrate reservoir still being poorly understood, improved knowledge has led to an order of magnitude drop of current estimates (Milkov, 2004). The highest estimates (e.g. 3×10^{18} m^3, Trofimuk et al., 1973) were based on the assumption that hydrates littered the entire floor of the deep ocean. However, improvements in our understanding of hydrate chemistry and sedimentology have revealed that they form only in a narrow range of depths (continental shelves), only at some locations within their potential depth range (10–30% of the gas hydrate saturation zone), and are typically found at low concentrations (0.9–1.5% by volume) at sites where they do occur. Recent estimates constrained by direct sampling suggest the global inventory lies between 1×10^{15} and 5×10^{15} m^3 (Milkov, 2004). This estimate, corresponding to 500–2500 GtC, is smaller than the 5000 GtC estimated for all other fossil fuel reserves, but substantially larger than the approximately 230 GtC estimated for other natural gas sources (USGS, 2000). These estimates are supported by an alternative approach of Buffett and Archer (2004), who used a mechanistic model to predict the distribution of methane hydrate in marine sediments, and used it to predict the sensitivity of the steady-state methane inventory to changes in the deep ocean. Their best estimate yields 3000 GtC in marine hydrate and another 1800 GtC in bubbles. They have refined this estimate with more detailed modelling. Archer et al. (2009) predict a combined total of methane hydrate and bubbles in the ocean today of between 1600 and 2000 GtC. Of interest they find that most of the hydrate in the model is in the Pacific Ocean, due to the lower oxygen levels that enhances the preservation of organic carbon. The evolution of our estimates of the global carrying capacity of methane hydrates is shown in Figure 11.4, which is an updated version of that presented in Marquardt et al. (2010).

In contrast the permafrost reservoir has not been modelled in detail and has been estimated at about 400 GtC in the Arctic (MacDonald, 1990) but no estimates have been made of possible Antarctic reservoirs. Despite the lowering of

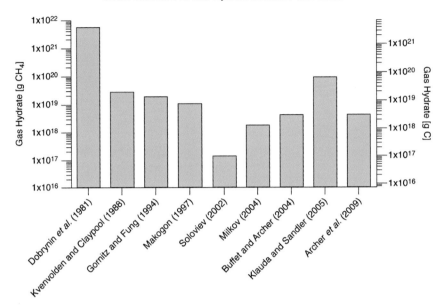

Global estimates of Gas Hydrate in marine sediments

Figure 11.4 Published estimates of global gas hydrate inventories since the early 1980s, based on an updated version of a draft diagram in Marquardt et al. (2010).

estimates over the last decade these are still extremely large amounts. For comparison total atmospheric carbon is around 760 GtC (IPCC, 2007).

Lower abundances of hydrates do not rule out their economic potential (Makogon, 1997; Milkov & Sassen, 2002) but a lower total volume and apparently low concentration at most sites do suggest that only a limited percentage of the deposits may provide an economically viable resource. Potential economic exploration of gas hydrates, however, raises two major issues:

1. Production and use of gas (methane) hydrate releases the greenhouse gas CO_2.
2. Production technology that is cost-effective, environmentally friendly and safe has yet to be developed, for either marine or permafrost gas hydrate.

Fossil fuel reservoirs are sufficient for at least the next century. Hence only a few countries – such as Japan, which does not have any significant deposits of oil and natural gas – are implementing sizeable programmes for an economic production of gas hydrate. Production of oceanic gas hydrate may seem most attractive because of the relatively large quantities available compared with permafrost regions; however, large-scale recovery of oceanic gas hydrates is unlikely to be achieved in the short term for several reasons:

• Low concentration of gas in the sediment, even though every m^3 of gas hydrate can produce 164 m^3 of methane, at atmospheric conditions, the highest concentrations of oceanic gas hydrates quoted are 17–20% by volume of the pore space, which equates to as little as 3% sediment volume.

- The conditions for production are far more complicated than in permafrost regions.
- The geohazards involved as well as the impact on the environment are difficult to assess.

Small-scale production of methane from permafrost gas hydrate already occurs at the Messoyakh field, in western Siberia (Krason, 2000) and has been trialled at Mallik, on the Mackenzie delta, northern Canada (Dallimore et al., 2005).

Formation and break down of gas hydrates

Clathrate/hydrates are not chemical compounds because the sequestered molecules are never bonded to the lattice. The formation and decomposition of hydrates are first-order phase transitions, not chemical reactions. Their detailed formation and decomposition mechanisms are still not well understood on a molecular level (Gao et al., 2005a, 2005b). Gas hydrate can form only if there are sufficient amounts of gas and water combined with high pressure and/or low temperatures. The required pressure and temperature for stability can be influenced by the type of gas incorporated into the hydrate. Hydrates of H_2S, CO_2 and higher hydrocarbon hydrates can exist at higher temperatures, whereas nitrogen hydrates and all hydrate formation in the presence of dissolved salts require lower temperatures. Naturally occurring gas hydrates can be based on various gases. Nitrogen hydrate has been inferred for parts of the Greenland ice sheet, whereas CO_2 hydrate has been suggested to exist on other planets. The most common type of gas hydrate on Earth is methane hydrate and the conditions required for its stability can occur in permafrost soil (Figure 11.5) and marine sediments (Figure 11.6).

Apart from adequate temperature and pressure conditions, gas hydrate formation is dependent on sufficient gas, primarily CH_4, being available. Most of the methane found in the oceans is produced by decomposition of organic components or by bacterial reduction of CO_2 in sediments. Sometimes it may also be a product of thermocatalytic alteration processes in deeper sediments. CH_4 production is highest at the continental margins. Here, high plankton productivity and a high amount continental in-wash yield large amounts of organic matter, which is the basis for gas production in the sediment. Gas hydrate is, therefore, found on any passive and active margins (Dale et al., 2008). However, there are also deposits in the Caspian Sea, the Black Sea, the Mediterranean, Lake Constance and Lake Baikal.

As the stability of gas hydrate is related to high pressure and relatively low temperatures, any change in these two parameters can increase or decrease the stability of the gas hydrate (Archer, 2007), e.g. if either the pressure is reduced or the temperature increased the gas hydrate will change phase from a solid to a gas and liquid. In the marine environment the ocean temperature can vary due to

Figure 11.5 Phase diagram for permafrost sediments (redrawn from Kvenvolden and Lorenson, 2001, and www.gashydrate.de), showing that the temperature gradients are considerably lower than in the ocean (see Figure 11.8), e.g. the temperature can be expected to change by 1.3°C per 100 m within the permafrost zone, compared with 2°C per 100 m in layers below the permafrost zone. The ambient temperature and the thickness of the frozen layer are of paramount importance for the stability of gas hydrate. If the permafrost base is located at a depth of 100 m or less (case 1), the physical conditions will not be adequate for formation of a gas hydrate. The situation is different in case 2, where the permafrost basis is located at greater depth. In polar regions, methane hydrate can occur at depths ranging from 150 m to 1650 m.

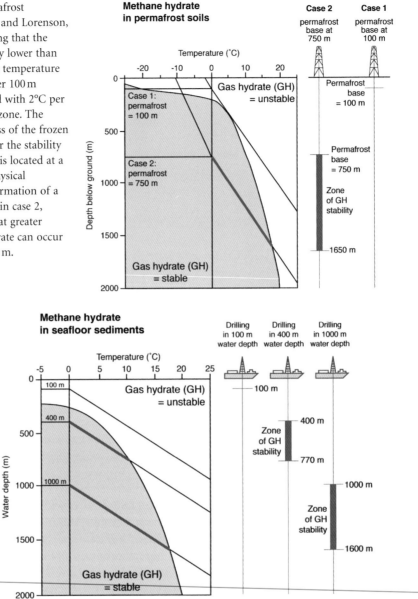

Figure 11.6 Phase diagram for ocean sediments (redrawn from Kvenvolden and Lorenson, 2001, and www.gashydrate.de), showing the physical conditions (temperature and pressure) required for the stability of methane hydrate in a marine environment. Assuming a constant temperature of 0°C, e.g. in polar regions, methane hydrate cannot be stable at a water depth of 100 m. It may occur in a seafloor, which is more than 400 m below sea level. The thickness of the hydrate zone will depend on the temperature gradient. However, with an increasing depth below the seafloor, temperatures get too high for formation of a gas hydrate, so that one can find free gas and water. Given an average temperature increase by 3°C per 100 m sediment depth, when drilling at a water depth of 300 m, we can expect to find a 300-m-thick hydrate layer. At 1000 m water depth, the layer will be 600 m thick. If, however, sediments are characterised by a stronger increase in temperature, which can be the case, for example, at active continental margins (4–6°C per 100 m depth), the hydrate zone will generally be thinner. Gas hydrate has been found in sediments up to 1100 m below the seafloor.

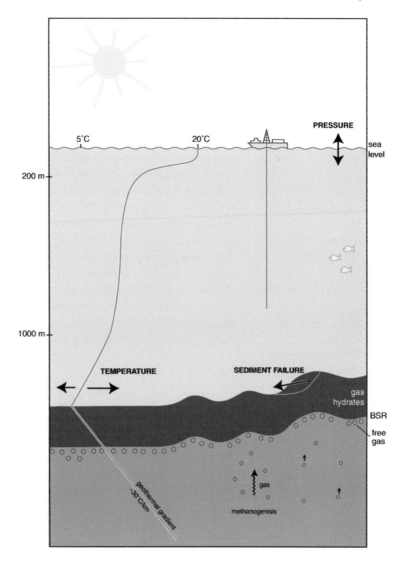

Figure 11.7 The two controls on the location of the gas hydrate layer, temperature and pressure. Temperature is controlled by the ocean bottom water temperature and the geothermal gradient at any given location. Pressure is controlled by sea level and sediment failures. Many gas hydrates reserves have been identified by their bottom-simulating reflector (BSR) which is the free gas trapped below the solid gas hydrate layer that shows up clearly on seismics and follows the sediment surface, not the structures, because the gas hydrate layer is controlled by pressure and temperature not lithology (from Maslin et al., 2010).

circulation changes and general changes in regional and global temperatures (Figure 11.7), e.g. during the last glacial period ocean temperatures were significantly lower so gas hydrates could form at shallower depth. Indeed it has been argued that warming at the end of the ice age did destabilise a large quantity of gas hydrates, leading to the observed peaks in atmospheric methane as seen in the ice core record (e.g. Kennett et al., 2003; Maslin & Thomas, 2003; Maslin et al., 2004). It has, therefore, been argued that warmer ocean temperatures predicted due to global warming will cause gas hydrate dissociation and may lead to a significant increase in atmospheric methane (see Figures 11.6 and 11.7).

The major influence on hydrate stability in permafrost regions is temperature. Regional temperatures can rise and fall, e.g. between ice age and non-ice age periods. Current and future global warming is a concern because its warm effects are thought to be largest in the high latitudes (IPCC, 2007). However, marine transgressions can also cause catastrophic release of gas hydrates from permafrost regions, as demonstrated for the Beaufort Sea by Paull et al. (2007). This is because the land temperature can be below −20°C, but salty oceanic water can be no colder than about −1.8°C before freezing (Maslin & Thomas, 2003). So if sea level rises and floods permafrost areas these will encounter a thermal shock of at least 20°C warming, which could lead to considerable break down of gas hydrates.

Marine gas hydrate stability can also be affected by pressure changes. Increased sea level will increase the hydrostatic pressure and stabilize gas hydrates to a deeper depth, although lowering sea level will reduce the pressure on the sediment causing gas hydrates to become unstable. Another way of removing pressure from underlying sediments is by marine sediment failure. When the overlying sediment column collapse occurs and sediment moves down slope, there is a reduction in weight of the sedimentary column; as upper layers have been removed, this decrease in pressure could cause gas hydrates to dissociate. This can be rapid because pressure effects can be transferred through the sediment quickly and thus can enlarge the sediment collapse as large quantities of hydrate turn to gas (Figure 11.8).

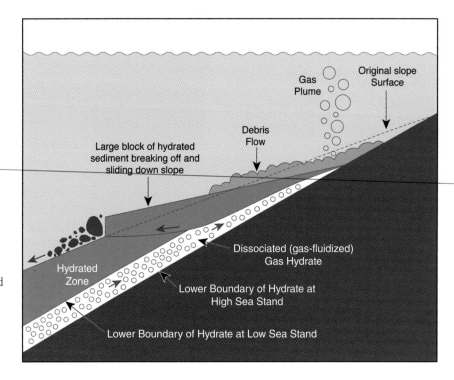

Figure 11.8 Champagne cork effect: what happens when a slope failure occurs above a gas hydrate layer and how large quantities of gas can be released. (Adapted from Kvenvolden and Claypool, 1988, and Maslin et al., 2010.)

Hydrates and past climate changes

Identification of large-scale release of methane gas from the collapse of gas hydrates deposits in the past may be postulated on the basis of two criteria. First is the identification of features associated with hydrate dissociation and gas release, large features such as pockmarks and seabed slumps as well as smaller scale features such as fluid venting chimneys (e.g. Haflidason et al., 2004; Paull et al., 2008). The second is changes in regional and global records of carbon isotopes. This is because hydrate methane has extremely depleted $\delta^{13}C$ values (−40 to −100‰ PDB [Pee Dee Belemnite standard], average ~−60‰ PDB, MacDonald, 1990; Dickens et al., 1997; Kvenvolden, 1998; Buffett, 2000; Maslin & Thomas, 2003) compared with average sea-water (~0‰ PDB). Three case studies are briefly reviewed below with evidence of large-scale methane hydrate release in the past.

Carbon isotope excursion at the Cretaceous–Tertiary boundary

One of the characteristic features of the Cretaceous–Tertiary (K/T) boundary extinction event is a marked but short-lived change or boundary excursion in the stable isotopic composition of carbon in carbonate sediments deposited at the time of the event (see references in Day & Maslin, 2005). Planktonic foraminifera $\delta^{13}C$ records indicate a significant shift to light or depleted values, meaning that the sediment is enriched in the light isotope ^{12}C. Similar, short-lived ^{12}C enrichments have been identified at several other turnover events in the history of life on Earth, such as the Palaeocene–Eocene Thermal Maximum or PETM (Dunkley Jones et al., 2010) and the Permian–Triassic mass extinction. The difference between the K/T boundary carbon isotope excursions and others such as the PETM should also be considered. At the K/T boundary, planktonic foraminifera indicate surface water ^{12}C enrichment, whereas the benthic foraminifera show an enrichment of the bottom water in ^{13}C. In contrast at the PETM both the surface and deep water become enriched in ^{12}C (e.g. Dunkley Jones et al., 2010), which has implications concerning speed of gas hydrate destabilisation and the effects of an impact on ocean productivity. Day and Maslin (2005) presented a plausible mechanism linking a large impact, such as Chicxulub, to significant continental sedimentary slope failure and gas hydrate release. Calculations of both the duration and intensity of syn- and post-impact-related seismicity suggest that there was a sufficient forcing to cause widespread sediment liquefaction and failure along tens of thousands of kilometres of continental slope. They have calculated the potential storage of gas hydrates based on known environmental conditions during the Cretaceous, which suggests that slope failures caused by the Chicxulub impact could have lead to the release of between 300 and 1300 GtC (best estimate ~700 GtC) of methane from the destabilization of gas hydrates. This would produce a global carbon isotopic excursion of between −0.5 and −2‰ (best estimate ~−1‰). This compares well with the observed change of −1‰ in the planktonic foraminifera records across the K/T boundary. This large release of methane may also account for the recently reconstructed very high atmospheric PCO_2 levels after the K/T boundary as our estimated gas hydrate releases could have increased atmospheric carbon dioxide by a maximum of 600–2300 ppm (best estimate ~1200 ppm or parts per million).

Paleocene/Eocene PETM

The PETM (~55.5 Ma) was first identified in the geological record by its expression in deep-sea sediment cores of the Southern Ocean as a rapid paired negative excursion in carbon and oxygen isotopes, coincident with a dramatic extinction event in benthic foraminifera (Kennett & Stott, 1991). First attributed to major ocean circulation changes (Kennett & Stott, 1991), for a number of years the magnitude and rate of the PETM carbon isotope excursion (CIE) proved resistant to explanation with a conventional understanding of atmosphere–ocean–biosphere carbon cycling. On the basis of relatively simple but compelling mass balance arguments, Dickens et al. (1995) first proposed a role for the massive global-scale dissociation of oceanic sediment-hosted methane hydrate deposits in generating such prominent negative CIEs in the geological record. These reservoirs of strongly ^{13}C-depleted carbon ($\delta^{13}C$ ~–60‰), with their inherent sensitivity to environmental temperature and pressure perturbations, have the potential to explain both the rapidity and the magnitude of global negative CIEs, such as observed across the PETM. The catastrophic methane hydrate destabilisation hypothesis has also gained some support from seismic and sedimentological evidence for large-scale PETM slope failure along the western margin of the Atlantic basin (Katz et al., 1999).

The last 15 years has seen a great deal of research focus on the PETM, largely driven by the potential analogues with modern anthropogenic climate change. This has investigated the primary sedimentary record of the PETM, including the magnitude of the CIE (e.g. Zachos et al., 2007), the refinement of age models for the duration of the event (e.g. Rohl et al., 2007), the associated climatic perturbation (e.g. Sluijs et al., 2006; Zachos et al., 2006), evidence for ocean circulation changes (e.g. Thomas et al., 2003; Nunes & Norris, 2006), the profound impacts on oceanic carbonate chemistry (e.g. Zachos et al., 2005; Zeebe & Zachos, 2007), and the effects of the event on benthic (e.g. Thomas & Shackleton, 1996), planktic (e.g. Gibbs et al., 2006) and terrestrial ecosystems (e.g. Gingerich 2006). It has also seen a number of climate and carbon cycle modelling studies attempting to integrate the available proxy data into a coherent view of earth system behaviour across the event (e.g. Bice & Marotzke, 2002; Panchuk et al., 2008; Zeebe et al., 2009). The current synthesis, although with significant continuing uncertainties, appears to consist of an approximately 4‰ negative CIE in the oceanic dissolved inorganic carbon pool, a global temperature anomaly of 5–6°C, a rapid (<10 ka) onset and total PETM duration of about 170 ka, and a carbon input within the range 4000–7000 GtC, with some contribution, but perhaps not the dominant one, from methane hydrates. For further discussion of the PETM see Lunt et al. (2010) and Dunkley Jones et al. (2010).

A forcing role for methane in the Quaternary glacial to interglacial cycles

The Quaternary glacial cycles are characterised by a saw-tooth pattern within which the methane and temperature records are in close agreement (e.g. Brook et al., 1996). This, among other evidence, has led various authors to suggest a role for methane hydrate release in the forcing of quaternary climate change (see Kennett et al., 2003 and references cited therein). Temperature and sea-level oscillations, associated with the changing climate signals, fit the model of hydrate phase shift and hence dissociation. However, whether methane hydrates have played a role in Quaternary climate change is a topic that has sparked much debate (e.g. Sowers, 2006; O'Hara, 2008).

Unfortunately this problem is extremely difficult to investigate because the main sources of evidence are subject to varying degrees of uncertainty and error. The ice-core methane record is subject to uncertainty over difference between ice age and gas age as well as the age distribution, as opposed to a discrete age, of a gas bubble (Spahni et al., 2003). Investigating past occurrence of events that may be related to hydrate dissociation and methane release, such as submarine slope failures and the occurrence of pockmarks, is an extremely expensive affair and accurate age constraint is difficult. Despite these difficulties recent work has focused on compiling databases of mass movement occurrence in order to assess the relationships with climate and the methane record (see Maslin et al., 2004; Owen et al., 2007). It has been noted previously that submarine mass movements appear to occur more frequently during periods of rapid climatic change (e.g. Maslin et al., 2004, 2005, 2006; Owen et al., 2007; Lee, 2009; Camerlenghi et al., 2010). The general consensus is that this is primarily due to changes in the sedimentary regime (with associated effects on preconditioning, trigger mechanisms and pore pressure) rather than a result of gas hydrate dissociation (see Masson et al., 2010). However, this does not rule out a role for mass movements in releasing methane, previously stored either within or beneath the hydrate layer, into the atmospheric system, nor does it preclude a role for gas hydrates in slope instability.

Figure 11.9 shows the occurrence of North Atlantic submarine mass movements plotted with Greenland Ice Sheet Project 2 (GISP2) atmospheric methane concentration for the last 45 ka. A key point to note is the lack of individual mass movements included within the figure: this is due to the difficulty of accurately dating these events. An attempt has been made to magnify any trends by grouping events into 3-ka periods. The Pre-Boreal methane peak coincides with the period with most observed mass movements, and the increased occurrence of mass movements at the glacial to interglacial transition is associated with the Bølling methane peak. However, due to dating uncertainties and the small sample number,

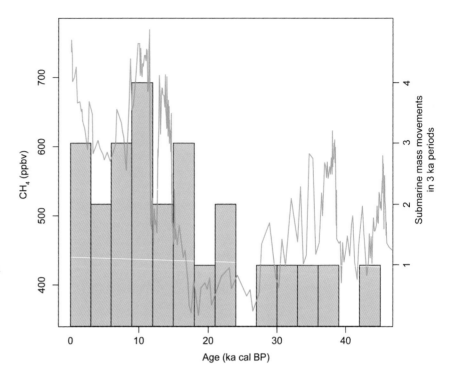

Figure 11.9 Greenland Ice Sheet Project 2 (GISP2) methane concentration (Brook et al., 1996) with North Atlantic submarine mass movement occurrence during the last 45 ka plotted as a histogram in 3 ka periods. (Based on data from Blunier T., Brook E.J. (2001) Timing of milennial-scale climate change in Antartica and Greenland during the last glacial period. Science 291:109–112.)

any meaningful statistical analysis of this data is impossible. It is possible to attempt to reconstruct the palaeo-methane budget by analysing the isotopic composition of ice-core methane records. Sowers (2006) investigates GISP2 δD-CH$_4$ at a stated temporal resolution of about 35 years. Methane hydrates appear to have a more distinctive δD signature than they do δ^{13}C, especially in comparison to global wetlands. Hydrates have a mean δ^{13}C value of −62.5‰, extremely similar to values of −62 and −56.8 to −58.9‰ for boreal and tropical wetlands, respectively; contrasting with this δD-CH$_4$ values are −190‰ and −360 to −380‰ for hydrates and wetlands, respectively (Whiticar & Schaefer, 2007).

GISP2 δD-CH$_4$ and submarine mass movement occurrence for the period 10–20 ka cal BP, are shown in Figure 11.10. During the glacial period the δD-CH$_4$ record is isotopically heavier, as would be expected with a greater role expected for geological, hydrate or biomass-burning methane sources. Progressing towards the Holocene the record is isotopically lighter as tropical and boreal wetlands assume a greater role. If methane hydrates were to release large volumes of methane gas into the atmospheric system the δD-CH$_4$ record would be expected to show a marked increase in the occurrence of ^{2}H, becoming isotopically heavier. We are in agreement with previous authors (e.g. Sowers, 2006; Whiticar & Schaefer, 2007) who state that there is no evidence of the so-called 'smoking gun', catastrophic hydrate dissociation. However, we also feel that the oscillating δD-CH$_4$ values observed at the Older Dryas–Bølling transition are intriguing and seem to

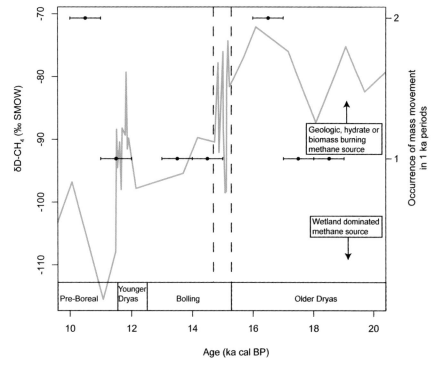

Figure 11.10 Greenland Ice Sheet Project 2 (GISP2) δD-CH$_4$ record (Sowers, 2006) with occurrence of North Atlantic submarine mass movements in 1 ka periods (adapted from Owen et al., 2007) for the period 10–20 ka cal BP, the last glacial to de-glacial transition.

indicate a complex period for the methane budget, with a potential for pulses of ^{2}H-enriched methane disrupting the gradual decrease in isotopic weight caused by the shift to wetland source-dominated atmospheric methane. Intriguingly, this period, 14.7–15.2 ka cal BP, is also a period within which one might expect to observe an increased occurrence of submarine slope failures associated with the deglaciation of mid- to high-latitude margins. The current data on submarine mass movement occurrence do not support this, as shown in Figure 11.9, with the greatest occurrence some 3 ka nearer the present. It is possible that our dataset is biased to more recent failures, which may obscure the record of older events, and it is also true that, on the whole, the dating of these events is not as accurate as we would wish. These two points lead us to the conclusion that, until we have more accurate dating of submarine mass movement occurrence, coupled with a larger database of events, it will not be possible to accurately assess their role as possible sources of methane.

Future global gas hydrate hazards

The future geohazard risks of gas hydrates can be divided into global and regional threats. In terms of global threats, as both marine and permafrost gas hydrate reservoirs are sensitive to environmental changes, they will be affected by global warming. Global warming prediction by the IPCC (2007) suggest that by year 2100 global mean surface temperature could rise by between 1.1 and 6.4°C (best estimate 4°C) and global mean sea level could rise at least 28–79 cm, more if the melting of Antarctica and Greenland accelerates. These predictions include the warming of both the oceans and the permafrost regions, which could cause significant dissociation of gas hydrates, releasing unknown amounts of methane into the atmosphere. As methane is a very strong greenhouse gas, 21 times more powerful than CO_2, gas hydrates have become one of the major concerns of global warming. We believe that our current best estimate is that there are 1600–2000 GtC (Archer et al., 2009) of methane stored beneath the ocean, and there may be another 400 GtC (MacDonald, 1990) in the Arctic permafrost regions compared with only 760 GtC in the atmosphere.

The general Intergovernmental Panel on Climate Change (IPCC, 2007) climate prediction quoted above masks regional difference, which may have a much greater impact on gas hydrate stability. First the temperature increase in the high latitudes is expected to be much larger than this average valve given. Liggins et al. (see Chapter 2) suggest that permafrost areas in the Northern Hemisphere could undergo up to a 12°C warming by 2100. This will have a significant detrimental effect on the stability of gas hydrates trapped in permafrost regions and will have important local affects, which are discussed in the next section. However, as we do not have a reliable estimate of the amount of methane stored in and under permafrost gas hydrates in the Arctic, and no estimate at all for the Antarctic, it is impossible to quantify the magnitude of any methane release.

Climate models also predict an increase in intermediate water temperature (IPCC, 2007). This is an important consideration because it is the intermediate ocean waters, between about 200 and 1500 m water depth, that bathes the sediment most likely to contain gas hydrates. This predicted warming of the intermediate ocean depth will tend to destabilise the marine gas hydrates. Although the small predicted future sea-level rise of up to 1 m may increase the hydrate stability slightly but in completely insufficient to counter the warming. However, to destabilise gas hydrates through increased temperatures, the increased warmth must penetrate through the gas hydrate layer to the phase transition at the hydrate–gas boundary at the base. Thermal diffusivity is a relatively slow process and related to the temperature gradient, hydrate depth and sediment composition (Clennell et al., 1999; Henry et al., 1999). It could, therefore, be possible for the destabilised gas hydrate at the base of the hydrate layer to become gas and migrate up into the stable zone and re-form hydrate. Archer et al. (2009) used models to estimate not only the gas hydrate inventory discussed above but also the possible future release. They also calculated that between 35 and 940 GtC of methane could escape due to global warming of 3°C. This is a huge range of uncertainty reflecting the problems discussed above. Archer et al. (2009) show that their estimate is highly dependent on the assumption of how much of the newly formed gas can escape the sediment into the water column and then into the atmosphere. They also predict that the maximum calculated methane release would add a maximum additional 0.5°C to global warming.

Permafrost regions

In the high latitude and high altitude areas permafrost exists where it is so cold that the ground is frozen solid to a great depth. During the summer months only the uppermost layer of the permafrost gets warm enough to melt, and this is the so-called active layer. In Sweden, active layer thickening at rates of 0.7–1.3 cm/year has been observed over the last 29 years and in the last decade rates have accelerated to 2.0 cm/year (Åkermann & Johansson, 2008). Future predictions suggest that there will be an increase in the thickness of the active layer of the permafrost or perhaps in some areas, the complete disappearance so called discontinuous permafrost over the next century (Lawrence et al., 2008). The thermal shock of this warming will destabilise gas hydrates trapped beneath the active layer. This widespread loss of permafrost and the loss of structural integrity due to dissolution of gas hydrates will produce a huge range of local problems, such as triggering erosion or subsidence, and changing hydrological processes. These changes in sediment cohesiveness may reduce the stability of slopes and thus increase incidence mass movement such as soil creep, landslides and avalanches. A more dynamic cryosphere may increase the natural hazards for people, structures and communication links. Already, buildings, roads, pipelines (such as the oil pipelines in Alaska) and communication links are being threatened. There is also the very remote possibility that methane could be explosively released and in the right circumstances would burn, posing a serious highly localised threat. Recent research carried out in the Siberian Arctic has shown that millions of tons of methane are already being released with concentrations in some regions reaching up to 100 times above normal (Shakhova et al., 2007, 2008).

Marine risks

Both the occurrence of gas hydrate deposits and predicted oceanic warming vary regionally so there will be particular areas that are more vulnerable to gas hydrate destabilisation. Moreover ocean temperature changes can be caused by variations in the location of key ocean currents. Westbrook et al. (2009) have shown large plumes of methane gas bubbles emanating from a seabed west of Svalbard. They suggest that this is due to the breakdown of gas hydrates caused by a shift and warming of the West Spitzbergen Current. These localised gas plumes could have safety implications for both shipping and marine oil/gas production. If sufficient methane is released in relatively shallow water the gas could produce negative buoyancy and cause boats/ship to founder. A more likely problem is the increased risk to oil and gas platforms in deeper water, i.e. between 200 and 1500 m water depth. As the localised increase in intermediate water temperatures could cause gas hydrate to break down, this may lead to significant sediment failure (see below) and/or out gassing. Both of these could be detrimental to the structural safety of the platforms and those working on them. Of course these risks will be significantly increased if economic exploitation of marine gas hydrate deposits is initiated.

Catastrophic submarine slope failures

When enclosed in sedimentary pore space, gas hydrate can act like cement, compacting and stabilising the seafloor (Grozic, 2010). The degree of strengthening is a function of temperature, strain rate, particle size, density and cage occupancy (Winters et al., 2004). However, if formed in deposits that are still unconsolidated, gas hydrate prevents the normal increase of compaction with increasing lithostatic pressure. If exposed to lower pressure and/or increased temperature, the gas hydrate decomposes. If this occurs, the compactness of the seafloor decreases, which may lead to submarine slope failures. Seismic, bathymetric and side-scan sonar mapping of the seafloor have shown that these mass movement deposits, of various sizes and morphologies, characterise all continental slopes, e.g. near slide scars one can often find seismic evidence of gas and fluid pathways up through the sediment column. The seafloor itself shows crater-like depressions, so-called pockmarks that indicate fluid or gas venting (e.g. Hustoft et al., 2009).

Sediments fail when the downslope force (shear stress) exceeds the resisting force (shear strength). Sediments deposited in a state where shear strength is the greater will be stable. However, an alteration of this state through either a reduction in shear strength or an increase in shear stress can lead to failure. At the upper shelf edge there are slopes of more than 4°. Submarine slope failures appear likely if such a slope moves out of equilibrium and into an unstable state. There are many means by which this can occur: one is gas hydrate dissociation. Stable sediments, strengthened at depth by a hydrate layer, may become unstable if the hydrate layer dissociates, weakening the sedimentary column and allowing either gravimetric forces, or a relatively minor trigger (such as a low-magnitude earthquake), to initiate slope failure. One important factor seems to be the expansion of the released gas, which increases with decreasing water depth, e.g. at 650 m water depth, the volume of released gas and water is almost three times the original gas hydrate volume. Gas hydrate decomposition at the upper shelf edge can result in an enormous pore pressure which leads to a massive loss of compactness whereas the large pore space makes the sediment highly deformable. This occurs only if there is no consistent venting of the methane gas. The free gas and hydrate layer that result in the BSRs, with their slope-parallel trend, are another potential preconditioning factor for submarine slope failure. When gas hydrate dissociates, they will be weak layers parallel to the slope.

There is some evidence that gas hydrate decomposition has contributed to the cause of a number of massive slides over the last 45 ka (Maslin et al., 2004; Owen et al., 2007; Lee, 2009; Maslin, 2009). Although it is extremely difficult to link hydrates directly to these geological slope failures, one such event is the Cape Fear slide, on the US Atlantic margin, for which methane hydrates, along with salt diapirism, are believed to have played a significant role (Hornbach et al., 2007). One of the most famous submarine slope failures in which gas hydrates are implicated is the Storegga slide off Norway, which has a known volume of 3000 km^3 and occurred about 8100 years ago (Bouriak et al., 2000; Bryn et al., 2003; Haflidason et al., 2004, 2005). This slide is know to have produced a tsunami impacting Norway, the Faeroes, Scotland and northern England, with a run-up of up to 20 m (Bondevik et al., 2005). More recently, the 1998 Sissano tsunami, which resulted in the loss of more than 2000 lives, may have also been caused by a submarine slope failure (Tappin et al., 2001).

Global warming via gas hydrate destabilisation may pose a potential threat to continental slope sediment stability. If significant parts of Greenland and Antarctica ice sheets melt (Hansen et al., 2008) the removal of the weight of ice from the continent initiates a slow movement upwards. This isostatic rebound can be seen in the British Isles, which are still recovering from the last ice age that ended 10,000 years ago. Scotland, on which the British ice sheet sat, is recovering at a rate of 3 mm/year whereas England is still sinking at a rate of 2 mm/year. This coastal recovery or isostatic rebound means that the relative sea level around the continental shelf reduces, removing the weight and thus pressure of the seawater on the marine sediment. In some cases pressure removal may be a more efficient way of destabilising gas hydrates than temperature increases, and hence there is a potential risk of slumping on the continental shelf around Greenland and Antarctic, which could lead to tsunami generation. There is already evidence that Arctic sea ice is in retreat (Wang & Overland, 2009; Boé et al., 2009) and the Greenland and Antarctic ice sheets are shrinking (e.g. Thomas, 2001; Wingham et al., 2006). Hence regional gas hydrate hazards must be considered and studied in more depth to assess the potential risks.

Conclusions

At the moment it is difficulty to assess the potential future geohazard represented by gas hydrates due to a lack of knowledge, e.g. we are unsure how much methane is stored in and below gas hydrates. For ocean sediments estimates range from 500 GtC to 10,000 GtC (best estimate 1600–2000 GtC) stored (see Figure 11.4), although our current estimates of gas hydrate storage in the Arctic region are very poor (~400 GtC) and non-existent for Antarctica. We do not know whether increased future temperatures could led to significant methane release from ocean sediments, because thermal penetration of marine sediments to the clathrate–gas interface could be slow enough to allow a new equilibrium to occur without any gas escaping. Even if methane gas does escape it is still unclear how much of this could be oxidised in the overlying ocean (Paull et al., 2003). Our best estimate by

Archer et al. (2009) using state-of-the-art modelling has a huge potential range from 35 GtC to 940 GtC. The destabilised gas hydrate reserves in permafrost areas seems more certain because climate models predict that high-latitude regions will be disproportionately affected by global warming, with temperature increases of up to 12°C predicted for much of North America and Northern Asia. But our current estimates of gas hydrate storage in the Arctic region are very poor and non-existent for Antarctica. Hence more research is required to (1) quantify the amount of methane stored in and below gas hydrates deposits and (2) increase our understanding of the possible limits of stability of these deposits, before we can successful evaluate the future risk that they pose to both the local and the global environment.

Acknowledgements

We would like to thank Dave Tappin, John Rees and Chris Rochelle for their excellent reviews. We would like to acknowledge the support of NERC QUEST dPETM and MethaneNet NE/H001964/1.

References

Abegg, F., Bohrmann, G., Freitag, J. & Kuhs, W. (2007) Fabric of gas hydrate in sediments from Hydrate Ridge-results from ODP Leg 204 samples. *Geo-Marine Letters* **27**, 269–277.

Abegg, F., Hohnberg, H-J., Pape, T., Bohrmann, G. & Freitag, J. (2008). Development and application of pressure-core-sampling systems for the investigation of gas – and gas-hydrate-bearing sediments. *Deep Sea Research* I **55**, 1590–1599.

Åkerman, H. J. & Johansson, M. (2008) Thawing Permafrost and Thicker Active Layers in Sub-arctic Sweden. *Permafrost and Periglacial Processes* **19**, 279–292.

Archer, D. (2007) Methane hydrate stability and anthropogenic climate change. *Biogeosciences* **4**, 521–544

Archer, D., Buffett, B. & Brovkin, V. (2009) Ocean methane hydrates as a slow tipping point in the global carbon cycle. *Proceedings of the National Academy of Sciences* **106**, 20596–20601.

Bice, K. L. & Marotzke, J. (2002) Could changing ocean circulation have destabilized methane hydrate at the Paleocene/Eocene boundary? *Paleoceanography* **17**, art. no. 1018.

Blunier, T. & Brook , E. J. (2001) Timimg of millenial-scale climate change in Antartica and greenland during the last glacial period. *Science* **291**, 109–112.

Boé, J., Hall, A. & Qu. X. (2009). September sea-ice cover in the Arctic Ocean projected to vanish by 2100. *Nature Geoscience* **2**, 341–343.

Bondevik, S., Lovholt, F., Harbitz, C., Mangerud, J., Dawson, A. & Svendsen, J. I. (2005) The Storegga Slide tsunami – comparing field observations with numerical simulations. *Marine and Petroleum Geology* **22**, 195–208.

Bouriak, S., Vanneste, M. & Saoutkine, A. (2000) Inferred gas hydrates and clay diapirs near the Storegga Slide on the southern edge of the Vøring plateau, offshore Norway. *Marine Geology* **163**, 125–148.

Brook, E. J., Sowers, T. & Orchardo, J. (1996) Rapid variations in atmospheric methane concentrations during the past 110,000 years. *Science* **273**, 1087–1091.

Bryn, P., Solheim, A., Berg, K., et al. (2003) The Storegga complex; repeated large scale sliding in response to climatic cyclicity. In: Locat, J. & Mienert, J. (eds), *Submarine mass movements and their consequences. Advances in Natural and Technological Hazards Research Series*. Dordrecht: Kluwer Academic Publishers, pp 215–222.

Buffett, B. (2000) Lathrate hydrates. *Annual Review of Earth and Planetary Sciences* **28**, 477–507.

Buffett, B. & Archer, D. (2004) Global inventory of methane clathrate: sensitivity to changes in the deep ocean. *Earth and Planetary Science Letters* **227**, 185–199

Camerlenghi, A., Urgeles, R. & Fantoni, L. (2010) A database on Submarine Landslides in the Mediterranean Sea. In: Mosher, D. C., Shipp, R. C., Moscardelli, L., et al. (eds), *Submarine Mass Movements and Their Consequences. Advances in Natural and Technological Hazards Research*, Vol. 28. Dordrecht: Springer, pp 503–513.

Clennell M. B., Hovland M., Booth J. S., Henry P. & Winters W. J. (1999) Formation of natural gas hydrates in marine sediments 1. Conceptual model of gas hydrate growth conditioned by host sediment properties. *Journal of Geophysical Research* **104**, 22985–23004.

Dale, A. W., Cappellen, P. V., Aguilera, D. R. & Regnier, P. (2008) Methane efflux from marine sediments in passive and active margins: Estimations from bioenergetic reaction–transport simulations. *Earth and Planetary Science Letters* **265**, 329–344.

Dallimore, S. R., Taylor, A. E., Wright, J. F., Nixon, F. M., Collett, T. S. & Uchida, T. (2005) Overview of the coring program for the JAPEX/JNOC/GSC et al. Mallik 5L-38 gas hydrate production research well. *Geological Survey of Canada Bulletin* **585**, 82.

Day, S. & Maslin M.A. (2005) Linking large impacts, gas hydrates and carbon isotope excursions through widespread sediment liquifaction and continental slope failure: The example of the K-T boundary: event. In: Kenkmann, T., Horz, F. & Deutsch, A. (eds), *Large Impacts III*. Special Paper 384. Geological Society of America, pp 239–258.

Dickens, G. R., O'Neil, J. R., Rea, D. K. & Owen, R. M. (1995) Dissociation of oceanic methane hydrate as a cause of the carbon isotope excursion at the end of the Paleocene. *Paleoceanography* **10**, 965–971

Dickens, G. R., Paull, C. K., Wallace, P. & ODP Leg 164 Scientific Party (1997) Direct measurements of in situ mathane quantities in a large gas-hydrate reservoir. *Nature* **385**, 426–428.

Dunkley Jones T., Ridgwell A., Lunt D. J., Maslin M. A., Schmidt, D. N. & Valdes, P. J. (2010) A Palaeogene perspective on climate sensitivity and methane hydrate instability. *Philosophical Transactions of the Royal Society of London A* **368**, 2395–2415

Gao, S., House, W. & Chapman, W.G. (2005a) NMR MRI Study of Gas Hydrate Mechanisms. *Journal of physical chemistry B* (American Chemical Society) **109**, 19090–19093.

Gao, S., Chapman, W.G., & House, W. (2005b) NMR and viscosity investigation of clathrate formation and dissociation. *Industrial & Engineering Chemistry Resistry* **44**: 7373–7379.

Gibbs, S. J., Bown, P. R., Sessa, J. A., Bralower, T. J. & Wilson, J. B. (2006) Nannoplankton Extinction and Origination across the Paleocene-Eocene Thermal Maximum. *Science* **314**, 1770–1773.

Gingerich, P. D. (2006) Environment and evolution through the Paleocene-Eocene thermal maximum. *Trends in Ecology and Evolution* **21**, 246–253.

Ginsburg G. D. & Soloviev V. A. (1998) *Submarine Gas Hydrates.* St Petersburg: VNIIOkeangeologia, p 220.

Grozic, J. L. H. (2010) Interplay between gas hydrates and submarine slope failure. In: Mosher, D. C., Shipp, R. C., Moscardelli, L., et al. (eds), *Submarine Mass Movements and Their Consequences. Advances in Natural and Technological Hazards Research*, vol. 28. Dordrecht: Springer, pp 11–30.

Haacke, R. R., Westbrook, G. K. & Riley, M. S. (2008) Controls on the formation and stability of gas hydrate-related bottom-simulating reflectors (BSRs): A case study from the west Svalbard continental slope. *Journal of Geophysical Research* **113**, B05104.

Harrison, W. E. & Curiale, J. A. (1982) Gas hydrates in sediments of holes 497 and 498A, Deep Sea Drilling Project Leg 67. *Initial Reports Deep Sea Drilling Project.* Washington DC: US Government Printing Office, pp 591–594.

Haq, B. (1998) Natural gas hydrates: searching for the long-term climate and slope stability records, In: Henriet, J-P. & Mienert, J. (eds), *Gas Hydrates: Relevance to world margin stability and climate change.* Geological Society Special Publication No. 137. London: Geological Society, pp 303–318.

Haflidason, H., Sejrup, H., Nygard, A., et al. (2004) The Storegga Slide: architecture, geometry and slide development. *Marine Geology* **213**, 201–234.

Haflidason, H., Lien, R., Sejrup, H. P., Forsberg, C. F. & Bryn, P. (2005) The dating and morphometry of the Storegga slide. *Marine and Petroleum Geology* **22**, 123–136.

Hansen, J., Sato, M., Kharecha, P., et al. (2008) Target atmospheric CO_2: Where should humanity aim? *The Open Atmospheric Science Journal* **2**, 217–231.

Henriet, J-P. & Mienert, J., eds (1998) *Gas Hydrates: Relevance to world margin stability and climate change.* Geological Society Special Publication No. 137. London: Geological Society, 338pp.

Henry, P., Thomas, M. & Clennell, M. B. (1999) Formation of natural gas hydrates in marine sediments 2: Thermodynamic calculations of stability conditions in porous sediments. *Journal of Geophysical Research* **104**, 23005–23020.

Hornbach, M. J., Lavier, L. L. & Ruppel, C. D. (2007) Triggering mechanism and tsunamogenic potential of the Cape Fear Slide complex, U.S. Atlantic margin. *Geochemistry Geophysics Geosystems* **8**, Q12008.

Hustoft, S., Bunz, S., Mienert, J. & Chand, S. (2009) Gas hydrate reservoir and active methane-venting province in sediments on <20 Ma young oceanic crust in the Fram Strait, offshore NW-Svalbard. *Earth and Planetary Science Letters* **284**, 12–24.

Intergovernmental Panel on Climate Change (IPCC) (2007) *The Scientific Basis, Contribution of Working Group I to the Fourth Assessment Report of the Intergovernmental Panel on Climate Change.* Cambridge: Cambridge University Press.

Katz, M. E., Pak, D. K., Dickens, G. R. & Miller, K. G. (1999) The source and fate of massive carbon input during the latest Paleocene thermal maximum. *Science* **286**, 1531–1533.

Kennett, J. P. & Stott, L. D. (1991) Abrupt deep-sea warming, palaeoceanographic changes and benthic extinctions at the end of the Paleocene. *Nature* **353**, 225–229.

Kennett, J., Cannariato, K. G., Hendy, I. L. & Behl, R. J. (2003) *Methane Hydrates in Quaternary Climate Change: The clathrate gun hypothesis.* Washington DC: American Geophysical Union, 216 pp.

Krason, J. (2000) Messoyakh gas field (W. Siberia) – A model for development of the methane hydrate deposits of Mackenzie Delta. *Annals of the New York Academy of Sciences* **912**, 173–188.

Kvenvolden, K. (1995) A review of the geochemistry of methane in natural gas hydrate. *Organic Geochemistry* **23**, 997–1008.

Kvenvolden, K. (1998) A primer on the geological occurrence of gas hydrate. In: Henriet, J-P. & Mienert, J. (eds), *Gas Hydrates: Relevance to world margin stability and climate change* Geological Society Special Publication No. 137. London: Geological Society, pp 9–30.

Kvenvolden, K. A. & Claypool, G. E. (1988). *Gas hydrates in oceanic sediment.* US Geological Survey, Open-File Rep. No. 88-216, 50pp.

Kvenvolden, K. A. & Lorenson, T. D. (2001) The global occurrence of natural gas hydrates. In: Paull, C. K. & Dillon, W. P. (eds), *Natural Gas Hydrates: Occurrence, distribution and detection.* AGU Geophysical Monograph 124. Washington DC: American Geophysical Union, pp 3–18.

Lawrence, D., Slater, A. G., Tomas, R. A., et al. (2008) Accelerated Arctic land warming and permafrost degradation during rapid sea ice loss. *Geophysical Research Letters* **35**, L11506.

Lorenson, T. D. & Kvenvolden, K. A. (2007) *A Global Inventory of Gas Hydrate.* US Geological Survey. Reston, VA, USA. Available at: walrus.wr.usgs.gov/globalhydrate/browse.pdf (accessed March 2011).

Lee, H. J. (2009) Timing of occurrence of large submarine landslides on the Atlantic Ocean margin. *Marine Geology* **264**, 53–64.

Lunt D. J., Valdes P. J., Dunkley Jones T., et al. (2010) CO_2-driven ocean circulation changes as an amplifier of Paleocene-Eocene thermal maximum hydrate destabilization. *Geology* **38**, 875–878.

MacDonald, G. J. (1990) Role of methane clathrates in past and future climates. *Climatic Change* **16**, 247–281.

Makogon, Y. F. (1997) *Hydrates of Hydrocarbons*. Tulsa, OK: Penn Well.

Marquardt, M., Hensen, C., Piñero, E., Wallmann, K. & Haeckel, M. (2010) A transfer function for the prediction of gas hydrate inventories in marine sediments. *Biogeosciences* **7**, 2925–2941.

Maslin M. A. (2009) Review of the timing and causes of the Amazon Fan Mass Transport and Avulsion Deposits during the latest Pleistocene. In: Kneller, B., Martinsen, O. J. & McCaffrey, B. (eds), *External Controls on Deep-Water Depositional Systems*. Society for Sedimentary Geology Special Publication No. 92. Tulsa, OK: SPEM, pp 133–144.

Maslin, M. A. & Thomas, E. (2003) Balancing the deglacial global carbon budget: the hydrate factor. *Quaternary Science Review* **22**, 1729–1736.

Maslin, M. A., Owen, M., Day, S. & Long, D. (2004) Linking continental slope failure to climate change: Testing the Clathrate Gun Hypothesis. *Geology* **32**, 53–56.

Maslin, M. A., Vilela, C., Mikkelsen, N. & Grootes, P. (2005) Causation of the Quaternary catastrophic failures of the Amazon Fan deduced from stratigraphy and benthic foraminiferal assemblages. *Quaternary Science Review* **24**, 2180–2193.

Maslin, M.A., Knutz, P.C. & Ramsay, T. (2006) Millennial-scale sea level control on avulsion events on the Amazon Fan. *Quaternary Science Review* **25**, 3338–3345.

Maslin, M., Owen, M., Betts, R., Day, S., Dunkley Jones, T. & Ridgwell, A. (2010) Gas hydrates: past and future geohazard? *Philosophical Transactions of the Royal Society of London A* **368**, 2369–2393

Masson, D. G., Wynn, R. B. & Talling, P. J. (2010) Large landslides on passive continental margins: processes, hypotheses and outstanding questions. In: Mosher, D. C., Shipp, R. C., Moscardelli, L., et al. (eds), *Submarine Mass Movements and Their Consequences. Advances in Natural and Technological Hazards Research*, Vol. 28, Dordrecht: Springer, pp 153–165.

Milkov, A. V. (2004) Global estimates of hydrate-bound gas in marine sediments: how much is really out there? *Earth Science Review* **66**, 183–197.

Milkov, A. V. & Sassen, R. (2002) Economic geology of offshore gas hydrate accumulations and provinces. *Marine Petroleum Geology* **19**, 1–11.

Nunes, F. & Norris, R. D. (2006) Abrupt reversal in ocean overturning during the Palaeocene/Eocene warm period. *Nature* **439**, 60–63.

O'Hara, K. D. (2008) A model for late Quaternary methane ice core signals: Wetlands versus a shallow marine source. *Geophysical Research Letters* **35**, L02712.

Owen, M., Day, S. & Maslin, M. (2007) Late Pleistocene submarine mass movements: occurrence and causes. *Quaternary Science Reviews* **26**, 958–978.

Panchuk, K., Ridgwell, A. & Kump, L. R. (2008) Sedimentary response to Paleocene-Eocene Thermal Maximum carbon release: A model-data comparison. *Geology* **36**, 315–318.

Paull, C. K. & Dillon W. P., eds (2001) *Natural Gas Hydrates: Occurrence, distribution and detection*. AGU Geophysical Monograph 124. p 315.

Paull, C. K., Brewer, P. G., Ussler, W., III, Peltzer, E. T., Rehder, G. & Clague, D. (2003) An experiment demonstrating that marine slumping is a mechanism to transfer methane from seafloor gas-hydrate deposits into the upper ocean and atmosphere. *Geo-Marine Letters* **22**, 198–203.

Paull, C. K., Ussler, W., III., Dallimore, S. R., et al. (2007) Origin of pingo-like features on the Beaufort Sea shelf and their possible relationship to decomposing methane gas hydrates. *Geophysical Research Letters* **34**, L01603.

Paull, C. K., Ussler, W., Holbrook, W. S., et al. (2008) Origin of pockmarks and chimney structures on the flanks of the Storegga Slide, offshore Norway. *Geo-Marine Letters* **28**, 43–51.

Rohl, U., Westerhold, T., Bralower, T. J. & Zachos, J. C. (2007) On the duration of the Paleocene-Eocene thermal maximum (PETM). *Geochemistry Geophysics Geosystems* **8**.

Sauter, E. J., Muyakshin, S. I., Charlou, J. L., et al. (2006) Methane discharge from a deep-sea submarine mud volcano into the upper water column by gas hydrate-coated methane bubbles. *Earth and Planetary Science Letters* **243**, 354–365.

Shakhova, N., Semiletov, I., Salyuk, A., Kosmach, D. & Bel'cheva N. (2007) Methane release on the Arctic East Siberian shelf. *Geophysical Research Abstracts* **9**, L01071

Shakhova, N., Semiletov, I., Salyuk, A. & Kosmach, D. (2008) Anomalies of methane in the atmosphere over the East Siberian shelf: Is there any sign of methane leakage from shallow shelf hydrates?, EGU General Assembly 2008. *Geophysical Research Abstracts* **10**, EGU2008-A-01526

Shipley, T. H. & Didyk, B. M. (1982) Occurrence of methane hydrates offshore southern Mexico. In: Watkins J. S., Moore J. C., et al. (eds), *Initial Reports, Deep Sea Drilling*. Washington DC: US Government Printing Office, pp 547–555.

Shipley T. H., Houston, M. H., Buffler, R. T., et al. (1979) Seismic reflection evidence for the widespread occurrence of possible gas-hydrate horizons on continental slopes and rises. *American Association of Petroleum Geologists Bulletin* **63**, 2204–2213.

Sloan, E. D. Jr (1998) *Clathrate Hydrates of Natural Gases*, 2nd edn. New York: Marcel Dekker Inc.

Sluijs, A., Schouten, S., Pagani, M., et al. (2006) Subtropical Arctic Ocean temperatures during the Palaeocene/Eocene thermal maximum. *Nature* **441**, 610–613.

Sowers, T. (2006) Late quaternary atmospheric CH_4 isotope record suggests marine clathrates are stable. *Science* **311**, 838–840.

Spahni, R., Schwander, J., Fluckiger, J., Stauffer, B., Chappellaz, J. & Raynaud, D. (2003) The attenuation of fast atmospheric CH_4 variations recorded in polar ice cores. *Geophysical Research Letters* **30**, art. no. 1571.

Stoll, R. D., Ewing, J. I. & Bryan, G. M. (1971) Anomalous wave velocities in sediments containing gas hydrates. *Journal of Geophysical Research* **76**, 2090–2094.

Tappin, D. R., Watts, P., McMurtry, G. M., Lafoy, Y. & Matsumoto, T. (2001) The Sissano, Papua New Guinea tsunami of July 1998 – offshore evidence on the source mechanism. *Marine Geology* **175**, 1–23.

Thomas, D. J., Bralower, T. J. & Jones, C. E. (2003) Neodymium isotopic reconstruction of late Paleocene-early Eocene thermohaline circulation. *Earth and Planetary Science Letters* **209**, 309–322.

Thomas, E. & Shackleton, N. J. (1996) The Palaeocene-Eocene benthic foraminiferal extinction and stable isotope anomalies. In: Knox, R. W. O. B., Corfield, R. M. & Dunay, R. E. (eds), *Global Perspective: Geochronology and the oceanic record*, vol. 101. London: Geological Society of London, pp 401–441.

Thomas, R.H. (2001) Remote sensing reveals shrinking Greenland Ice Sheet. *Eos Transactions AGU* **82**, 369–373.

Trofimuk, A. A., Cherskiy, N. V. & Tsarev, V. P. (1973) Accumulation of natural gases in zones of hydrate – formation in the hydrosphere [in Russian]. *Doklady Akademii Nauk SSSR* **212**, 931–934.

Trofimuk A. A., Cherskiy, N. V. & Tsarev, V. P. (1975) The biogenic methane resources in the oceans. *Doklady Academii Nauk SSSR* **225**, 936–943.

Trofimuk A. A., Cherskiy, N. V. & Tsarev, V. P. (1979) The gas-hydrate sources of hydrocarbons. *Priroda* **1**, 18–27.

Tucholke, B. E., Bryan, G. M. & Ewing, J. I. (1977) Gas-hydrate horizons detected in seismic-profiler data from the western North Atlantic. *American Association of Petroleum Geologists Bulletin* **61**, 698–707.

Vasil'ev, V. G., Makogon, Y. F., Trebin, F. A., Trofimuk A. A. & Cherskiy, N. V. (1970) *The property of natural gases to occur in the Earth crust in a solid state and to form gas hydrate deposits. Otkrytiya v SSSR* 1968–1969, 15–17.

von Stackelberg, M. & Müller, H. M. (1954) *Zeitschrift für Elektrochemie* **58**, 16–83.

USGS World Energy Assessment Team (2000) *US Geological Survey world petroleum assessment 2000 – description and results.* USGS Digital Data Series DDS-60.

Wang, M. & Overland, J. E. (2009) A sea ice free summer Arctic within 30 years? *Geophysical Research Letters* **36**, L07502.

Westbrook, G. K., Thatcher, K. E., Rohling, E. J., et al. (2009) Escape of methane gas from the seabed along the West Spitsbergen continental margin. *Geophysical Research Letters* **36**, L15608.

Whiticar, M. & Schaefer, H. (2007) Constraining past global tropospheric methane budgets with carbon and hydrogen isotope ratios in ice. *Philosophical Transactions of the Royal Society of London A* **365**, 1793–1828.

Wingham, D. J., Shepherd, A. P., Muir, A. S. & Marshall, G. J. (2006) Mass balance of the Antarctic Ice Sheet. *Philosophical Transactions of the Royal Society of London A* **364**, 1627–1635.

Winters, W. J., Pecher, I. A., Waite, W. F. & Mason, D. H. (2004) Physical properties and rock physics models of sediment containing natural and laboratory-formed methane gas hydrate. *American Mineralogist* **89**, 1221–1227.

Wood, W. T., Gettrust, J. F., Chapman, N.R., Spence, G. D. & Hyndman, R. D. (2002), Decreased stability of methane hydrates in marine sediments owing to phase-boundary roughness. *Nature* **420**, 656–660.

Yefremova, A. G. & Zhizhchenko, B. P. (1974) Occurrence of crystal hydrates of gas in sediments of modern marine basins. *Doklady Akademii Nauk SSSR* **214**, 1179–1181.

Zachos, J. C., Rohl, U., Schellenberg, S. A., et al. (2005) Rapid acidification of the ocean during the Paleocene-Eocene thermal maximum. *Science* **308**, 1611–1615.

Zachos, J. C., Schouten, S., Bohaty, S., et al. (2006) Extreme warming of mid-latitude coastal ocean during the Paleocene-Eocene Thermal Maximum: Inferences from TEX86 and isotope data. *Geology* **34**, 737–740.

Zachos, J. C., Bohaty, S. M., John, C. M., McCarren, H., Kelly, D. C. & Nielsen, T. (2007) The Palaeocene-Eocene carbon isotope excursion: constraints from individual shell planktonic foraminifer records. *Philosophical Transactions of the Royal Society A* **365**, 1829–1842.

Zeebe, R. E. & Zachos, J. C. (2007) Reversed deep-sea carbonate ion basin gradient during Paleocene-Eocene thermal maximum. *Paleoceanography* **22**.

Zeebe, R. E., Zachos, J. C. & Dickens, G. R. (2009) Carbon dioxide forcing alone insufficient to explain Palaeocene-Eocene Thermal Maximum warming. *Nature Geoscience* **2**, 576–580.

12 Methane hydrate instability: a view from the Palaeogene

Tom Dunkley Jones[1], Ruža F. Ivanović[2], Andrew Ridgwell[3],
Daniel J. Lunt[2], Mark A. Maslin[4], Paul J. Valdes[2] and Rachel Flecker[2]

[1]School of Geography, Earth and Environmental Sciences, University of Birmingham, Birmingham, UK
[2]School of Geographical Sciences, University of Bristol, Bristol, UK
[3]Department of Geography, Bristol University, Bristol, UK
[4]Department of Geography, University College London, London, UK

Summary

The Palaeocene–Eocene Thermal Maximum (PETM), a rapid global warming event and carbon-cycle perturbation of the early Paleogene, provides a unique test of climate and carbon cycle models as well as our understanding of sedimentary methane hydrate stability, albeit under conditions very different from the modern. The principal expression of the PETM in the geological record is a large and rapid negative excursion in the carbon isotopic composition of carbonates and organic matter from both marine and terrestrial environments. Palaeo-temperature proxy data from across the PETM indicate a coincident increase in global surface temperatures of approximately 5–6°C. Reliable estimates of atmospheric CO_2 changes and global warming through past transient climate events can provide an important test of the climate sensitivities reproduced by state-of-the-art atmosphere–ocean general circulation models (AOGCMs). Here we synthesise the available carbon-cycle model estimates of the magnitude of the carbon input to the ocean–atmosphere–biosphere system, and the consequent atmospheric pCO_2 perturbation, through the PETM. We also review the theoretical mass balance arguments and available sedimentary evidence for the role of massive methane hydrate dissociation in this event. The plausible range of carbon mass input, approximately 4000–7000 PgC (10^{15} g carbon), strongly suggests a major alternative source of carbon in addition to any contribution from methane hydrates. We find that the potential range of PETM atmospheric pCO_2 increase, combined with proxy estimates of the PETM temperature anomaly, do not necessarily imply climate sensitivities beyond the range of state-of-the-art climate models.

Introduction

The Palaeocene/Eocene Thermal Maximum ('PETM', ~55.5 × 10[6] years or 55.5 Ma) is the most prominent transient global warming event recorded in the Cenozoic geological record. PETM palaeo-temperature proxy data from some localities indicate a rapid surface ocean warming of approximately 5–9°C within 1–10 ka (Thomas et al., 2002; Sluijs et al., 2007). Near synchronous with this warming are prominent negative carbon isotopic excursions recorded in both marine carbonates (−3.5 to −4‰) (Zachos et al., 2007) and terrestrial carbon (−5 to −6‰) (Bowen et al., 2004), indicative of a massive release of isotopically light carbon to the ocean–atmosphere system (Dickens et al., 1995). The PETM offers a unique opportunity to investigate the effects of an anthropogenic-scale (thousands of giga-tonnes) release of greenhouse gases on the Earth's climate, ocean–atmosphere chemistry and biota, although with a release rate at least an order of magnitude slower than that achieved by the modern combustion of fossil fuels and deforestation (Zeebe et al., 2009). There were also major differences in the background climate state of the early Palaeogene that preclude a direct comparison with the modern system. These include significantly higher surface- and deep-ocean temperatures, the absence of polar ice sheets (Zachos et al., 2008), as well as a distinct palaeo-geography, ocean chemistry and biosphere (see Kump et al., 2009 for a review).

Despite these caveats, the PETM is a pertinent geological analogue for future global change and can be used to test the behaviour of climate models – such as their simulation of global climate sensitivity – under conditions of extreme warming (Zeebe et al., 2009). To assess (past) climate sensitivity – which is generally defined as the equilibrium surface warming in response to a doubling of CO_2 (and/or the equivalent radiative increase from a different greenhouse gas) – both atmospheric CO_2 and the corresponding warming response must be known. In the case of the PETM there remains considerable uncertainty over: (1) the source and magnitude of carbon release; (2) the size of the PETM atmospheric CO_2 perturbation; and (3) the magnitude and spatial pattern of the associated climate response. The potential involvement of a massive destabilisation of methane gas hydrates, driven by a progressive warming of oceanic intermediate waters (Dickens et al., 1995), also has a particular relevance within a discussion of the climatic forcing of major geohazards and is a further point of considerable current uncertainty regarding our understanding of the PETM. This chapter thus critically reviews two aspects of PETM research that have a particular relevance to modern climate change and the associated climate-forced geohazards: first, the evidence for a global-scale release of methane hydrate at the PETM and, second, the magnitude of global climate sensitivity that can be inferred from the PETM climate–carbon cycle perturbation.

The involvement of methane hydrates within the PETM climate–carbon cycle perturbation was first proposed – based on mass–balance arguments – to explain the magnitude of the observed PETM carbon isotope excursion (CIE) (Dickens et al., 1995). Subsequently some seismic and sedimentological evidence has been published indicating large-scale slope failure along the western margin of the Atlantic basin close to the Palaeocene/Eocene boundary, which may have been associated with the catastrophic dissociation of methane hydrate (Katz et al., 1999, 2001). We review these arguments in the light of more recent estimates of the PETM CIE and the size of the PETM carbon input. We also present a new global review of slump features recorded in Palaeocene to middle Eocene deep-sea drilling programme cores, with a particular focus on the western Atlantic continental margin. The second part of this chapter, concerning the quantification of PETM climate sensitivity, reviews the available estimates of both the atmospheric CO_2 concentration (partial pressure) (pCO_2) perturbation and global average surface warming across the event. At present, there is no reliable proxy estimate of pCO_2 change across the PETM. Instead, quantification of the pCO_2 increase (ΔpCO_2) depends on carbon-cycle model estimates of the size of the PETM carbon pulse and the time-dependent partitioning of CO_2 among the atmosphere, ocean and, eventually, carbonate sediments. These model estimates, in turn, rely on observations from the geological record of the intensity and extent of carbonate dissolution across the PETM, as well as the magnitude of the CIE. There are multiple assumptions made throughout this process, as well as inherent variation in the response of different carbon-cycle models. In addition, although direct proxy data are available for the extent of PETM surface warming at multiple locations, the use of these data as an estimate of global surface warming requires care (Dunkley Jones et al., 2010).

The PETM and methane hydrates

The magnitude and rapidity of the PETM CIE is a critical constraint on the mass and source of carbon injected into the exogenic carbon reservoir during the PETM. With reliable estimates of the mean size of the PETM CIE in the exogenic carbon reservoir, the total mass of this reservoir and an independent estimate of the mass of carbon input, it would be possible to determine the carbon isotopic composition, and hence nature, of the injected carbon using a relatively simple mass balance. Unfortunately there remains some uncertainty in the magnitude of the PETM CIE, with significant variations in the size of $\delta^{13}C$ excursions recorded in different substrates and across marine/terrestrial environments, and as yet no consistent estimate of the mass of carbon input. To the first order, there appear to be three distinct and generally consistent sets of CIE records: (1) those from deep-sea benthic foraminifera, which should record the $\delta^{13}C$ evolution of deep-ocean dissolved inorganic carbon (DIC); (2) those from well-preserved planktonic foraminifera, recording the $\delta^{13}C$ of surface ocean DIC; and (3) terrestrial records, largely controlled by the $\delta^{13}C$ of terrestrial vegetation and hence recording the $\delta^{13}C$ evolution of atmospheric CO_2. The critical question is which of these records most accurately records the pattern and magnitude of the global carbon isotope excursion.

Benthic foraminiferal $\delta^{13}C$ records from deep-ocean cores typically indicate a CIE on the order of −2 to −3‰ (Nunes & Norris, 2006). The magnitude of the CIE in these pelagic successions is, however, clearly influenced by the extent of carbonate dissolution and the ecological exclusion (through extinction) of calcareous benthic foraminifera across the onset to peak PETM (Thomas, 2007; McCarren et al., 2008). These two effects will both tend to cause the loss of peak negative $\delta^{13}C$ values to an extent that is difficult to assess at any single location. Across the Walvis Ridge depth transect, however, there is a clear relation between palaeo-depth, and hence dissolution intensity, and CIE magnitude: the CIE recorded in benthic foraminifera declines from −3.5‰ to −1.5‰ between sites with palaeo-depths of approximately 1500m to 3600m respectively (McCarren et al., 2008). This effect partly accounts for the discrepancy between early estimates of the CIE on the order of −2 to −3‰ (Dickens et al., 1995) and more recent estimates of −4 to −6‰ (Pagani et al., 2006; Zachos et al., 2007). To overcome the problem of deep-ocean carbonate dissolution, attention has recently turned to the PETM records from shelf environments. The high sedimentation rates and shallow water depths at these locations combine to provide higher-resolution records, particularly of the onset of the PETM, that are less impacted by carbonate dissolution. The planktonic foraminiferal (i.e. surface ocean DIC) $\delta^{13}C$ records of the PETM in these locations indicate a CIE of approximately −4‰ (Zachos et al., 2006; Handley et al., 2008; John et al., 2008), a magnitude that is consistent with the largest excursion values recorded in planktonic foraminifera from pelagic sections, such as Ocean Drilling Program (ODP) Site 690, with minimal dissolution (Thomas et al., 2002; Zachos et al., 2007).

Although the magnitude of the CIE recorded in carbonates from the most complete and well-preserved marine sections appear to be converging on approximately −4‰, there remains a persistent discrepancy between this value and the CIE magnitude of approximately −6‰ recorded across both organic and inorganic carbon phases from the terrestrial biosphere (Bowen et al., 2004; Pagani et al., 2006; Handley et al., 2008). A range of perturbations in the terrestrial environment has been proposed as mechanisms for CIE amplification on land, including changes in soil cycling rates, relative humidity and vegetation (Bowen et al., 2004). None of these has yet provided a satisfactory solution to the marine-terrestrial CIE discrepancy. It should be noted that invoking large (~20%) increases in relative humidity as a comprehensive (i.e. global) explanation of the terrestrial CIE amplification (Bowen et al., 2004) is inconsistent with both the hydrological response of global climate models (Pagani et al., 2006) and much palaeo-hydrological proxy data (Wing et al., 2005; Kraus & Riggins, 2007). Given the continuing uncertainty surrounding the controls on the terrestrial CIE, the emerging consensus on the size of the CIE in surface ocean DIC, and the dominance of the oceanic carbon pool relative to the terrestrial biosphere and atmospheric carbon (Figure 12.1), we make the conservative assumption that the global exogenic carbon reservoir CIE is in the order of −4‰.

Carbon exchanges rapidly between the Earth's ocean, atmosphere and biomass reservoirs, which on 10^4-year timescales can be considered as a single reservoir

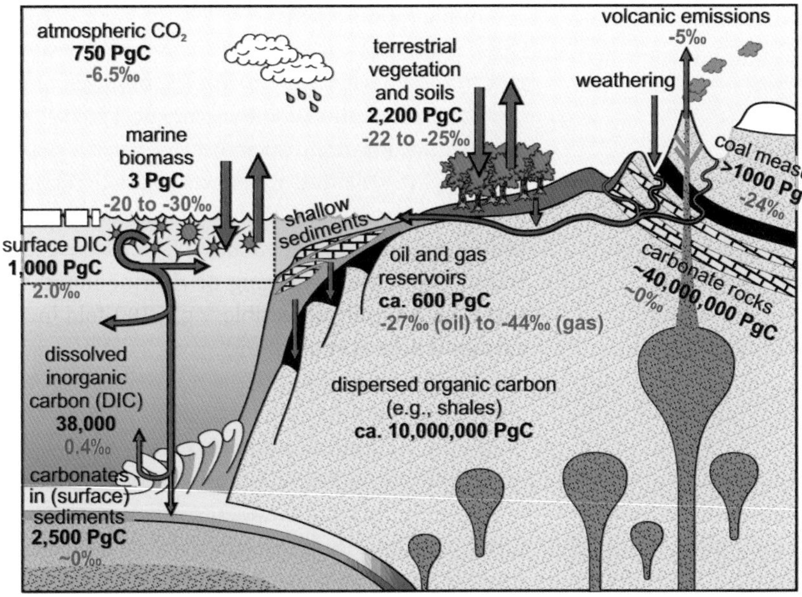

Figure 12.1 Schematic model of the major exogenic carbon reservoirs within the modern Earth system. (Modified from Ridgwell, A and Edwrads, U (2007). Geological Carbon Sinks. 74–97. Wallingford: CAB International with permission.)

with a mean carbon isotopic composition driven by the rate of external inputs or removal of carbon (Dickens, 2003). Some deep-ocean records suggest a stratigraphic lag between the onset of the CIE recorded in mixed layer-dwelling foraminifera and the onset in deeper-dwelling thermocline species (Thomas et al., 2002; Zachos et al., 2007). The approximately 5 ka duration of this inferred top-down propagation of the $\delta^{13}C$ signal from the surface ocean to the thermocline is, however, contrary to current understanding of ocean physics and circulation, even allowing for enhanced warming-induced stratification. In particular, the recorded $\delta^{13}C$ transition at both surface and thermocline depths is abrupt with no evidence of intermediate values (Thomas et al., 2002), whereas the signal expected from a protracted propagation of a signal by microscale mixing and diffusion would be significantly 'smeared'. Thus, the existence of a secondary, species and/or habitat specific control on the recorded $\delta^{13}C$ patterns seems a more likely explanation, although one that has yet to be understood. Notwithstanding that possibility, there is widespread evidence that the light carbon released at the onset of the PETM was relatively rapidly cycled throughout the whole ocean–atmosphere–biosphere system, with dramatic falls in carbon isotope values recorded across terrestrial, surface-ocean and deep-marine environments within approximately 10 ka (Bowen et al., 2004; Nunes & Norris, 2006; Zachos et al., 2007).

The size of the PETM CIE thus provides a critical constraint on the source of carbon entering the exogenic reservoir (Dickens et al., 1995). Any given $\delta^{13}C$ excursion in the total exogenic carbon reservoir can be simply related to the quantity of mass input and $\delta^{13}C$ signature of the external carbon source:

$$(M_T + M_R)(\delta^{13}C_{T'}) \approx (M_R)(\delta^{13}C_R) + (M_T)(\delta^{13}C_T) \qquad (1)$$

where M_T and M_R are the masses of carbon in the exogenic reservoir before input and the mass of the carbon input, and $\delta^{13}C_R$, $\delta^{13}C_T$, $\delta^{13}C_{T'}$ are the carbon isotopic compositions of the carbon input and the exogenic carbon reservoir before and after mass input. Taking into account the large mass of the total exogenic carbon reservoir, to produce a −4.0‰ PETM $\delta^{13}C$ excursion within about 10 ka, the fluxes from any of the conventional long-term (background) carbon sources, such as volcanic activity, weathering or riverine input, would have to increase at a rate that is geologically unfeasible, e.g. a 100-fold increase in the rate of volcanic outgassing (Dickens et al., 1995).

Today, by the burning of fossil fuels and deforestation (Prentice et al., 2001), and during the glacial–interglacial cycles of the late Neogene (Kohfeld & Ridgwell, 2009), carbon isotopic changes in the atmosphere and ocean are driven primarily through the release of isotopically depleted CO_2 from organic matter, with $\delta^{13}C$ signatures of about −26 for biospheric carbon (Prentice et al., 2001) and about −28‰ for fossil fuels (Andres et al., 1993). The total mass of carbon necessary to produce the PETM CIE from such a source is, however, problematic. Using estimates from the modern system to produce a −2 to −3‰ excursion in the ocean–atmosphere carbon reservoir would require the loss of between 75 and 90% of the total organic carbon reservoir (Dickens et al., 1995). To obtain a −4‰ magnitude of excursion requires approximately 8000 PgC for a −22 to −24‰ source (Figure 12.2) – a greater loss of organic carbon than exists in the modern carbon cycle. Proposals for the PETM carbon release involving organic matter – the desiccation of shallow epicontinental basins (Higgins & Schrag, 2006) and the global-scale combustion of terrestrial peat and coal (Kurtz et al., 2003) – hence invoke a larger reservoir before the PETM than exists today. Other potential sources exist for carbon of this approximate isotopic composition, such as that released from the impact of a carbon-rich meteorite (Kent et al., 2003) and, more plausibly, as thermogenic methane released during volcanic activity in the North Atlantic (Svensen et al., 2004).

The difficulties faced in reconciling the isotopic composition of either a volcanic (~−5‰) or organic matter (~−22 to −24‰) carbon source with the magnitude of the PETM CIE imply a significant additional contribution from an extremely ^{13}C-depleted carbon reservoir. In addition, this reservoir has to be outside the exchangeable ocean–atmosphere–terrestrial carbon cycle and yet be released into the ocean–atmosphere system on timescales of 1–10 ka. The leading contender for this source of light carbon remains methane gas hydrates (Dickens et al., 1995). In the modern system methane hydrates are stable under relatively high-pressure and/or low-temperature systems, notably in high-latitude permafrost and along continental margins (Maslin et al., 2010). Hydrates within continental margin sedimentary prisms are typically associated with both free and dissolved methane gas trapped beneath the hydrate layer (Kvenvolden, 1993). The methane present in these deposits, sourced from either the microbial fermentation of organic matter or thermal cracking, typically has a $\delta^{13}C$ signature of the order

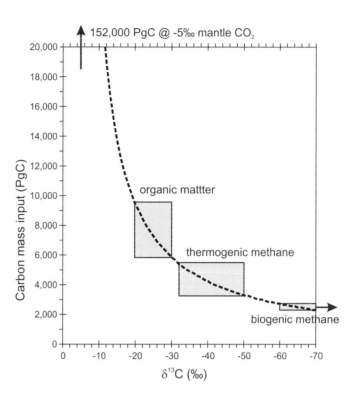

Figure 12.2 The carbon mass input required to produce a −4‰ carbon isotope excursion within an exogenic carbon reservoir of 38,000 PgC, with carbon sources of varying isotopic composition. Assumes that a mean isotopic composition of the ocean and atmosphere before carbon input is 0‰; estimated isotopic values of major sources based upon Maslin & Thomas, 2003.

of −60‰ (Kvenvolden, 1993). The pressure–temperature–salinity-defined region within a sediment column in which gas hydrates are theoretically stable is termed the 'gas hydrate stability zone' (GHSZ), whereas the actual occupancy of the GHSZ by methane hydrates is determined by the supply of methane, itself dependent on organic carbon burial rates and local redox and fluid flow conditions (see also the review of methane gas hydrates in Chapter 11). Due to significantly warmer oceanic intermediate and bottom water temperatures in the Palaeocene, it is estimated that the upper limit of the Palaeocene GHSZ was located in considerably deeper waters (~900 m) compared with modern systems (~250–500 m) (Dickens, 2003). Despite this, a major decrease in pressure, or more likely an increase in temperature at sub-thermocline to intermediate water depths, and its propagation down through the sediment geotherm, could have triggered the dissociation of significant portions of a Palaeocene gas hydrate reservoir. In the initial presentation of this hypothesis, Dickens et al. (1995) calculated that a 4°C warming on background latest Palaeocene bottom water temperatures of 11°C would have depressed the top of the GHSZ from approximately 900 m to approximately 1400 m water depth with an associated dissociation of 1000–2000 Pg methane carbon.

Although the transfer pathways for dissociated methane gas from the GHSZ into the exchangeable carbon reservoir are still poorly understood, the two most likely routes are via diffusion into the water column or ebullition into the atmosphere during sediment failure (Kvenvolden, 1993; Paull et al., 2003). Once released, this methane would have been rapidly oxidized (<10 years) to carbon dioxide in the oceans (Valentine et al., 2001) and atmosphere (Ehhalt et al., 2001), which in turn would be relatively rapidly exchanged between the surface and deep ocean, the atmosphere and terrestrial biomass. The dissociation of substantial volumes of gas hydrate-hosted methane over approximately 10^4 years would have both increased the total mass of the exchangeable carbon reservoir and produced the observed decrease in its mean isotopic composition (Dickens et al., 1995).

A critical part of a warming-driven methane hydrate dissociation hypothesis is that intermediate water warming, as the trigger for hydrate destabilisation, should lead the observed CIE. There is now some evidence for a precursor warming, at least in surface waters, immediately before the onset of the CIE observed in high-resolution continental shelf records of the North Atlantic (Sluijs ct al., 2007). This finding needs to be substantiated by additional records, particularly from intermediate water depths. One plausible scenario is that this initial warming is itself a response to rapid emissions of CO_2 from latest Palaeocene explosive volcanism in either the North Atlantic (Storey et al., 2007) or the Caribbean (Bralower et al., 1997). Recent climate modelling studies, configured to PETM-like boundary conditions, have also shown a strong, non-linear warming response of Atlantic intermediate waters to such a CO_2 forcing (Lunt et al., 2010), providing an amplification of warming in a hydrate-critical region.

Sedimentological evidence for Palaeogene mass movements

On the basis of seismic stratigraphy and sedimentological evidence of slumping in PETM cores from the western North Atlantic margin ODP Site 1051B, Katz et al. (1999, 2001) proposed that major submarine mass movement could have been associated with the dissociation of continental slope methane hydrates at the PETM. The finding that Palaeogene Atlantic intermediate waters were particularly sensitive to atmospheric CO_2 forcing (Lunt et al., 2010) makes this a critical region to investigate for further evidence of continental slope failure and methane hydrate release. Here we present new data from a complete sedimentological review of all early Palaeogene (Palaeocene to middle Eocene; 65–40 Ma) deep-sea sediment cores recovered during the Deep Sea Drilling Project (DSDP) and ODP. The purpose of this review was to determine the frequency and timing of sediment failures and slumps recorded in deep-sea sediment cores and their relationship to the PETM and other early Palaeogene hyperthermal events.

The location of DSDP and ODP sites, which traditionally have targeted deep-sea calcareous oozes for the purposes of palaeo-oceanographic studies, introduces a primary bias into this review when searching for evidence of continental slope failure. In fact the only sites with both good recovery of the early Palaeogene time

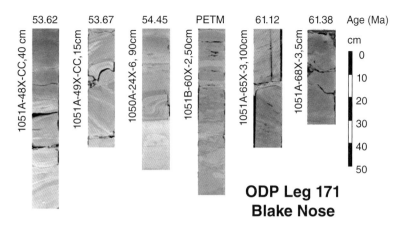

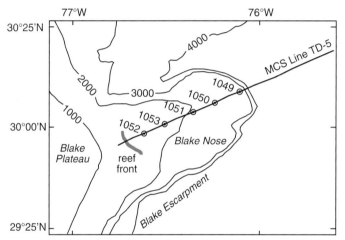

Figure 12.3 Core images of sediment deformation and slump features identified in the Ocean Drilling Program (ODP) Leg 171B sites in the age interval 40–65 Ma (top). ODP Leg 171 site locations across the Blake Nose depth transect (bottom). (Modified from Norris et al., 1998. Proc. ODP, Init. Repts., 171B: College Station, Texas with permission.)

interval (65–40 Ma) and repeated evidence of soft sediment deformation and slumping were situated on the western Atlantic continental margin, drilled during ODP Legs 171 and 207 (Figures 12.3 and 12.4). It is notable that both of these legs were located on the flanks of topographic salients, Blake Nose and Demerara Rise, which protrude outwards from the western Atlantic continental margin (Figures 12.3 and 12.4). The strategy for these legs was to drill a series of sites on depth transects along the slopes of these topographic highs, in both cases spanning approximately some 1300 m of vertical depth: Blake Nose (Leg 171) sites range from 1300 m to 2600 m water depth (Norris et al., 1998); Demerara Rise (Leg 207) sites range from 1900 m to 3200 m water depth (Erbacher et al., 2004).

Evidence for sediment slumping directly coincident with the PETM from Blake Nose Site 1051B has already been documented by Katz et al. (1999). In a review of all Palaeocene to middle Eocene (65–40 Ma) cores from Leg 171, five additional intervals of slumping were identified based on original shipboard visual core

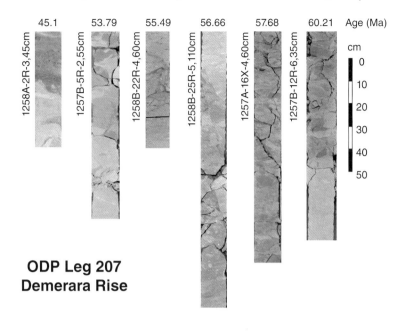

ODP Leg 207
Demerara Rise

Figure 12.4 Core images of sediment deformation and slump features identified in Ocean Drilling Program (ODP) Leg 207 sites in the age interval 40–65 Ma (top). ODP Leg 207 site locations across the Demerara Rise depth transect (bottom). (Modified from Erbacher et al. 2004. Proceedings of the Ocean Drilling Programme, Initial Reports, 207. College Station, Texas with permission.)

descriptions verified by the examination of core photographs (Figure 12.3). Tentative descriptions of slumping or deformation in the shipboard core descriptions, as well as intervals that lacked clear evidence of deformation in core photographs, were excluded from this analysis. Slumped intervals were dated based on linear interpolation from the nearest biostratigraphic tie points (Table 12.1); the estimated uncertainty for the great majority of these dates are less than 100 ka. All the identified slump features are found exclusively at two sites on the lower slope of Blake Nose, Site 1051 at 1900 m water depth and Site 1050 at 2300 m water depth (Figure 12.3).

The same analysis was applied to sites from Demerara Rise, where six slump features were identified within the 65–40 Ma time interval. As with Blake Nose, all of these slumps were found in the two deeper, lower slope sites, Sites 1257 and 1258, at water depths of 2951 m and 3192 m respectively. The ages of the slumps identified from Legs 171 and 207 are plotted alongside the composite benthic foraminiferal stable isotope records of Zachos et al. (2001) on Figure 12.5. Care is needed in interpreting this initial review of ODP sedimentology data, but the results are intriguing. Although the PETM interval is coincident with slumping on Blake Nose, there is no 'smoking gun' of massive, coordinated slope failure associated with the PETM. The PETM, expressed as a characteristic clay layer, was recovered at all of the Demerara Rise sites and is not marked by evidence of slumping in any of these cores (Erbacher et al., 2004). It is apparent, however, that

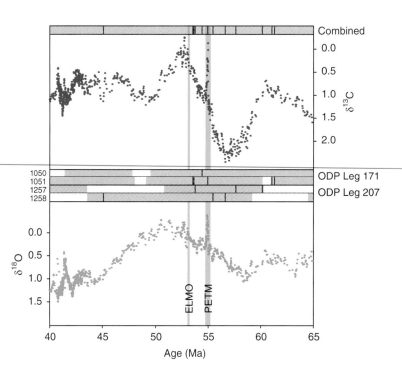

Figure 12.5 Ages of sediment slumps identified in Ocean Drilling Program (ODP) Legs 171B and 207 within the interval 40–65 Ma against the global compilation of deep-sea benthic foraminiferal carbon (top, red) and oxygen (bottom, blue) stable isotope data of Zachos et al. (2001). Ages of individual slumps are indicated by black vertical lines; grey boxes indicate the age of recovered sediments at each of the sites. Top line shows the compiled records of sediment slumps from ODP Legs 171B and 207.

Table 12.1 Palaeocene–Eocene Thermal Maximum (PETM) submarine mass movement data from ocean core analysis with drilling leg, site, hole, core, depth, description of feature evidencing mass movement, biostratigraphic age estimates and reference

Site	Core-section, interval (cm)	Depth (mbsf)[a]	Description	Datum above	Datum age (Ma)[b]	Depth (mbsf)	Datum below	Datum age (Ma)[b]	Depth (mbsf)	Interpolated age of slump (Ma)	Reference
Leg 171B											
1051A	48X-CC, 40	457.18	Cross- and convolute bedding	B *D. iodoensis*	53.11	433.40	B *T. orthostylus*	53.66	458.90	53.62	Norris et al. (1998)
1051A	49X-CC, 15	460.28	Sometimes slumped sediment	B *T. orthostylus*	53.66	458.90	T *T. bramlettei*	53.71	463.70	53.67	Norris et al. (1998)
1050A	24X-6, 90	226.80	Slump fold	B *T. orthostylus*	53.66	208.18	T *F. tympaniformis*	54.71	233.01	54.45	Norris et al. (1998)
1051B	60X-2, 50	512.90		PETM	55.00						
1051A	65X-3, 100	595.80	Convolute bedding and possible slump	B *Ch. Bidens*	61.15	596.90	B *P. uncinata* (PF)	61.37	606.47	61.12	Norris et al. (1998)
1051A	68X-3, 5	623.75	Cross-lamination and convolute bedding	B *Ch. Bidens*	61.15	606.47	B *P. uncinata* (PF)	61.37	623.04	61.38	Norris et al. (1998)

(Continued)

Table 12.1 (*Continued*)

Site	Core-section, interval (cm)	Depth (mbsf)[a]	Description	Datum above	Datum age (Ma)[b]	Depth (mbsf)	Datum below	Datum age (Ma)[b]	Depth (mbsf)	Interpolated age of slump (Ma)	Reference
Leg 207											
1257A	16X-4, 60	136.20	Debris flow and possible slump flow	B *D. multiradiatus*	56.10	116.32	B *D. mohleri*	57.57	134.85	57.68	Erbacher et al. (2004)
1257B	5R-2, 55	71.05	Slump	B *T. orthostylus*	53.64	70.11	T *F. tympaniformis*	54.71	76.99	53.79	Erbacher et al. (2004)
1257B	12R-6, 35	144.06	Several debris flows and slumps	B *D. mohleri*	57.57	122.30	B *F. ullii*	60.31	144.90	60.21	Erbacher et al. (2004)
1258A	2R-3, 45	8.55	Possible slump	Presence *N. fulgens*						43.4–46.8	Erbacher et al. (2004)
		Depth (mcd)				Depth (mcd)					
1258B	22R-4, 60	222.50	Possible stylolites or slumping	T. *Fasciculithus*	54.71	196.82	B *D. mohleri*	57.57	244.37	55.49	Erbacher et al. (2004)
1258B	25R-5, 110	262.10	Slump clast breccia	B *D. mohleri*	57.57	244.37	B *Cr. tenuis*	64.5	271.42	56.66	Erbacher et al. (2004)

[a] Depth of the midpoint of the slumped/disturbed interval. All depths are in meters below sea floor (mbsf) except for site 1258B which is correlated back into data from 1258A using the Site 1258 composite depth scale (mcd).

[b] Calcareous nannofossil datum ages are from Agnini et al. (2007) except for B *Cr. tenuis* which is taken from Erbacher et al. (2004); planktic foraminifera datum ages are from Berggren and Pearson (2005).

All data are calcareous nannofossils except those marked 'PF'. Data are denoted by B, bottom (lowest/first occurrence) and T, top (highest/last occurrence).

although core recovery from both legs covers most of the study interval (65–40 Ma), the majority of slumps cluster on the long-term late Paleocene–early Eocene deep-sea warming trend recorded in benthic foraminifera $\delta^{18}O$ values. This pattern requires further documentation, but if it is supported indicates a potential link between long-term climatic change and continental slope sediment dynamics. Whether this is related to sediment-hosted methane production and release is also open to question.

Magnitude of PETM carbon release

Obtaining a reliable estimate of the total mass input of carbon into the exogenic carbon cycle is critical to understanding the size and rate of carbon fluxes between reservoirs, the magnitude of the perturbation in atmospheric CO_2 and the response of the climate system through the PETM. There are two key constraints on the mass of carbon input at the PETM: the size of the PETM CIE and the degree of sedimentary carbonate dissolution in the deep oceans due to CO_2-driven carbonate under-saturation. Some theoretical limits on the specific contribution of methane gas hydrates to the PETM event may also be achieved by the predictive modelling of the latest Palaeocene GHSZ and its occupancy by methane hydrates. The massive release of carbon into the ocean/atmosphere system over approximately 10^4 years, the majority being absorbed by the ocean, would have caused significant reductions in both the pH and carbonate ion concentration of the deep ocean (Broecker et al., 1971, 1993). This in turn, would have increased the dissolution of biogenic carbonates (primarily calcareous nannofossil liths and foraminiferal tests) in the surface sediments of the deep ocean:

$$CO_2 + H_2O + CaCO_3 = 2HCO_3^- + Ca^{2+}$$

One observable consequence of this is a marked shoaling of both the lysocline and carbonate compensation depth (CCD) across the global oceans (Kump et al., 2009; Ridgwell & Zeebe, 2005). The magnitude of this dissolution event observed in the geological record is fundamentally controlled by the mass and the rate of carbon input to the ocean–atmosphere system and, within the constraints imposed by the size of the PETM CIE, can be used to estimate the magnitude of the carbon input. In reality the extent of carbonate dissolution at any given location is also locally regulated by primary carbonate production rates, local deep-water chemistry and the rate of mixing/bioturbation at the sediment surface, which all confound a simple solution to this problem (Ridgwell, 2007). The reservoir of carbonate in deep-sea sediments available for reaction is also finite (Archer et al., 1997) and the relationship between $CaCO_3$ reacted (chemically eroded) and CO_2 released becomes increasingly non-linear at CO_2 releases >4000 PgC (Goodwin & Ridgwell, 2010), requiring global carbon-cycle models to interpret observations.

To date there have been three key studies that aim to quantify the magnitude of the PETM carbon pulse using the extent of deep-ocean carbonate dissolution. The first of these is based on detailed weight percentage carbonate and $\delta^{13}C$ analyses of deep-ocean sediment cores from five locations along an approximately 2-km depth transect on Walvis Ridge in the South Atlantic (Zachos et al., 2005). These sites effectively provide a vertical profile of carbonate ion concentrations and PETM CCD shoaling through the deep ocean. In addition, the close proximity of all five sites greatly reduces potential biases between these records due to variations in carbonate export productivity and terrigenous particle flux (Zachos et al., 2005). Based on correlations between these sites and the orbital age model at ODP Site 690, the CCD at Walvis Ridge shoaled by more than 2 km in a few thousand years (Zachos et al., 2005). By a comparison with numerical models of anthropogenic-forced ocean acidification and CCD shoaling (Archer et al., 1997), Zachos et al. (2005) suggest that the PETM carbon pulse required to produce such a dramatic CCD change would have been approximately 4500 PgC.

Two subsequent numerical modelling studies used both the Walvis Ridge data and the pattern of carbonate sedimentation across the other major ocean basins, to explicitly simulate the carbon-cycle perturbation through the PETM (Panchuk et al., 2008; Zeebe et al., 2009). The first of these model studies used the Earth system model GENIE-1 (Grid ENabled Integrated Earth system model), in what was the first pre-Quaternary application of a three-dimensional ocean model with full biogeochemical cycling and a spatially resolved coupled sediment model (Panchuk et al., 2008). Configured with an equal area grid of resolution of 36 × 36 (comprising 10° increments in longitude but uniform in sine of latitude, giving 3.2° latitudinal increments at the equator increasing to 19.2° in the highest latitude band) and early Eocene bathymetry and palaeo-geography (Figure 12.6), GENIE was used to simulate a series of carbon input scenarios representative of the diversity of potential carbon sources. These ranged from purely biogenic methane with an isotopic composition of −60‰ to mantle-derived volcanic CO_2 with a $\delta^{13}C$ signature of −5‰. In each case the carbon mass input was sufficient to instantaneously decrease the isotopic composition of the combined ocean and atmosphere carbon reservoir by 4‰ and was added uniformly across atmospheric grid cells over 10 ka. The explicit production of a synthetic sediment column within the GENIE model, with a composition that is a function of model bottom-water chemistry and calcium carbonate/organic carbon export production, allows for a direct comparison with the calcium carbonate content and carbon isotopic composition of available deep-sea PETM sediment cores. The sediment model also includes a representation of bioturbation, the intensity of which exerts a critical control on both the shape and the magnitude of the PETM dissolution and carbon isotope signals (Ridgwell, 2007). This approach allows a direct comparison between the observed and modelled patterns of global marine carbonate sedimentation for both pre-PETM and PETM intervals (Figure 12.6).

The more recent PETM carbon-cycle modelling of Zeebe et al. (2009) used the LOSCAR (Long-term Ocean-atmosphere Sediment CArbon cycle Reservoir) box model. LOSCAR uses a three-box (surface, intermediate and deep) representation

Figure 12.6 Pre-Palaeocene–Eocene Thermal Maximum (PETM) model (CaCO₃ wt%) of Panchuk et al. (2008: Sedementary respinse to Paleocene-Eocene Thermal Maximum Carbon release. Geology 36, 315–318) plotted in map view (top) and versus depth (bottom), both with observations superimposed. Carbonate variation with depth (below) are plotted for the data-rich regions delineated in the top panel by white boxes: (a) central equatorial Pacific Ocean, (b) Walvis Ridge and (c) southern Indian Ocean. Bottom right panel shows the pre-PETM CCD reconstructions of Zeebe et al. (2009). (Adapted by permission from MacMillan Publishers Ltd: Nature Geoscience 2, 576–580, copyright 2009.)

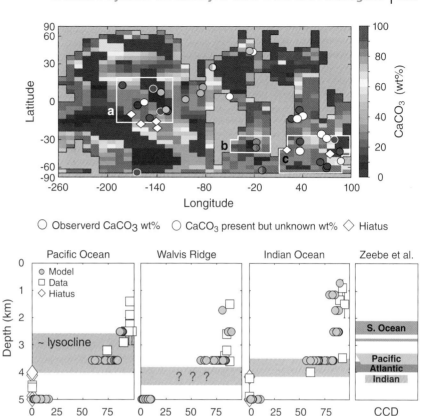

of each of the major ocean basins. Before the PETM, deep water formation was prescribed to occur in the Southern Ocean and bottom water temperatures for pre- and peak-PETM were set at 12°C and 16°C respectively. As in the study of Panchuk et al. (2008), carbonate dissolution records were used to constrain the mass of PETM carbon input, in this case based on comparison between modelled and reconstructed PETM CCD shoaling in the major ocean basins. In order to fit the observed pattern of a relatively prolonged interval of PETM dissolution and CIE, the best-fit scenario consisted of an initial rapid (~5 ka) carbon input of about 3000 PgC followed by an extended (~50 ka) release of 1480 PgC (Figure 12.7). Combined with the authors' assumed surface ocean CIE of −3‰, this carbon release scenario indicates a source isotopic signature of about 50‰, which is consistent with a significant contribution from biogenic methane hydrates (Zeebe et al., 2009). It is worth noting that, assuming a larger CIE, e.g. the −4‰ suggested herein, implies an even more ^{13}C-depleted carbon source, unless the magnitude of carbon release also increases.

A crucial part of all PETM modelling studies is determining an appropriate initial 'latest Palaeocene' pre-PETM climate and ocean chemistry. In the approach

Figure 12.7 Comparison of time-series results from the GENIE (left: Panchuk et al., 2008) and LOSCAR (right: Zeebe et al., 2009) models. The GENIE model shows two carbon input scenarios of 2200 PgC and 6800 PgC. The LOSCAR model shows the best-fit carbon input scenario. From top to bottom: (1) carbon input in giga-tonnes of carbon per year; (2) atmospheric pCO$_2$ perturbation – for each carbon input scenario in GENIE three curves are shown: (a) without bioturbation, (b) with bioturbation, (c) with bioturbation and a temperature dependent weathering feedback; (3) modelled CaCO$_3$ deposition in the south Atlantic at locations and depths comparable with observed CaCO$_3$ deposition across the Walvis Ridge depth transect (Zachos et al., 2005). GENIE data are from the 6800 PgC input scenario with no bioturbation. (Adapted by permission from MacMillan Publishers Ltd: Nature Geoscience 2, 578–580, copyright 2009.)

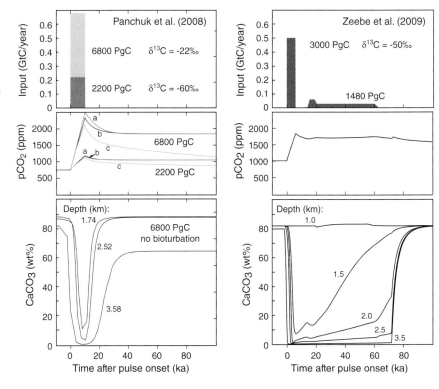

of Panchuk et al. (2008), GENIE-1 is started with a pre-PETM pCO$_2$ of 750 ppm (parts per million), which produces modelled deep-ocean temperatures of about 7°C. This is warmer than modern but cooler than the approximate 12°C indicated by benthic foraminiferal δ^{18}O proxy data (T. Dunkley Jones, D. J. Lunt, P. J. Valdes, J. Tindall, A. Ridgwell, M. Maslin, D. N. Schmidt and A. Sluijs unpublished data). The other key unknowns in the pre-PETM model configuration are global ocean alkalinity and the particulate inorganic to organic carbon rain ratio (PIC:POC ratio). In the absence of suitable proxy data, Panchuk et al. (2008) constrain these variables by running an ensemble of pre-PETM simulations across a range of weathering rates (ocean alkalinity) and spatially uniform PIC:POC ratios. Using the observed pattern of pre-PETM deep-sea carbonate sedimentation, Panchuk et al. (2008) suggest a pre-PETM CCD of approximately 3500–4000 m in the central equatorial Pacific, Walvis Ridge and southern Indian Ocean (see Figure 12.6). These depths appear to be broadly consistent with the CCD reconstructions of Zeebe et al. (2009) and, more importantly, within the limits of the data, reproduce the lysocline depth observed at these three data-rich regions (see Figure 12.6). The best-fit model calibration to the observed pattern of carbonate deposition requires a PIC:POC ratio of 0.20, which is somewhat higher than estimates of the modern global average of 0.06–0.14 (Sarmiento et al., 2002; Ridgwell et al., 2007), and a global weathering rate of 35×10^{15} mol HCO$_3^-$/ka to balance the rate of deep-sea calcium carbonate sedimentation. The modelled deep-ocean circulation, and associated $\delta^{13}C$ gradients are consistent with available pre-PETM benthic foraminiferal $\delta^{13}C$ data (Nunes & Norris, 2006), with decreasing $\delta^{13}C$ from the South to the North Atlantic and from the Indian to the Pacific Ocean, indicative of a dominant Southern Ocean deep water source.

In Panchuk et al. (2008), the carbon release scenario that was most consistent with the observed pattern of deep-sea carbonate sedimentation at peak-PETM conditions consists of a pulse of 6800 PgC with bioturbation in the Pacific and a cessation of bioturbation in the Atlantic (Panchuk et al., 2008). The basin-to-basin variation in bioturbation is required to reproduce the differential dissolution between the Atlantic and the Pacific observed during the event. Carbonate accumulation drops to zero during peak PETM conditions at even the shallowest of the Walvis Ridge sites in the South Atlantic (i.e. CCD <1500 m), which can be compared with the 20–50 wt% carbonate preserved at 2500 m water depth on Shatsky Rise in the central equatorial Pacific (i.e. CCD >2500 m). Again it should be noted that these best-fit scenarios are consistent with the PETM CCD reconstructions of Zeebe et al. (2009), with the only discrepancy being the presence of a small amount of modelled carbonate below their reconstructed CCD in the equatorial Pacific even with a pulse size of 6800 PgC. The absence of bioturbation in the Atlantic increases the intensity of dissolution in surface sediments by removing the resupply of 'old' carbonate from deeper in the sediment column, hence reducing the size of carbon pulse required to produce a zero carbonate layer. There is evidence for such a reduction or cessation of bioturbation in the Atlantic, with the preservation of fine laminations in both equatorial and South Atlantic PETM sediment cores (Erbacher et al., 2004; Zachos et al., 2004). In contrast, there is clear evidence for continued sediment mixing through the onset of the PETM at shallower sites on Shatsky Rise and Alison Guyot in the Pacific, as well as in the Southern Ocean (Kelly et al., 1998; Bralower et al., 2002; Thomas et al., 2002).

An alternative cause of enhanced dissolution in the Atlantic is an altered deep-ocean circulation, with a major reduction in Southern Ocean deep-water formation and its subsequent ventilation of the South Atlantic (Zeebe & Zachos, 2007). During the carbon release experiments of Panchuk et al. (2008), the ocean circulation in GENIE does not respond to warming by a reversal or weakening in the Southern Ocean deep-water source. This is consistent with previous ocean modelling studies that only induced a reversed circulation through an explicit modification of the hydrological cycle and an altered North Atlantic bathymetry (Bice & Marotzke, 2002). More recent, fully coupled AOGCMs have, however, reproduced a significant slowdown in Southern Ocean deep-water formation in response to greenhouse gas-forced warming alone (Lunt et al., 2010). This slowdown is more consistent with the available proxy data (Nunes & Norris, 2006; Chun et al., 2010) and may help to explain some of the basin-to-basin dissolution variability (Zeebe & Zachos, 2007). Any circulation change does not, however, preclude the importance of changes in bioturbation intensity. Indeed, these two effects may represent a coherent change in environmental conditions, with a slowdown in South

Atlantic deep-ocean circulation causing a reduction in bottom water ventilation, dysoxia and the cessation of *in situ* biotic activity (Chun et al., 2010).

Although both of these modelling studies indicate a PETM carbon input at least double the initial estimates of 1500–2500 PgC, which were based purely on the CIE and an assumed methane hydrate source (Dickens et al., 1995), there are some discrepancies between both the model approaches and their final estimates of carbon release. As Zeebe et al. (2009) point out, if a pre-PETM model configuration has an unrealistically deep CCD, this increases the mass of carbonate material available for dissolution at the ocean floor during the PETM acidification event. This in turn will produce a model overestimate of the mass of carbon input required to reproduce the observed carbonate dissolution during the PETM itself – effectively it makes the available oceanic carbonate buffer too large. Although this is an implied criticism of the GENIE model configuration of Panchuk et al. (2008), and their proposed pulse size of 6800 PgC, there is no apparent mismatch between their modelled pattern of carbonate sedimentation, lysocline and CCD and the available observations (see Figure 12.6). As noted above, however, the PIC:POC ratios and biocarbonate weathering fluxes required in the pre-PETM GENIE calibration are higher than the modern and there may be scope for further tuning of the non-carbonate detrital flux rates (currently a globally uniform value of 0.18 g/cm per year) to lower total ocean alkalinity and yet retain a similar pattern of observed weight percentage carbonate sedimentation. There is, however, at least a similar uncertainty involved in the production and use of the CCD reconstructions of Zeebe et al. (2009) as a tuning target for modelled pre- and PETM ocean alkalinity. The accurate reconstruction of the CCD, and subsequent inference of deep-ocean carbonate chemistry, is not straightforward even from modern core-top sediments (Broecker, 2008). One advantage of the synthetic sediment cores produced by GENIE is that the model response is translated into a spatially resolved pattern of sediment compositions that can be directly compared to the carbonate content of actual sediment cores. As noted above, this allows detailed model-data comparisons of the depth-dependent pattern of carbonate sedimentation, and hence changes in both the lysocline and CCD, in data-rich regions of the ocean (see Figure 12.6).

After the discrepancy in total carbon input (6800 PgC of Panchuk et al., 2008, versus the 4480 PgC of Zeebe et al., 2009), the second principal difference in the models concerns the cause of the observed pattern of variation in dissolution intensity between the Atlantic and Pacific basins (Zeebe & Zachos, 2007). This differential is not reproducible in the models without some secondary control on the pattern of carbonate dissolution. GENIE runs with and without bioturbation globally suggest that this might be achieved by a cessation of bioturbation at peak PETM conditions in the Atlantic while continuing to allow Pacific sediments to be actively mixed (Panchuk et al., 2008), although this scenario has not yet been explicitly tested. In LOSCAR, this differential dissolution pattern was achieved by prescribing a weakening of Southern Ocean deep-water formation and an increase in the formation of North Pacific deep water during the PETM (Zeebe et al., 2009). As discussed above, these two mechanisms both have some support from available sediment core and proxy data, and are probably both important aspects of the PETM perturbation.

PETM climate sensitivity

One of the critical conclusions made at the end of the modelling study of Zeebe et al. (2009) was that a carbon input of 4480 PgC, together with the associated model rise in atmospheric pCO_2 from 1000 to 1700 ppmv (Figure 12.7), implied a climate sensitivity for the PETM considerably above the range of values produced by the Intergovernmental panel on Climate Change (IPCC) climate model ensemble for the modern climate system. This conclusion relies on two pieces of information: first their modelled change in atmospheric pCO_2 and second their estimate of PETM surface warming of between 5 and 9°C (Zeebe et al., 2009). In the absence of any reliable proxy data for either the absolute levels or relative changes in atmospheric pCO_2 through the PETM, the only means of assessing the behaviour of the earth system is through the comparison of PETM modelling studies. Within the LOSCAR model runs of Zeebe et al. (2009), a set of sensitivity studies indicates that, for a given carbon pulse, the relative increase in pCO_2 through the PETM is relatively insensitive to baseline pre-PETM atmospheric pCO_2 (within the range 500–1500 ppmv). This suggests that the carbon pulse size is the critical factor controlling the relative pCO_2 increase. It is, however, difficult to make a direct comparison between the results of GENIE and LOSCAR because not only does the initial atmospheric pCO_2 (and state of global carbon cycling) differ, but so too do the tested scenarios for CO_2 release, in both magnitude and duration (Figure 12.7).

By assessing and contrasting multiple model studies (here: Panchuk et al., 2008 and Zeebe et al., 2009), and with the addition of future carbon-cycle model results, we stand to gain a better understanding of the range of possible pCO_2 variations across the PETM. These carbon-cycle model estimates of PETM pCO_2 can then be compared against climate model simulations and the available palaeotemperature data to reduce uncertainty and target future proxy data collection and model refinements. In this vein, it is worth comparing the modelled pCO_2 changes with a suite of early Eocene configurations (0.4% reduction in solar constant compared with modern; early Eocene topography and bathymetry; ice-free land surface; uniform vegetation) of HadCM3L, one of the UK Met Office's fully coupled AOGCMs (Lunt et al., 2010). Model runs have been integrated for >2000 years at one, two, four and six times the pre-industrial pCO_2 levels (280 ppmv), providing a reasonable coverage of potential pre- and PETM conditions (range 280–1680 ppmv). A critical finding of this ensemble is that climate sensitivity is relatively constant over this range of pCO_2, at approximately 3.6°C (Figure 12.8). This value can then be used to estimate the global average temperature changes implied by the estimated PETM pCO_2 increases of Panchuk et al. (2008) and Zeebe et al. (2009) (Figure 12.8); these are 5.5° and 3°C respectively.

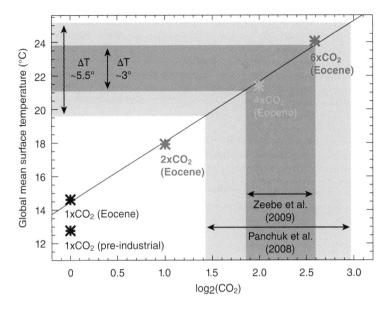

Figure 12.8 Mean annual surface air temperature as a function of CO_2 forcing across a suite of HadCM3L general circulation model (GCM) simulations (Lunt et al., 2010). CO_2 forcing is expressed as $\log_2$ of the CO_2 concentration normalised to the pre-industrial level of 280 ppmv. Also plotted are the pCO_2 perturbations of the best-fit carbon release scenarios of Panchuk et al. (2008) and Zeebe et al. (2009), and the associated estimates of mean annual surface air temperature anomaly predicted for these scenarios by the HadCM3L simulations.

Finally, we should consider the accuracy and implications of the available surface temperature proxy data. Although Zeebe et al. (2009) use the widely quoted range of PETM surface warming of 5–9°C, it is not clear how consistently these proxy data have been interpreted and how representative the specific palaeo-geographical locations of these data are of global surface warming. A new compilation of the available geochemical palaeo-temperature proxy data (Dunkley Jones et al., 2010), which employs a consistent methodology to obtain representative pre-PETM and peak PETM mean values, as well as using a uniform set of palaeo-temperature calculations and calibrations across the available sites. Given that many of these proxy data are subject to issues of stratigraphic completeness and preservational bias, there are only two sites that indicate warming in excess of 6°C, one from the shelf environment of Wilson Lake and the other from the Southern Ocean ODP Site 690. Although these are arguably two of the better-preserved PETM records, they are also in environments that may be subject to locally enhanced warming, due to either shallow water depths or high-latitude amplification of the PETM warming signal. Furthermore, particularly at ODP Site 690, there is a considerable range of variation between proxy measurements on different planktonic foraminifera species, with the average being approximately 6°C. It should be clear from the limited geographical spread, quality and quantity of available PETM palaeo-temperature proxy data that we are some way from the reliable estimate of global mean surface warming necessary to constrain PETM climate sensitivity with confidence. At this stage a conservative interpretation of the proxy data would indicate a pattern of global surface warming in the range 5–6°C rather than the 5–9°C quoted by Zeebe et al. (2009). This is also consistent with available estimates of warming from the terrestrial record: 3–7°C from carbonate soil nodule $\delta^{18}O$ (Bowen et al., 2001; Koch et al., 2003); 4–6°C from $\delta^{18}O$ of biogenic phosphate (Fricke et al., 1998; Fricke & Wing, 2004) and ~5°C from leaf margin analysis (Wing et al., 2005).

The challenge to our current understanding of climate sensitivity presented by Zeebe et al. (2009) thus comes from their estimate of the PETM carbon pulse size, which is at the lower range of the plausible carbon cycle models of the PETM, combined with an interpretation of the temperature anomaly that leans towards the higher end of available proxy data. An alternative scenario, using the PETM carbon pulse modelled by Panchuk et al. (2008), together with a more conservative interpretation of the palaeo-temperature proxy data as representing a 5–6°C warming, places the PETM climatic perturbation within the range of state-of-the art climate model simulations.

Concluding remarks

Although a great deal is now known about the rate, timings and global nature of the PETM event, critical uncertainties still remain about the cause of the carbon-cycle perturbation and the details of the climate system response. On consideration of the size and global expression of the negative carbon isotopic excursion, it still appears likely that a large-scale release of methane hydrate contributed substantially to the massive input of carbon to the exogenic reservoir. With improved estimates of the mass of PETM carbon input, now within the range of about 4000–7000 PgC, it is difficult to ascribe all of this carbon to a methane hydrate source. Even in the modern system, the best estimates of the hydrate reservoir are approximately 500–2500 PgC (Milkov, 2004) and, in the significantly warmer Palaeogene world, the size of the continental slope hydrate stability zone should have been reduced not increased. One possibility currently being discussed is the role of polar permafrost hydrates in Early Palaeogene hyperthermal events. Extracting clear evidence of such releases from the geological record is extremely challenging. That methane hydrates may have operated as a positive feedback mechanism to rapid global warming, even in the warmer ocean/smaller hydrate reservoir Palaeogene world, leaves open the question of what role hydrate-bound carbon will play in the coming centuries of earth history.

Acknowledgements

We are grateful for the constructive comments of two anonymous reviewers. This work was supported by Natural Environment Research Council (NERC) grant *The Dynamics of the Paleocene/Eocene Thermal Maximum*, part of the QUEST research programme.

References

Agnini, C., Fornaciari, E., Raffi, I., Rio, D., Röhl, U. & Westerhold, T. (2007) High-resolution nannofossil biochronology of middle Paleocene to early Eocene at ODP Site 1262: Implications for calcareous nannoplankton evolution. *Marine Micropaleontology* **64**: 215–248.

Andres, R. J., Marland, G., Boden, T. & Bischof, S. (1993) Carbon dioxide emissions from fossil fuel combustion and cement manufacture, 1751–1991, and an estimate of their isotopic composition and latitudinal distribution. In: Wigley, T. M. L. & Schimel, D. S. (eds), *The Carbon Cycle*. Cambridge: Cambridge University Press, pp 53–62.

Archer, D., Kheshgi, H. & Maier-Reimer, E. (1997) Multiple timescales for neutralization of fossil fuel CO_2. *Geophysical Research Letters* **24**, 405–408.

Berggren, W. & Pearson, P. (2005) A revised tropical to subtropical Paleogene planktonic foraminiferal zonation. *Journal of Foraminiferal Research* **35**: 279–298.

Bice, K. L. & Marotzke, J. (2002) Could changing ocean circulation have destabilized methane hydrate at the Paleocene/Eocene boundary? *Paleoceanography* **17**, 8.1–8.13.

Bowen, G. J., Koch, P. L., Gingerich, P. D., Norris, R. D., Bains, S. & Corfield, R. M. (2001) Refined isotope stratigraphy across the continental Paleocene-Eocene boundary at Polecat Bench in the northern Bighorn Basin. In: Gingerich, P. D. (ed.), *Paleocene-Eocene Stratigraphy and Biotic Change in the Bighorn and Clarks Fork Basins, Wyoming*, vol. 33. University of Michigan, pp 73–88.

Bowen, G. J., Beerling, D. J., Koch, P. L., Zachos, J. C. & Quattlebaum, T. (2004) A humid climate state during the Palaeocene/Eocene thermal maximum. *Nature* **432**, 495–499.

Bralower, T. J., Thomas, D. J., Zachos, J. C., et al. (1997) High-resolution records of the late Paleocene thermal maximum and circum-Caribbean volcanism: Is there a causal link? *Geology* **25**, 963–966.

Bralower, T. J., Premoli Silva, I., Malone, M. J., et al. (2002) *Proceedings of the Ocean Drilling Program, Initial Reports 198*. College Station, TX: Ocean Drilling Program.

Broecker, W. S. (2008) A need to improve reconstructions of the fluctuations in the calcite compensation depth over the course of the Cenozoic. *Paleoceanography* **23**, PA1204.

Broecker, W. S., Li, Y-H. & Peng, T. H. (1971) Carbon dioxide – Man's unseen artifact. In: Hood, D. W. (ed.), *Impingement of Man on the Oceans*. New York: John Wiley, pp 287–324.

Broecker, W. S., Lao, Y., Klas, M., et al. (1993) A search for an early Holocene $CaCO_3$ preservation event. *Paleoceanography* **8**, 333–339.

Chun, C. O. J., Delaney, M. L. & J. C. Zachos. (2010) Paleoredox changes across the Paleocene–Eocene thermal maximum, Walvis Ridge (ODP Sites 1262, 1263, and 1266): Evidence from Mn and U enrichment factors. *Paleoceanography* **25**, PA4202.

Dickens, G. R. (2003) Rethinking the global carbon cycle with a large, dynamic and microbially mediated gas hydrate capacitor. *Earth and Planetary Science Letters* **213**, 169–183.

Dickens, G. R., O'Neil, J. R., Rea, D. K. & Owen, R. M. (1995) Dissociation of oceanic methane hydrate as a cause of the carbon isotope excursion at the end of the Paleocene. *Paleoceanography* **10**, 965–971.

Dunkley Jones, T., Ridgwell, A., et al. (2010) A Paleogene perspective on climate sensitivity and methane hydrate instability. *Philosophical Transactions of the Royal Society of London A* **368**: 2395–2415.

Ehhalt, D., Prather, M., Dentener, F., et al. (2001) Atmospheric chemistry and greenhouse gases. In: Houghton, J. T., Ding, Y., Griggs, D. J., et al. (eds), *Climate Change 2001: The Scientific Basis. Contribution of Working Group I to the Third Assessment Report of the Intergovernmental Panel on Climate Change.* Cambridge: Cambridge University Press, pp. 239–287.

Erbacher, J., Mosher, D. C., Malone, M. J., et al. (2004) *Proceedings of the Ocean Drilling Program, Initial Reports, 207.* College Station, TX: Ocean Drilling Program.

Fricke, H. C. & Wing, S. L. (2004) Oxygen isotope and paleobotanical estimates of temperature and $\delta^{18}O$-latitude gradients over North America during the early Eocene. *American Journal of Science* **304**, 612–635.

Fricke, H. C., Clyde, W. C., O'Neil, J. R. & Gingerich, P. D. (1998) Evidence for rapid climate change in North America during the latest Paleocene Thermal Maximum: Oxygen isotope compositions of biogenic phosphate from the Bighorn Basin (Wyoming). *Earth and Planetary Science Letters* **160**, 193–208.

Goodwin, P. & Ridgwell, A. (2010) Ocean-atmosphere partitioning of anthropogenic carbon dioxide on multi-millenial timescales. *Global Biogeochemical Cycles* **24**, GB2014.

Handley, L., Pearson, P. N., McMillan, I. K. & Pancost, R. D. (2008) Large terrestrial and marine carbon and hydrogen isotope excursions in a new Paleocene/Eocene boundary section from Tanzania. *Earth and Planetary Science Letters* **275**, 17–25.

Higgins, J. A. & Schrag, D. P. (2006) Beyond methane: Towards a theory for the Paleocene-Eocene Thermal Maximum. *Earth and Planetary Science Letters* **245**, 523–537.

John, C. M., Bohaty, S. M., Zachos, J. C., et al. (2008) North American continental margin records of the Paleocene-Eocene thermal maximum: Implications for global carbon and hydrological cycling. *Paleoceanography* **23**.

Katz, M. E., Pak, D. K., Dickens, G. R. & Miller, K. G. (1999) The Source and Fate of Massive Carbon Input During the Latest Paleocene Thermal Maximum. *Science* **286**, 1531–1533.

Katz, M. E., Cramer, B. S., Mountain, G. S., Katz, S. & Miller, K. G. (2001) Uncorking the bottle: What triggered the Paleocene/Eocene thermal maximum methane release? *Paleoceanography* **16**, 549–562.

Kelly, D. C., Bralower, T. J. & Zachos, J. C. (1998) Evolutionary consequences of the latest Paleocene thermal maximum for tropical planktonic foraminifera. *Palaeogeography, Palaeoclimatology, Palaeoecology* **141**, 139–161.

Kent, D. V., Cramer, B. S., Lanci, L., Wang, D., Wright, J. D. & Van der Voo, R. (2003) A case for a comet impact trigger for the Paleocene/Eocene thermal maximum and carbon isotope excursion. *Earth and Planetary Science Letters* **211**, 13–26.

Koch, P. L., Clyde, W. C., Hepple, R. P., Fogel, M. L., Wing, S. L. & Zachos, J. C. (2003) Carbon and oxygen isotope records from paleosols spanning the Paleocene-Eocene boundary, Bighorn Basin, Wyoming. In: Wing, S. L., Gingerich, P. D., Schmitz, B. & Thomas, E. (eds), *Causes and Consequences of Globally Warm Climates in the Early Paleogene*, vol. 369. Geological Society of America, pp 49–64.

Kohfeld, K. E. & Ridgwell, A. (2009) Glacial-interglacial variability in atmospheric CO_2. In: Le Quéré, C. & Saltzman, E. S. (eds), *Surface Ocean – Lower Atmosphere Processes*, vol. 187. Washington DC: American Geophysical Union, pp 251–286.

Kraus, M. J. & Riggins, S. (2007) Transient drying during the Paleocene-Eocene Thermal Maximum (PETM): Analysis of paleosols in the bighorn basin, Wyoming. *Palaeogeography Palaeoclimatology Palaeoecology* **245**, 444–461.

Kump, L. R., Bralower, T. J. & Ridgwell, A. (2009) Ocean acidification in deep time. *Oceanography* **22**, 94–107.

Kurtz, A. C., Kump, L. R., Arthur, M. A., Zachos, J. C. & Paytan, A. (2003) Early Cenozoic decoupling of the global carbon and sulfur cycles. *Paleoceanography* **18**, 1090.

Kvenvolden, K. A. (1993) Gas hydrates – geological perspective and global change. *Reviews of Geophysics* **31**, 173–187.

Lunt, D. J., Valdes, P. J, Dunkley Jones, T., et al. (2010) CO_2-driven ocean circulation changes as an amplifier of Paleocene-Eocene thermal maximum hydrate destabilization. *Geology* **38**, 875–878.

Maslin, M. A. & Thomas, E. (2003) Balancing the deglacial global carbon budget: the hydrate factor. *Quaternary Science Reviews* **22**, 1729–1736.

Maslin, M.A., Owen, M., Betts, R., Day, S., Dunkley Jones, T. & Ridgwell, A. (2010) Gas hydrates: Past and future geohazard? *Philosophical Transactions of the Royal Society of London A* **368**: 2369–2393.

McCarren, H., Thomas, E., Hasegawa, T., Röhl, U. & Zachos, J. C. (2008) Depth dependency of the Paleocene-Eocene carbon isotope excursion: Paired benthic and terrestrial biomarker records (Ocean Drilling Program Leg 208, Walvis Ridge). *Geochemistry Geophysics Geosystems* **9**, Q10008.

Milkov, A. V. (2004) Global estimates of hydrate-bound gas in marine sediments: how much is really out there? *Earth-Science Reviews* **66**, 183–197.

Norris, R.D., Kroon, D., Klaus, A., et al. (1998) *Proceedings of the Ocean Drilling Program, Initial Reports, 171B*. College Station, TX: Ocean Drilling Program.

Nunes, F. & Norris, R. D. (2006) Abrupt reversal in ocean overturning during the Palaeocene/Eocene warm period. *Nature* **439**, 60–63.

Pagani, M., Pedentchouk, N., Huber, M., et al., Expedition 302 Scientists (2006) Arctic hydrology during global warming at the Palaeocene/Eocene thermal maximum. *Nature* **442**, 671–675.

Panchuk, K., Ridgwell, A. & Kump, L. R. (2008) Sedimentary response to Paleocene-Eocene Thermal Maximum carbon release: A model-data comparison. *Geology* **36**, 315–318.

Paull, C. K., Brewer, P. G., Ussler, W., Peltzer, E. T., Rehder, G. & Clague, D. (2003) An experiment demonstrating that marine slumping is a mechanism to transfer methane from seafloor gas-hydrate deposits into the upper ocean and atmosphere. *Geo-Marine Letters* **22**, 198–203.

Prentice, I. C., Farquhar, G. D., Fasham, M. J. R., et al. (2001) The carbon cycle and atmospheric carbon dioxide. In: Houghton, J. T., Ding, Y., Griggs, D. J., et al. (eds), *Climate Change 2001: The Scientific Basis*. Contribution of Working Group I to the Third Assessment Report of the Intergovernmental Panel on Climate Change. Cambridge: Cambridge University Press, pp 183–237.

Ridgwell, A. (2007) Interpreting transient carbonate compensation depth changes by marine sediment core modeling. *Paleoceanography* **22**, PA4102.

Ridgwell, A. & Zeebe, R. E. (2005) The role of the global carbonate cycle in the regulation and evolution of the Earth system. *Earth and Planetary Science Letters* **234**, 299–315.

Ridgwell, A. & Edwards, U. (2007) Geological carbon sinks. In: Reay, D., Hewitt, C. N., Smith, K. & Grace, J. (eds), *Greenhouse Gas Sinks*. Wallingford: CAB International, pp 74–97.

Ridgwell, A. J., Hargreaves, J. C., Edwards, N. R., et al. (2007) Marine geochemical data assimilation in an efficient Earth system model of global biogeochemical cycling. *Biogeosciences* **4**, 87–104.

Sarmiento, J. L., Dunne, J. P., Gnanadesikan, A., Key, R. M., Matsumoto, K. & Slater, R. (2002) A new estimate of the $CaCO_3$ to organic carbon export ratio. *Global Biogeochemical Cycles* **16**, 1107.

Sluijs, A., Schouten, S., Pagani, M., et al., the Expedition 302 Scientists (2006) Subtropical Arctic ocean temperatures during the Palaeocene/Eocene thermal maximum. *Nature* **441**, 610–613.

Sluijs, A., Brinkhuis, H., Schouten, S., et al. (2007) Environmental precursors to rapid light carbon injection at the Palaeocene/Eocene boundary. *Nature* **450**, 1218–1221.

Storey, M., Duncan, R. A. & Swisher, C. C. (2007) Paleocene-Eocene thermal maximum and the opening of the northeast Atlantic. *Science* **316**, 587–589.

Svensen, H., Planke, S., Malthe-Sorenssen, A., et al. (2004) Release of methane from a volcanic basin as a mechanism for initial Eocene global warming. *Nature* **429**, 542–545.

Thomas, E. (2007) Cenozoic mass extinctions in the deep sea: What disturbs the largest habitat on Earth? In: Monechi, S., Coccioni, R. & Rampino, M. (eds), *Large Ecosystem Perturbations: Causes and consequences*, vol. 424. Geological Society of America, pp 1–23.

Thomas, D. J., Zachos, J. C., Bralower, T. J., Thomas, E. & Bohaty, S. (2002) Warming the fuel for the fire: Evidence for the thermal dissociation of methane hydrate during the Paleocene-Eocene thermal maximum. *Geology* **30**, 1067–1070.

Valentine, D. L., Blanton, D. C., Reeburgh, W. S. & Kastner, M. (2001) Water column methane oxidation adjacent to an area of active hydrate dissociation. *Geochemica et Cosmochemica Acta* **65**, 2633–2640.

Wing, S. L., Harrington, G. J., Smith, F. A., Bloch, J. I., Boyer, D. M. & Freeman, K. H. (2005) Transient floral change and rapid global warming at the Paleocene-Eocene boundary. *Science* **310**, 993–996.

Zachos, J., Pagani, M., Sloan, L., Thomas, E. & Billups, K. (2001) Trends, rhythms, and aberrations in global climate 65 Ma to present. *Science* **292**, 686–693.

Zachos, J. C., Kroon, D., Blum, P., et al. (2004) *Proceedings of the Ocean Drilling Program, Initial Reports, 208*. College Station, TX: Ocean Drilling Program.

Zachos, J. C., Röhl, U., Schellenberg, S. A., et al. (2005) Rapid acidification of the ocean during the Paleocene-Eocene thermal maximum. *Science* **308**, 1611–1615.

Zachos, J. C., Schouten, S., Bohaty, S., et al. (2006) Extreme warming of mid-latitude coastal ocean during the Paleocene-Eocene Thermal Maximum: Inferences from TEX86 and isotope data. *Geology* **34**, 737–740.

Zachos, J. C., Bohaty, S. M., John, C. M., McCarren, H., Kelly, D. C. & Nielsen, T. (2007) The Palaeocene-Eocene carbon isotope excursion: constraints from individual shell planktonic foraminifer records. *Philosophical Transactions of the Royal Society of London A* **365**, 1829–1842.

Zachos, J. C., Dickens, G. R. & Zeebe, R. E. (2008) An early Cenozoic perspective on greenhouse warming and carbon-cycle dynamics. *Nature* **451**, 279–283.

Zeebe, R. E. & Zachos, J. C. (2007) Reversed deep-sea carbonate ion basin gradient during Paleocene-Eocene thermal maximum. *Paleoceanography* **22**.

Zeebe, R. E., Zachos, J. C. & Dickens, G. R. (2009) Carbon dioxide forcing alone insufficient to explain Palaeocene-Eocene Thermal Maximum warming. *Nature Geoscience* **2**, 576–580.

Index

Climate Forcing of Geological Hazards, First Edition. Edited by Bill McGuire and Mark Maslin.
© 2013 The Royal Society and John Wiley & Sons, Ltd. Published 2013 by John Wiley & Sons, Ltd.